The
Methyl Bromide
Issue

JOHN WILEY & SONS SERIES ON AGROCHEMICALS AND PLANT PROTECTION

Agrochemicals and Plant Protection

Volume 1

The
Methyl Bromide
Issue

Edited by

C. H. Bell, N. Price and B. Chakrabarti
Central Science Laboratory, UK Ministry of Agriculture, Fisheries and Food, UK

JOHN WILEY & SONS
Chichester · New York · Brisbane · Toronto · Singapore

Other Wiley Editorial Offices

John Wiley & Sons, Inc., 605 Third Avenue,
New York, NY 10158-0012, USA

Jacaranda Wiley Ltd, 33 Park Road, Milton,
Queensland 4064, Australia

John Wiley & Sons (Canada) Ltd, 22 Worcester Road,
Rexdale, Ontario M9W 1L1, Canada

John Wiley & Sons (Asia) Pte Ltd, 2 Clementi Loop #02-01,
Jin Xing Distripark, Singapore 0512

Library of Congress Cataloging-in-Publication Data

The methyl bromide issue/edited by C. H. Bell, N. Price, B. Chakrabarti.
    p.   cm. – (Agrochemicals and plant protection ; v. 1)
  Includes bibliographical references and index.
  ISBN 0-471-95521-3 (hardcover ; alk. paper)
    1. Bromomethane.   2. Bromomethane–Environmental aspects.
I. Bell, C. H.   II. Price, N.   III. Chakrabarti, B.   IV. Series.
SB952.B75M48 1995
632′.94–dc20   95-42367
                                                                    CIP

British Library Cataloguing in Publication Data

A catalogue record for this book is available from the British Library

ISBN 0 471 95521 3

Typeset in 10/12pt Times by Keytec Typesetting Ltd, Bridport, Dorset, UK
Printed and bound in Great Britain by Biddles Ltd, Guildford, Surrey
This book is printed on acid-free paper responsibly manufactured from sustainable forestation,
for which at least two trees are planted for each one used for paper production.

Contributors to Volume 1

P. C. ANNIS — CSIRO Stored Grain Laboratory, Canberra ACT, Australia

C. H. BELL — Central Science Laboratory, MAFF, London Road, Slough, Berkshire, SL3 7HJ, UK

J. H. BUTLER — Climate Monitoring and Diagnostics Laboratory, National Oceans and Atmospheric Administration, 325 Broadway, Boulder 80303, Colorado, USA

B. CHAKRABARTI — Central Science Laboratory, MAFF, London Road, Slough, Berkshire, SL3 7HJ, UK

L. KLEIN — Dead Sea Bromine Group, PO Box 180, Beer Shava 84101, Israel

O. MACDONALD — Central Science Laboratory, MAFF, Hatching Green, Harpenden, Herts, UK

M. MILLER — Environmental Policy Analyst, PO Box 665, Napier, New Zealand

N. PRICE — Central Science Laboratory, MAFF, London Road, Slough, Berkshire, SL3 7HJ, UK

C. REICHMUTH — Federal Biological Research Institute for Agriculture and Forestry, Institute for Stored Product Protection, Komigin-Luise-Str 19, Berlin D-14196, Germany

J. M. RODRIGUEZ — Atmospheric and Environmental Research, Cambridge, Massachusetts, USA

C. J. WATERFORD — CSIRO Stored Grain Laboratory, Canberra ACT, Australia

Contents

Series Scope and Aims

The production of food and fibre for an increasing world population continues to be of critical importance as we approach the 21st century. However, interest in the protection of our environment and concerns over the use of chemicals have increased rapidly and have tended to receive more attention in recent years. As a result, controversy over the use of agrochemicals to increase the quality and quantity of food produced worldwide has become more polarised. Research on alternatives to agrochemicals has also gained momentum. With this in mind, the new 'Wiley Series in Agrochemicals and Plant Protection' has been launched to bring together current scientific and regulatory knowledge and perspectives on all aspects of the use of chemicals and biotechnology in agriculture. This new series has evolved from the well established '*Progress in Pesticide Biochemistry and Toxicology*' edited by Hutson and Roberts, published in 1980, and will have considerably wider scope.

SCOPE

The 'Wiley Series in Agrochemicals and Plant Protection' will continue to focus on topical issues covered by Hutson and Roberts, notably environmental and toxicological aspects of the use of agrochemicals. However, the scope of the new series will encompass all subject areas of importance including pesticide discovery and use, agricultural biotechnology, integrated pest management (IPM) and integrated crop management (ICM), mode of action, resistance, structure-activity relationships, analytical and formulation technology and bioremediation. It will also aim to bring together scientific, regulatory and public perception issues relating to the use of chemicals in agriculture. Opportunities will also be sought to publish monographs on specific classes of agrochemicals (for example, the triazole fungicides and the sulphonylurea herbicides).

This important new series will complement the international Wiley journals '*Pesticide Science*' and '*Journal of Science Food and Agriculture*' as well as the existing volumes of '*Progress in Pesticide Biochemistry and Toxicology*'.

EDITORIAL BOARD

The Editors of 'Wiley Series on Agrochemicals and Plant Protection' is Dr Terry Roberts (JSCI, UK) and Dr Junshi Miyamoto (Sumitomo, Japan). In addition, an international Editorial Advisory Board has been appointed initially comprising:

Professor Fritz Fuhr (KFA Julich, Germany)
Dr David Hutson (Falmouth, UK)
Dr Philip Kearney (USDA, USA)
Professor Don Mackay (University of Toronto, Canada)

who together will provide an invaluable source of knowledge on current issues with the discovery and use of chemicals and biotechnology in agriculture.

Preface

The proposal that CFCs contribute to depletion of the stratospheric ozone layer, the subsequent scientific and political activity and the arrival at international agreements to put in place phase-out schedules, proved to be a long drawn out and painful process. Indeed, there is still much activity in this field. The same ozone depleting proposal, levelled against the minor-use chemical methyl bromide might be thought to have been an insignificant issue in comparison. This has not proved to be the case, and a fierce debate has raged since 1991 on the relative merits of scientific, political, social, environmental and economic arguments for and against the production and use of this chemical.

Methyl bromide is a naturally occurring compound but is also made by man, both intentionally and unintentionally. The vast majority of the intentionally man-made methyl bromide is used as a fumigant in agriculture, horticulture and the preservation of structures and structural materials. The unintentional production sources include biomass burning and leaded petrol usage. The fate of the natural and man-made chemical is the focus of much of the debate and this provides one contrast with the CFC situation. Other contentious issues include: the lifetime of methyl bromide in the upper atmosphere; its ozone depletion potential, the opportunities available for its replacement; its impact on developing countries' economies, and the impact on overall atmospheric bromine levels of removing only man-made methyl bromide. In addition, there is the powerful argument that we have a responsibility to do all that we can to protect our planet, its environment and the health of its inhabitants.

Against this background, this book does not set out to convince the reader of a predetermined viewpoint. Its purpose is to set out as much of the scientific debate as is possible to date and to let the reader weigh up the available evidence. We hope that the book covers the major relevant fields of science including; agriculture, atmospheric chemistry, oceanography, environmental sciences, chemistry, biology and toxicology, as well as including two chapters on potential alternatives to methyl bromide.

At the time of writing, several important meetings are on the horizon, which will determine the fate of methyl bromide and the likely future of fumigation technology. It is hoped that this book will inform those participating in this debate and that it will come to be regarded as a useful tool for all interested in the related disciplines.

Acknowledgements

The editors would like to thank all the authors for their dedication and, (reasonably) prompt delivery of manuscripts, Miss Carol Wakefield for acting as editorial co-ordinator and keeping the editors organised, Mr Jonathan Stein for his support in provision of library services at CSL, Miss Alison Simms for her sterling efforts in organising the reference lists of many of the chapters, and Mr Kevin Norman of the CSL Pesticides Group at CSL for advice on certain aspects of methyl bromide analysis. We thank Dr Terry Corbitt for constructing and collating the index.

We would also like to express our gratitude to the senior management of CSL for allowing and encouraging us to carry out this project.

1

Methyl Bromide in Perspective

N. PRICE

Central Science Laboratory, MAFF, Slough, UK

1.1 INTRODUCTION

Until 1991 methyl bromide was a relatively unheard of chemical. True, within the fields of agriculture and horticulture it had become a well used

C.H. Bell, N. Price and B. Chakrabarti: The Methyl Bromide Issue and important fumigant for the control of weed, insect, nematode, fungal, bacterial and viral pests and diseases. Yet it remained a low volume, high value product, requiring, in general, trained professional application. The discovery in 1991 of the ability of methyl bromide to interact with ozone and thus potentially to contribute to global ozone depletion has elevated methyl bromide to the headlines and caused a flurry of political and scientific activity, together with no small measure of controversy. Over the years methyl bromide has been used in a variety of applications from chemical synthesis to fire retardation, but its pest and disease control applications have become an industry of some size.

Methyl bromide, CH_3Br, is a simple alkyl halide, with properties that fit into the very narrow band of requirements of physical, chemical and toxicological characteristics needed for a successful fumigant. This chapter will explore the properties of methyl bromide, (MB) together with its preparation, analysis, toxicology and mode of action. A number of the elements touched upon briefly in this chapter will be elaborated upon in later sections of the book.

2.1 CHEMISTRY

2.1.1 PREPARATION

Methyl bromide is readily prepared by refluxing methanol with excess constant-boiling hydrobromic acid in the presence of small amounts of sulphuric acid. Alternatively, heating methanol with potassium bromide in an excess of concentrated sulphuric acid also gives a good yield of methyl bromide.

$$CH_3OH + KBr + H_2SO_4 = CH_3Br + KHSO_4 + H_2O \qquad (1)$$

2.1.2 PROPERTIES AND REACTIONS

Methyl bromide has the empirical formula CH_3Br and a relative molar mass of 94.95. Alternative names for methyl bromide are monobromomethane, MBX, and, more recently, as a result of greater familiarity, MB. Trade names include, Bromofume, Desbrom, Haltox, MBR-2, Metabrom, Methyl-brom, Methyl-o-gas, Sobrom 9B, Terr-o-gas 100, Bromogaz, Selfume, Brom-o-sol, and Bromopic. It is a colourless gas at temperatures above 3.5 °C and at low concentrations has no noticeable odour. At high concentrations it is said to have a sickly sweet or musty odour and a burning taste. The lack of odour has disadvantages when MB is used as a fumigant and this has sometimes been overcome by the inclusion of a warning gas, such as chloropicrin, at concentrations of about 2%.

Methyl bromide freezes at $-93°C$ and has a latent heat of vapourisation of 61.52 cal/g. Its solubility in water at 25 °C is 13.4 g/l.

Methyl bromide is a powerful solvent of many organic materials including many plastics and natural rubber. Polyethylene, polypropylene and poly-tetrafluoroethylene are only slighly affected by liquid methyl bromide and not at all in the concentrations of gas used for fumigation. Pure methyl bromide is not corrosive to metals but in the liquid state will react with aluminium. In this reaction methyl aluminium bromide is formed, which, in the presence of oxygen ignites spontaneously.

Mixtures of methyl bromide in air are non-inflammable except in the range 9–20% in air, and methyl bromide has been used successfully in fire extinguishers.

Methyl bromide and other monoalkyl halides undergo a number of reactions which make them important chemical intermediates in a range of organic synthesis pathways, and for this reason it is pertinent briefly to examine some of the general reactions of alkyl halides.

Alkyl halides can be hydrolysed to alcohols. Although this reaction is slow in water alone, hydrolysis by boiling in alkalis or in a boiling aqueous suspension of silver oxide is rapid.

$$CH_3Br + KOH = CH_3OH + KBr \qquad (2)$$

This reaction represents the reverse reaction to that shown above for the preparation of methyl bromide and is a nucleophilic substitution.

Alkyl halides are reduced by nascent hydrogen giving rise to the alkane.

$$CH_3Br + 2[H] = CH_4 + HBr \qquad (3)$$

Formation of higher alkanes can occur via the Wurz reaction:

$$2CH_3Br + 2Na = C_2H_6 + 2NaBr \qquad (4)$$

However, when heated with certain metal alloys, alkyl halides can form organometallic compounds such as tetra-alkyl lead compounds. These latter compounds along with the alkyl halide 1,2-dibromoethane are used as fuel additives. The dibromoethane prevents emission of lead by forming volatile lead halides, but also gives rise to the formation, in the exhaust gases, of methyl bromide as the most abundant organobromine component in the exhaust, (Baumann, Heumann and Fresenius 1992). Indeed in the UK alone it has been calculated that 46 tonnes of methyl bromide per annum could be emitted to the atmosphere from vehicle exhaust (Chemistry and Industry 1993).

Organometallic complexes which can function as Grignard reagents are also formed when liquid methyl bromide reacts with magnesium in the presence of ether. Grignard reagents can be used in a wide range of synthetic processes including the production of hydrocarbons, alcohols, ethers, ketones, carboxylic acids, esters, and amines.

Primary alkyl halides give good yields of amines when heated with ethanolic ammonia under pressure and when heated with aqueous ethanolic potassium cyanide, alkyl halides react to produce alkyl cyanides.

$$CH_3Br + KCN = CH_3CN + KBr \qquad (5)$$

Alkyl halides can also form thioalcohols, thioethers, and sulphonates, by reaction with alcoholic solutions of sulphides, mercaptides and with sulphites respectively. They can also be used in the Friedel–Crafts reaction, reacting, for example, with benzene to form toluene.

2.1.3 REACTIONS WITH ORGANIC MATTER

MB can react with a range of organic materials, its primary reaction being methylation. This can give rise to problems if some types of material are allowed, inadvertantly to come into contact with the gas. Indeed the Manual of Fumigation for Insect Control, (Bond 1984), gives a list of materials which it is advised should not come into contact with MB. In view of the discussion in this section of the reactions of MB with organic chemical groupings, it is of interest to reproduce Bond's full listing below:

- iodised salt, stabilised with hyposulphite
- certain baking sodas, salt blocks used for cattle licks or other foods containing reactive sulphur compounds
- full fat soya flour
- sponge rubber
- foam rubber, as used in rug padding, pillows, cushions and mattresses
- rubber stamps and similar forms of reclaimed rubber
- furs, horsehair and pillows (especially feather pillows)
- leather goods, particularly white kid or any other leather goods tanned with a sulphur process
- woollens, especially angora; some adverse effects have been noted on woollen socks, sweaters and yarn
- viscose rayons made by a process that uses carbon disulphide
- cinder blocks or mixtures of mortar; mixed concrete occasionally picks up odours
- charcoal, which not only becomes contaminated but sorbs great quantities of MB
- paper that has been cured by a sulphide finishing process, and silver polishing papers
- photographic chemicals, not including cameras or film

- rug padding, cellophane, vinyl
- any other materials that may contain reactive sulphur compounds.

Methylation is the major reaction of interest in the interaction of methyl bromide with organic and living materials. Methylation appears to occur with the SH, carboxyl, and amino groups of enzymes and other proteins, and with the nucleotide bases of nucleic acids. A common factor in many of the materials in the list above is the presence of sulphur, and since S-methylation appears to occur readily with MB it is not surprising that these materials may suffer detrimental effects. Indeed the structural protein keratin, (the primary component of hair, fur, feather, wool, and skin) is stabilised at the molecular level by disulphide cross-links formed between adjacent cysteine residues in the polypeptide chain.

Blackburn and Phillips (1941) found that wool treated with methyl bromide underwent primary O- and N-methylation. Collagen and gelatin treated with MB also underwent methylation which appeared to be primarily esterification of free carboxyl groups, (Blackburn, Consden and Phillips 1944). The SH group of proteins is also a major target of MB methylation and bisulphited wool treated with MB showed methylation of the thiol groups (Blackburn, Consden and Phillips 1944). However, most studies on biochemical effects of methyl bromide are on the products of its most common use, fumigation. Thus Winteringham and Harrison (1946) investigated the decomposition and consequential reactions of methyl bromide during fumigation and wheat flour. The greater rate of decomposition of methyl bromide in low moisture flow samples compared to high moisture samples, led to the conclusion that hydrolysis was not the major route of degradation. Indeed in the case of gluten treated with methyl bromide, production of methanol by hydrolysis of the MB accounted for less than 10% of the reaction products (Winteringham 1955).

Winteringham and Barnes (1955), conducted a number of pioneering studies using the then relatively new technique of radioactive 'tracing'. Methyl bromide was prepared labelled with trace amounts of ^{14}Carbon. In wheat flour treated with this $^{14}CH_3Br$ the reaction of MB with gluten accounted for about 80% of the decomposition of MB. N-methyl derivatives of the proteins accounted for 50% of the reaction products whilst dimethyl sulphonium and thiomethoxyl derivatives (S-methylation), accounted for 30% and 10% respectively and methoxy derivatives (O-methylation), accounted for 10%. Using $^{14}CH_3Br$, Bridges (1955) used paper and starch column chromatography to isolate and identify the N-methyl reaction products of MB with gluten. He concluded that the reaction between MB and the nitrogen groups of wheat proteins was primarily due to methylation of the imidazole rings in the histidine residues of the protein. He claimed that this reaction accounted for 75% of the total N-methylation whilst a further 10% was due to methylation of the ε-amino group of lysine. Subsequently many N-methyl derivatives of plant proteins have been found to result from MB treatment (Winteringham 1977).

A number of studies were carried out in the 1970s and 1980s at the CSL laboratories, on the reaction of protein components of various commodities with methyl bromide. Using electrophoresis and isoelectric focusing protein banding patterns were found to be altered in MB treated wheat compared to untreated. A number of protein bands were shown to change their staining density in a dose-dependent manner. When the gels were stained for esterase activity, it was found that some of the protein perturbation effects were due to changes in isoelectric point of a number of esterase enzymes, presumably due to methylation of these enzymes. Again the appearance of esterases with altered P_is was related to the dose of MB used. Similar effects were shown with proteins from treated barley, lupins, peanuts and navy beans. This technique at one time was proposed as a diagnostic test for fumigation of wheat with MB but suffered from lack of sensitivity. Studies using $^{14}CH_3Br$ treatment of wheat followed by enzymic degradation, hydrolysis and HPLC analysis confirmed that N-methylation of amino acids was the primary reaction with histidine, lysine and arginine, (all basic amino acids) being the most susceptible to attack. Greater amounts of N-methyl derivatives were formed at higher pHs.

At high exposures to MB the germination and growth of seeds may be retarded. However, with dry seeds and normal fumigation dosages, most seeds are insensitive to damage by MB. Indeed Powell (1975) showed that most seeds tested out of 40 varieties of vegetables, cereals, fodder and grasses, were unaffected by MB treatments of $200-400 \, \text{mg} \, l^{-1} h$ at a range of moisture contents. Repeated fumigation of seeds is not recommended since cumulative dosages can adversely affect viability.

With regard to growing plants, such as nursery stock, which is treated with MB, it has been estimated that almost 95% are tolerant to doses of the gas that will kill pests infesting them (Bond 1984).

The methylating effect of MB on nucleic acid has been investigated. Djalali-Behzad et al (1981) incubated DNA with $^{14}CH_3Br$ in-vitro and also isolated mouse spleen cells and red blood cells were given the same treatment. Methylation of the nitrogenous nucleotide bases of DNA was found to occur, and methylation of the 7-N position of guanine was used as a marker. DNA methylation in mice treated with MB by intraperitoneal injection or by inhalation was found to be low compared to the in-vitro experiments. Liver methylation, both of DNA and protein was also low, possibly surprising in view of the liver's crucial function in detoxication of xenobiotics. Despite the preponderance of N-methyl protein derivatives following MB treatment, reaction with SH groups in proteins has been seen as important, particularly as an explanation for the toxic action of MB.

Toennies and Kolb (1945) and Lewis (1948) used solutions of sulphur-containing amino acids to investigate the effect of MB. It was shown that MB reacts with methionine to form alkyl sulphonium salts. In addition, sulphydryl methylation of cysteine and glutathione, was found to be directly proportional to the concentration of Br^- in solution. Glutathione is a tripeptide, γ-L-glutamyl-L-cysteinylglycine, which is known to play a vital role in the detoxication of xenobiotics, so its methylation by MB may be a key factor in MB action and detoxication. Indeed cytotoxicity of MB to cultured mammalian cells is diminished by the addition of reduced glutathione, (GSH), (Nishimura et al 1980).

Using a target pest of methyl bromide, the granary weevil, Starrat and Bond (1981) further studied the role of glutathione in MB action. They found that MB-resistant weevils had higher levels of bodily GSH than their susceptible counterparts. Following treatment of the weevils with $^{14}CH_3Br$ they found that the primary radioactive metabolites were S-methyl glutathione and S-methyl cysteine, and that susceptible insects incorporated higher levels of radioactivity into their tissues. The implication of this finding is that GSH is 'mopping up' MB and the higher levels of GSH in the resistant insects offer greater protection against the toxicant.

Glutathione detoxifies a range of xenobiotics by conjugating their leaving groups to a range of biochemical substrates and to itself. The reactions are mediated by enzymes called glutathione S-transferases, (GST). Some in-vivo studies on rats have attempted to elucidate the role of glutathione in MB toxicity and detoxication. Davenport et al (1992) showed that in rats which had inhaled MB, glutathione was depleted and brain GST was inhibited. Thomas and Morgan (1988) proposed that MB, through its action on GSH, could interfere with the metabolism of leukotrienes, prostaglandins and primary carbohydrate metabolism. Depletion of GSH in rats by pretreatment with buthionine sulphoximine increased MB toxicity, indicating that GSH is a route of detoxication of MB in-vivo.

As already discussed, MB reacts with the sulphur-containing amino acid cysteine. When this reaction takes place with the cysteine residues in red blood cell haemoglobin, the presence of the resulting S-methyl cysteine conjugate can be used as a diagnostic tool for MB exposure. These adducts have a life span of only a few months so no cumulative history of exposure can be detected but recent and suspected exposures can be checked out.

2.1.4 MODE OF ACTION

As can be seen from the foregoing, the reactions of MB with organic and living material are not highly specific. As such it has proved difficult for researchers to pinpoint a specific reaction in living organisms which gives rise to the toxicity of MB. Such a vital reaction is often called the 'toxic lesion' or the 'biochemical lesion'. As an additional complication, the fact that MB is used against such a wide range of pest and disease organisms, of very different metabolism and biochemistry also indicates that there may not be a single toxic lesion, or if there is it must be fundamental to the life process.

MB is highly soluble in lipids. Lipids form a major constituent of the membranes that surround and protect living cells and their organelles. It has been suggested that this lipophilicity leads to the dissolution of MB in membranes resulting in general chaotropic effects. These effects may be on the vital selective permeability properties of membranes or by methylation of the proteins which are embedded in or traverse the membrane.

Some of the most sensitive membraneous systems are those associated with nerves in animals. Whilst this is clearly not the site of the biochemical lesion in lower organisms such as bacteria, fungi and viruses, it is an attractive hypothesis in organisms with nervous systems, though it has not been explored in terms of conventional neurophysiological techniques.

The suggestion that MB may act primarily on the CNS was supported by Honma *et al* (1982, 1983) who demonstrated that the norepinephrine content of the hypothalamus and cortex was depleted on exposure to MB. Brain amino acid levels and metabolism were altered and it was suggested that alterations in metabolisms of the biogenic amine catecholamine were a factor in MB-induced neurotoxicity.

By virtue, one suspects, of being more amenable to experimentation, the reaction of MB with SH groups in enzymes has received more attention as a possible mode of action (Lewis 1948).

Loveday and Winteringham (1951) showed that methylation of protein SH groups did indeed occur in insects, and there was some evidence to show that the reaction was irreversible *in-vivo*. Classic SH inhibitors such as *iodoacetate* exert their effect by disrupting the process of glycolysis, causing irreversible depletion of ATP by inhibiting the enzyme triose phosphate dehydrogenase. ATP, (adenosine triphosphate) is produced primarily in the mitochondria of cells by a complex process known as oxidative phosphorylation, and is the major source of energy for living cells.

Using houseflies as a target animal, Winteringham, Hellyer and McKay (1958), concluded that inhibition of SH enzymes as a result of methylation by MB was unlikely to be the sole cause of MB's toxicity to the fly. Immobilisation of flies exposed to MB was associated with a reversible breakdown of ATP. Flies which recovered from initial known-down by MB were found to have levels of ATP that had returned to normal. To spoil this elegant explanation of the toxic effect of MB, it was found that flies which, having temporarily recovered and subsequently died did not have depleted levels of ATP. Bond (1956) suggested that methyl bromide inhibits succinate dehydrogenase, another SH-dependent enzyme involved in the generation of ATP. He proposed that, in the early stages of poisoning by MB this would lead to a stimulation of glycolysis, but in later stages both this process and the oxidative synthesis of ATP would be inhibited.

Whilst the toxic lesion or lesions of MB remain unknown, it appears clear that the effect is chemical rather than a physical narcotic or anoxic effect, since the concentration of MB needed to kill 50% of a population of the granary weevil indicated that the thermodynamic activity was very low, (Hayes, 1963). From the large number of enzymes which depend on SH groups for stabilisation of their active sites, it is clear that MB is likely to cause a range of detrimental effects, though an irreversible depletion of ATP would certainly prove lethal.

3.1 RESIDUES AND ANALYSIS

3.1.1 RESIDUES

A consequence of the reaction between a chemical and a commodity is the presence of a residue. This combined with the need to measure concentrations of MB in air have led to a considerable body of work on the analysis of MB and its reaction products, (residues) in a range of materials. It is not the intention of this chapter to give a comprehensive review of the literature on this subject, but to introduce some of the issues concerning residues and the methodologies used to assess them.

Following the fumigation of a commodity with MB, most of the gas is desorbed and diffuses away quickly. Indeed it is this aeration of MB which is now causing concern as will be evident in later sections of this book. Gaseous methyl bromide does not normally present a residue problem in the commodity, but, as seen above, the reaction of methyl bromide with SH, NH_2, COOH and other functional groups of plant and animal origin must give rise to some reaction products. The 'bound' reaction product will be the methylated derivative of the protein or other component of the commodity whilst the 'free' reaction product, (or leaving group), will be the bromide ion Br^-, which, because of its reactivity is normally recovered as inorganic bromide.

In analysing foodstuffs for inorganic bromide, a complicating factor is the natural occurrence of bromide in many foodstuffs. Under normal circumstances, the very small amounts of residual bromide in foodstuffs fumigated with MB presents no problem when considered alongside normal consumption levels and the presence of naturally occurring bromides. For example, it would require about 135 kg of MB fumigated apples to provide the average medicinal dose of bromide salt, (in Bond 1984). Since residues in fumigated commodities which are high in lipids (high fat content) such as nuts and cheese, tend to be higher, occasionally problems of taint or odour can occur, especially if the nuts are subsequently roasted. This results from the reaction of MB with the sulphur-containing amino acid methionine, the methylated product then breaking down to release the odoriferous compound dimethyl sulphide. As to the fate of the methylated residue from MB treatment, a considerable number of studies have addressed the toxicological significance of this 'bound' residue. As already seen, treatment with MB results in widespread methylation of protein amino acids. There does not appear to be any appreciable loss of essential amino acids from foodstuffs treated with MB, (Winteringham 1955). MB has also been shown to react to some extent with the vitamins of the B group, this also appears to be of no toxicological or nutritional consequence, since under normal fumigation conditions, no loss of vitamins has been detected (Clegg and Lewis 1953). In general the formation, in a fumigated food commodity of the main methylation products, (1-methyl histidine, S-methyl cysteine, and a range of O-methyl products) does not appear to pose a health hazard. In addition S-methyl methionine is a naturally-occurring component of some foodstuffs and is nutritionally equivalent to its parent amino acid, whilst 1-methyl histidine is a secondary metabolite of vertebrate metabolism, and both S-methyl cysteine and O-methyl derivatives are readily metabolised.

3.1.2 ANALYSIS

Three applications of analytical techniques are relevant to the fumigation of commodities or soil with MB. These are; the approximate determination of the concentration of MB in air, needed for an estimation of the safety of fumigation operations; the accurate determination of gaseous MB useful in establishing and confirming dosage schedules; and the quantitative determination of the residues discussed above.

Halide leak detector lamps have been commonly used for monitoring purposes. They are also used in the detection of freon-type refrigerants and commonly comprise a propane-fuelled flame which plays on to a copper ring. In the presence of halide-containing gas, the flame burns with a green or greenish-blue hue. As indicated above, the device is not specific to MB and cannot be used if other halide-containing gases might be present. Similarly, because of the naked flame, this type of leak detector cannot be used in the vicinity of inflammable materials. Other fuels, such as methanol, or paraffin, have been used in this type of detector.

In general, due to the variations in responses of individual lamps, and the change in response as the copper ring gets dirty, readings below 30 ppm are likely to be unreliable, and since the lowering of the long term occupational exposure standard to 5 ppm, halide detectors are no longer used in practice. Instead electronic leak detectors have become available for this purpose.

Gas detector tubes provide a common 'field analysis' method. These are small glass tubes filled with an indicator chemical into which a known volume of air is drawn via a hand-operated or mechanical pump. Those for MB also contain a pre-tube in which is a strong oxidising agent. As the air is drawn in, any MB present will react with the oxidising agent to form bromine which then reacts with the indicator causing a change in colour. Since the air is drawn in at one end of the tube, the length of the indicator in the tube which has changed colour is indicative of the concentration of MB in the air. Tubes are calibrated by the manufacturer, and thus provide a degree of *quantitation*.

Portable thermal conductivity meters can be used to analyse MB concentrations above $1\,\mathrm{gm}^{-3}$ (250 ppm) up to $300\,\mathrm{gm}^{-3}$. When a constant current is passed through a wire filament, the equilibrium temperature of the filament and hence its electrical resistance, is a function of the thermal conductivity of the surrounding gas. A change in the composition of the gas will alter its thermal conductivity and the subsequent change in electrical resistance can be measured. Instruments using this principle have been constructed (Phillips and Bulger 1953, Heseltine, Pearson and Wainman 1958) and calibrated with various concentrations of MB. An unknown concentration of MB in air can be determined from the meter reading using the calibration.

More accurate still is the use of infra-red gas analysis (IRGA). MB like many other gases absorbs light in the infra-red region of the spectrum. The pattern of this absorption is specific and thus spectrometers can be calibrated accurately to measure MB-air concentrations. The equipment is relatively expensive but does provide sensitive, accurate and quantitative data. The MIRAN is one such instrument which can monitor atmospheric MB as low as 2.3 ppm. Some doubt still exists as to the stability of such instruments when used under difficult conditions in the field.

At the top of the sensitivity and accuracy scale is gas-chromatography, (GC). This is the chosen laboratory method for the analysis of MB and has been adapted for use in the field. A portable GC for use in the field is described by Bond (1984) with sensitivity down to 0.01 ppm. Photoionisation detectors and argon ionisation GC detectors have been used for this purpose though the former suffers from being non-specific for MB, especially in the presence of other halohydrocarbons. In the absence of other materials, electron capture detection allows reliable and routine GC analysis of air samples.

Analysis of MB in water also relies heavily on GC. Various methods have been used including 'purge and trap', and headspace analysis, (see later).

With the advent of 'Bench Top' mass spectrometers, GC-MS is sometimes used as a confirmatory technique, though this is a laboratory-based method.

Analysis of MB in soil, in common with that from commodities, relies on the extraction of samples with solvent prior to GC analysis.

Despite the volatility of MB, evidence exists for the persistence of unchanged MB residues in commodities following fumigation, as well as for the reaction products discussed above. This is particularly true for commodities with high fat content such as nuts and cheese, presumably as a result of the lipohilic MB dissolving in these fatty materials. Most of the published methods for analysing MB residues in food commodities rely on the determination of total bromide. This will include organic bromides, inorganic bromide, (including naturally occurring) and, where the extraction method lends itself, parent MB. These methods almost exclusively use gas chromatography as the analytical tool, though the method of preparation of the sample varies. Initially, cold solvent extraction followed by gas chromatography, (GC) was a favoured method. MB is extracted from the commodity with solvent, (acetone has been commonly used) and may be analysed directly be GC using a flame ionisation detector. In a more specific variation, the extract is reacted with sodium iodide, any unchanged MB reacting to form methyl iodide which can then be analysed by GC, (Fairall and Scudamore 1980).

More recently, headspace GC analysis has been developed for analysis of MB residues (Scudamore 1988).

This technique involves the blending of a commodity with water or water mixed with miscible solvent, followed by equilibration of the extract in a sealed vial fitted with a septum. The headspace gas above the extract is sampled by gas-tight syringe, either manually or as part of an automatic apparatus and the vapour injected into a GC. Determination of the extracted residues is accomplished usually using electron capture detection GC. The amount of MB in the vapour phase is quantifed by carrying out calibration extraction of known amounts of MB. There have been many refinements to this technique and these combined with automation have brought the limits of detection down to the parts per billion level (Daft 1992). The technique has been applied to a wide range of commodities, including fruits, herbs, spices, cheese, bread, cereal products, meats, vegetables, and edible oils (Daft 1993).

Direct headspace analysis of native, unextracted foodstuff is also possible, (Norman 1991). The method has been automated and found to be suitable for detecting low levels of MB in a variety of commodities. There are problems with this direct method, in that the high temperatures needed to release MB from the commodity into the headspace, induce reaction between MB and the organic components of the food. The method may nevertheless be useful as a qualitative screening tool.

4.1 TOXICOLOGY

4.1.1 TOXICITY TO LABORATORY ANIMALS

It has already been established that MB is an effective biocide for a wide range of pest and disease organisms. It follows that it is also highly toxic to a wide range of life forms. As far as target species is concerned, much has been written concerning the toxicity of methyl bromide to insects, mites, fungi, bacteria, viruses and nematodes. This forms the basis of Chapter 4 and thus here we will concentrate on the toxicity of MB to non-target species and the related toxicology.

The normal route of MB exposure is by inhalation, though dermal contact can sometimes occur. In laboratory experiments with animals, oral toxicities are also usually determined for comparative purposes. MB is lethal to rabbits orally at dosages above about $60\,mg\,kg^{-1}$ body weight, though accurate data is hard to find. Guinea pigs exposed to MB vapour suffer various effects depending on the combination of time and concentration used. This introduces the principle of concentration-time products, which is vital to understanding the toxic effects of gaseous toxicants. In dealing with solid or liquid poisons, the compound is administered in known amounts to experimental animals and the LD_{50} (the amount needed to kill 50% of the test population) is determined. With gas-phase poisons the time of exposure to the gas is as important, (and sometimes more so) than the concentration of gas in air. As a general rule the exposure time and the concentration of gas contribute equally to the toxic effect and thus, for example a concentration of MB of $1\,mg\,l^{-1}$ in air administered to a test animal for 5 hours would show the same toxicological effects as $0.5\,mg\,l^{-1}$ over 10 hours. Thus toxicity of gases is often expressed as the concentration $\times$ time or CT product.

Early studies showed that pigs that were briefly exposed to high concentrations of MB died more quickly than those exposed to low concentrations over several hours. In the latter case mortality was delayed and followed symptoms of delayed weakness, rapid pulse, lung irritation and pneumonia. However the logarithmic relationship of time and concentration was linear for similar symptomatic outcomes, underlining the CT principle discussed above. This was also confirmed with rats and rabbits all of which were killed by a CT product of $2\,mg\,l^{-1}h$. When animals were exposed to MB for 8 hours a day for 5 days a week over a 6 month period they were able to tolerate concentrations of the same order of magnitude to that for a single 8 hour exposure. For rats and guinea pigs this was $0.025\,mg\,l^{-1}$ (about 25% of the single 8 hour maximum), and for monkeys and rabbits, $0.013\,mg\,l^{-1}$ and $0.0065\,mg\,l^{-1}$ respectively. After 6 months of this treatment all animals showed normal growth, no symptoms and no histopathological changes. At higher concentrations symptoms of increased activity, muscular tremors and paralysis of the extremities occurred, though these disappeared on removal from exposure. More recent studies, (Miyagawa cited in Alexeev and Kilgore, 1983), put the CTP for 100% mortality at $3.6-5.5\,\text{mg}\,\text{l}^{-1}\text{h}$ for rats. From this more recent data in which a series of mortality levels was studied, a 24 hour LC_{50} can be estimated for rats at $0.05\,\text{mg}\,\text{l}^{-1}$ (13 ppm). For fish, LC_{50}s of $4.18\,\text{mg}\,\text{l}^{-1}$ and $4.68\,\text{mg}\,\text{l}^{-1}$ for 96 hours were estimated for bluegill sunfish (freshwater) and tidewater silversides, (salt water) respectively, (Dawson et al 1977).

Kato, Morinobu and Ishizu (1986) determined the 4 hour LC50 for Sprague–Dawley rats as 780 ppm in air, whilst Zwart (1988) found a 1 hour LC50 of 1876 ppm.

Inhalation of up to 120 ppm MB for 13 weeks caused no mortality (Drew, Haber and Tice 1984) or little mortality (Haber et al 1988), in laboratory rats. The high mortality caused by a few days exposure at 159 ppm shows either the steep dose response curve of MB or the difficulty in obtaining consistent data with this chemical.

Recent studies (Kaneda et al 1993), reported that the highest practical residue remaining after deliberate overtreatment of rat diets with MB was $500\,\text{mg}\,\text{kg}^{-1}$. They fed rats with an MB-treated diet containing 80 200 and $500\,\text{mg}\,\text{kg}^{-1}$ residues of total bromine for two complete generations (18 weeks per generation). No effects were found at the two lower levels which represented 1.5 and 5 times the FAO-recommended level of bromine intake. At the $500\,\text{mg}\,\text{kg}^{-1}$ level there was a reduction in feeding of the F1 and F2 females but no pathological or physiological changes. It was concluded that residues of 200 and $500\,\text{mg}\,\text{kg}^{-1}$ of total bromine in diets treated with MB were, respectively, the no-observed-effect level, (NOEL) and the minimum toxic level for rats.

The normal route of excretion of inhaled MB is via exhaled air with CO_2, but ingested MB is excreted primarily in the urine. This has been confirmed using ^{14}C radiolabelled MB, with 43% of orally administered radiolabel appearing in the urine of treated rats over a 24 hour period (Medinsky et al 1984), whereas 47% of an inhaled dose of radiolabelled MB administered to rats was recovered as [14]CO_2, (Bond et al 1985), MB appears to be rapidly metabolised in the tissues and eliminated from the body.

4.1.2 TOXICITY TO HUMANS

Most of the information relating to MB toxicity comes from accidental cases of human exposure, and many have been fatal. High concentrations of MB can produce rapid unconsciousness and death. Although symptoms of high concentration exposure may resemble anaesthesia, the action is chemical and at most dosages the symptoms are many, varied and often delayed.

Symptoms, as well as recovery or death may all be delayed by a matter of minutes, hours or even days. The cause of death in delayed cases is usually circulatory failure. The most common symptoms are; general malaise, headache, visual disturbances, nausea and vomiting. These are accompanied by a variety of central nervous system disorders including; numbness, ataxia, tremor, myoclonus, abnormalities in the electroencephalogram, agitation, personality change, coma and convulsions. Death is usually the result of pulmonary oedema leading to respiratory failure or cardiovascular collapse. In non-fatal cases, recovery may take several weeks and result in permanent disability.

By the very nature of the human cases reported, concentrations of MB were not measured and estimates based on available information may be misleading. However, it seems clear that death usually resulted from acute exposures to very high concentrations or from chronic repeated low level exposure which culminated in a somewhat higher level exposure. From a number of fatal cases, estimates of concentrations to which the victim was exposed range from $6.2 \, \text{mg} \, \text{l}^{-1}$ to $231 \, \text{mg} \, \text{l}^{-1}$, and the times of exposure range from 1.5 hours to 20 hours, (cited in Alexeev and Kilgore, 1983). One common factor is that deaths resulting from exposure of humans to MB follow exposure periods of at least 1.5 hours and are usually somewhat longer. Non-fatal incidents have been reported with exposures estimated up to $230 \, \text{mg} \, \text{l}^{-1}$ for 20 minutes, (Alexeev and Kilgore 1983).

4.1.3 PATHOLOGY

The literature on the acute and chronic effects of MB on animals and humans is extensive, and has been reviewed comprehensively by Alexeev and Kilgore (1983), and more recently by the World Health Organisation, (IPCS). For this reason comments here will be restricted to studies of humans accidentally poisoned with MB. The symptoms reported above are consistent with pathological findings on autopsy. These include pulmonary oedema, bronchopneumonia, congestion, and haemorrhaging. Sometimes pathology includes direct effects on the gastro-intestinal tract such as stomach congestion, sub-mucous haemorrhaging and sloughing off of the squamous epithelium of stomach, small and large intestine. It is not known whether such effects are caused directly by MB or as a result of disruption of the central nervous system.

Effects on the kidney have been reported with effects on urea production and blood or albumen in the urine. There is some evidence to suggest that the effects on the kidney are directly attributed to MB, with reports of enlarged and inflamed kidneys.

Detailed investigation of the glomeruli and tubules have revealed a range of, often conflicting abnormalities including constriction, dilation, degenerative changes and necrosis. Many obscure, varied and bizarre neurological effects of MB poisoning have been reported, including epilepsy, polyneuropathy, and even conditions which have been mistaken by physicians for psychological disfunction. Indeed a number of physiological manifestations normally associated with psychological disorders are accompaniments to MB intoxication including mood swings, loss of memory, mental confusion, anorexia, and vertigo.

More specific neural effects are often produced in acute poisoning cases. These include speech impairment, blurred vision, temporary blindness, limb and muscle twitching and tremors. Autopsy examination of brain tissue reveals multiple minute haemorrhages, spongy necrosis, and breakdown of the myelin sheath.

In chronic cases of MB exposure, loss of initative, loss of libido, inability to tolerate beer, hallucinations, confusion, euphoria, personality changes, irritability, psychosis, and neurosis have all been reported. Autopsy usually reveals a range of lesions in the brain, though to what extent these are due directly to MB or as a result of many and various CNS effects, is unknown.

4.1.4 CYTOTOXICITY, MUTAGENICITY

Studies on the effect of MB on germ cells have shown that inhalation of the gas reduces sperm motility in rats, though no effects on the reproductive cells of females was found (Morrissey *et al* 1988). Testicular degeneration was accompanied by the sloughing of spermatocytes and the formation of abnormal cells, (Eustis *et al* 1988). A number of studies have failed to discover any teratogenic effects of MB. As discussed earlier MB reacts with DNA and adducts have been isolated following exposure of laboratory animals to the gas. DNA adducts have been isolated from liver, lung and stomach and have been identified as 3-methylguanine, 7-methylguanine and 0^6-methylguanine.

MB has been tested in a range of mutagenicity test systems, with highly variable results. Simmon *et al* (1971), assessed the mutagenicity of MB using the Ames test. Although MB did produce mutant *Salmonella typhimurium* it was less potent than other common halocarbons. In tests with barley kernels, the mutagenic potential of MB was found to be about 800 times less than that of ethylene dibromide, (a misnomer for 1,2,dibromoethane), itself once used as a fumigant.

In tests with other strains of *S. typhimurium* MB does not show up as mutagenic. A positive result has been reported with *E. coli* WP2, (Moriya *et al* 1983), but not with *E. coli* Sd-4 (Djalali-Behzad *et al* 1981). The table below indicates the variability of mutagenicity tests using MB.

5.1 SAFETY AND MEDICAL TREATMENT

By the very nature of pest and disease controlling chemicals, their biocidal activities can cause safety problems for operators and bystanders. In recent years there has been a move towards the development of more specific pesticides, such as insect and plant growth regulators which interact with specific biochemical processes not present in higher animals. No gaseous chemicals have been discovered which have such desirable properties and all fumigants are toxic to a wide range of organisms. As shown in earlier sections of this chapter, MB is no exception to this and as such, in most countries and for most applications, MB is allowed to be used only by trained and licensed operators.

During a fumigation it is important to know whether gas is escaping into the nearby environment, whether workers are being exposed and what to do if they are. Threshold limit values (TLV) were introduced to provide guidance on the level of gas concentration above which it is not safe to expose humans for a normal full working day. The TLV (Occupational Exposure Standard or OEL in UK) is normally expressed as a concentration of gas in parts per million (ppm) in air. The TLV is set at a value of safety on the assumption that all workers would be exposed repeatedly for a full working day. This of course is rarely likely to occur but represents a margin of safety built in to the TLV. The data used to set TLVs are the best available at any time taken from industrial, medical and research sources. From time to time, as new information becomes available, TLVs are modified in the light of new knowledge. The Manual of Fumigation for Insect Control gives two types of TLV value. The Time Weight Average, (TLV-TWA), which is the time-weighted average concentration for a normal 8 hour working day or a 40 hour working week to which all workers may be exposed, day after day without adverse effects. The Short Term Exposure Limit, (TLV-STEL), is the maximum concentration to which workers should be exposed for a period of 15 minutes without suffering from initial symptoms of poisoning. The TLV-TWA for MB is 5 ppm and the TLV-STEL is 15 ppm. At the concentrations likely to be encountered in practice MB has no odour. This means that the first-line safety factor for humans, that of smelling the gas is not effective. For this reason MB has often been formulated with chloropicrin, which is a lacrymatory gas which acts as a warning agent for the MB in the mixture.

There are a number of general safety measures which should be taken when MB, (or any fumigation) is being undertaken. These include ensuring that no one works alone, the whereabouts of all personnel is known at all times, and that a first-aid kit and expertise in its use is available. The use of communication aids such as mobile telephones or radio-communicators and medical contact cards are all sensible precautions. A range of preventative measures are taken as part of normal good fumigation practice. Sealing of

Table 1.1. Mutagenicity tests with methyl bromide

Test	Test system concentration	Dose level, activation presence/absence	Metabolic	Response	Reference
Reverse mutation	*Salmonella typhimurium* TA 100	0.02–0.2% (in desiccators)	–	Positive	Simmon *et al* (1977)
Reverse mutation	*Salmonella typhimurium* TA 100, 1533 TA 98, 1537, 1538	500–5000 mg/m³	+/–	Positive Negative Positive	Moriya *et al* (1983)
Reverse mutation	*Escherichia coli* WP 2 hcr *Salmonella typhimurium* TA100 TA98	950–19000 mg/m³ (plate)	+/–	Positive at 19000 mg/m³ Negative	Kramers *et al* (1985a)
SOS-umu (modified Ames)	*Salmonella typhimurium* TA 1535/pSK 1002	1.5 litre/min for 30 min	–	Negative	Ong *et al* (1987)
Forward mutations Streptomycin resistance	*Klebsiella pneumoniae* ur⁻ pro⁻	950–19000 mg/m³		Positive at 4750 mg/m³	Kramers *et al* (1985a)
Forward mutation	*Escherichia coli* Sd 4	0.5–6 mM.h		1 mutation/10⁸ surviving bacteria mM.h	Djalali-Behzad *et al* (1981)
Forward mutation	L5178Y	0.03–30 mg/litre	–	Positive	Kramers *et al* (1985a)
Sex-linked recessive lethal	*Drosophila melanogaster*	78 or 272 mg/m³ (5 h)		Negative	McGregor (1981)
Sex-linked recessive	*Drosophila melanogaster*	750 mg/m³ (6 h) 375 mg/m³ (5 × 6 h) 200 mg/m³ (15 × 6 h)		Negative Positive Positive	Kramers *et al* (1985a, b)
Sister chromatide exchanges (SCE)	Phytohaemagglutinin-stimulated human whole blood	4.3%		Increased SCE frequency from 10 to 16.84/cell after 100 sec cultures	Tucker *et al* (1985)

Assay	Test system	Dose/concentration	+/−	Result	Reference
SCE	Cultured human lymphocytes	10^{-4}–10^{-6} M	+/−	Positive	Garry et al (1990)
SCE	Bone marrow cells of exposed B6C3F$_1$ mice	0, 47, 97, 195 389, 778 mg/m^3 (6 h/day; 5 d/week; 14 days) Same dose range and exposure		Dose response observed; higher in females than males Negative	NTP (1992)
Unscheduled DNA synthesis	Human diploid fibroblasts (human embryonic intestinal cells)	Up to 70%/3 h	+/−	Negative	McGregor (1981)
Unscheduled DNA synthesis	SPF male Wistar rats, primary liver cells	10–30 mg/litre		Negative	Kramers et al (1985a)
Cell transformation	Syrian hamster embryo cells	3890–31 120 mg/m (2 to 20 h in sealed chambers)		Negative	Hatch et al (1983)
Chromosomal aberrations	Male and female CD (ex Sprague–Dawley) rat bone marrow cells	78 or 272 mg/m^3 single (7 h or 7 h/d; 5 days)		Negative	McGregor (1981)
Micronucleus	Male and female BDF1 mice	600, 778, 1011, 1314 and 1712 mg/m^3 (6 h/d; 5 d/week; 14 days)		Polychromatic erythro-cytes with micronuclei in the bone marrow increased 10-fold in males (778 mg/m^3) and 6-fold in females (600 mg/m^3); in peripheral blood increased 32-fold in males (778 mg/m^3) and 3-fold in females (9600 mg/m^3)	Ikawa et al (1986)

continued overleaf

Table 1.1. (*continued*)

Test	Test system concentration	Dose level, activation presence/absence	Metabolic	Response	Reference
Micronucleus	Male and female F344 rats	600, 778, 1011, 1314 and 1712 mg/m^3 (6 h/d; 6 d/week; 14 days)		Polychromatic erythrocytes with micronuclei in the bone marrow increased 10-fold in males and 3-fold in females at 1314 mg/m^3	Ikawa *et al* (1986)
Micronucleus	In peripheral erythrocytes of B6C3F$_1$ mice	0, 47, 97, 195, 389, 778 mg/m^3 (6 h/day; 5 d/week; 14 days) Same dose range and exposure routine (13 weeks)		Elevated responses over entire dose range with greatest response at 389 and 778 mg/m^3 in females; males less responsive Negative	NTP (1992)
Dominant lethal	Rats, male CD (Sprague–Dawley)	78 or 272 mg/m^3 (7 h/day; 5 days)		Negative	McGregor (1981)

the fumigation area, gas monitoring, leak testing, dosage calculation and airing time calculation are all part of this process and will not be dealt with in more detail here.

The wearing of protective clothing, and specifically, respiratory protection is vital to the safe use of MB. Because of the liquid state of MB, its low boiling point and its reactivity with rubber, protective clothing made of rubber, (e.g., gloves) should be avoided.

Respirators should be only those approved for the specific use and should be regularly maintained. Respirators which cover the entire face are required and used with an organic vapour canister. This type is usually packed with activated charcoal. Self-contained respirators, in which the operative breathes compressed air supplied by tube from a remote cylinder are also available. The use of gas monitoring equipment is essential, and the types of equipment used were discussed in Section 3.1.2 on page 10. This is important for a number of reasons, not the least being that canister type respirators will not be effective above certain concentrations stated by the manufacturer, and it must be known if these concentrations are being exceeded. In addition, the capacity of the canister must be known and the likely duration of their effectiveness in a given concentration of MB. This should be calculated from the known concentration of MB being applied and an assumed high breathing rate. Tables are available to assist this estimation and one such is given in Bond (1984) and reproduced below.

Canisters should be discarded after use or if they appear damaged, blocked or out of date. They should be used only once because MB may continue to diffuse through the activated charcoal. Contact of liquid MB with skin causes severe blisters due to the low boiling point, which rapidly extracts heat from the body producing a high temperature gradient and resulting 'burning'. Gloves, bandages or any other item which could cause retention of MB in contact with the skin should not be worn. Any liquid MB that is spilled on the skin will evaporate quickly and the affected area should be washed with soap and water immediately.

There is no known antidote to MB although compounds which protect SH groups may alleviate the effects of MB (Winteringham 1955). Symptoms of

Table 1.2. Maximum lifetime of OV type canister for methyl bromide

MB concentration (g/m^3)	Maximum time (minutes)
0–16	60
16–32	30
32–48	22
48–64	15

exposure are likely to be delayed and thus no specific procedures are available to bring about rapid recovery. Anyone in the vicinity of a MB fumigation who suffers any of the following symptoms with 48 hours should receive medical attention immediately: nausea, vomiting, dizziness, fatigue, headache, double or blurred vision, loss of appetite, abdominal pain, impaired or slurred speech, mental confusion or convulsions. Medical treatment may involve the administration of anti-emetic drugs to reduce vomiting and nausea. Tracheostomy may be required if the airway cannot be kept clear, otherwise oxygen can relieve respiratory difficulties. The many and varied neurological manifestations mentioned earlier make it difficult to prescribe specific medical intervention but barbiturates or diazepam may be used against nervous excitability. Survival over the first few days will usually indicate eventual recovery. In essence, because of the wide range of MB intoxication symptoms, treatment has to be symptomatic.

6.1 GLOBAL PRODUCTION AND USES

Methyl bromide has been used as a fumigant to control pests and diseases since before the second world war. It was first used to control pests of horticultural produce and pests of stored grain (de Francolini 1935a, 1936b, Shepard and Buzicky 1939). Later it became established as a treatment for structures (Hill and Border 1953) and more recently as a fumigant against pests and diseases carried in soil (Kohn 1962, Kempton and Maw 1973).

Today, about 80% of its total usage as a fumigant is on soil. Between 1984 and 1992 the worldwide use of methyl bromide increased by 60% to over 70 thousand tonnes per annum, largely as a result of soil treatments in the open. Throughout this period uses other than on soil have tended to remain constant (Table 1.3).

Methyl bromide is used throughout the world, with the USA being the major producer and user (Table 1.4).

There are six companies reporting production of methyl bromide, (61 724 metric tons in 1990), primarily in USA and Western Europe though production in Asia, China and Eastern Europe is known to be significant. Estimates from these regions are shown in Table 1.5.

7.1 RECENT EVENTS

In recent years there have been reports (SORG 1990, 1991, WMO 1990, 1992) drawing attention to the potential ozone depleting properties of methyl bromide, and environmental organisations, particularly in the USA, have called for a rapid phasing out of the fumigant (Friends of the Earth

Table 1.3. Estimated global sales of methyl bromide (tonnes) by use sector

Year	Soil	Post harvest	Structural	Chemical intermediates	Total sales*
1984	30 408	9001	2166	3997	45 572
1985	33 976	7533	2257	4507	48 273
1986	36 090	8332	2029	4004	50 455
1987	41 349	8708	2923	2710	55 690
1988	45 131	8028	3647	3804	60 610
1989	47 542	8919	3613	2496	62 570
1990	51 306	8411	3234	3693	66 644
1991	55 079	10 290	1817	4071	71 257
1992	57 407	9855	2264	2648	72 174

Data source: Methyl Bromide Global Coalition (1994).
*Use on perishables not included: 6537 tonnes estimated in 1992 (UNEP, 1995).

Table 1.4. Sales of methyl bromide (tonnes) by region including chemical feedstock, but excluding China, India and the CIS

Year	North America	South America	Europe	North Africa	Africa	Asia
1984	19 659	1389	11 364	183	1595	10 687
1985	20 062	1503	14 414	45	1975	9743
1986	20 410	1774	13 870	380	2205	11 278
1987	23 004	1820	15 359	385	1751	12 816
1988	24 848	2058	17 478	277	1582	13 555
1989	26 083	1701	16 952	618	2075	14 386
1990	28 101	1621	19 119	432	1838	14 605
1991	30 909	2068	17 447	1058	2093	16 843
1992	29 466	2300	18 521	1363	1697	16 944

Data source: Methyl Bromide Global Coalition (1994).

Table 1.5. Production (tonnes) of methyl bromide in the former USSR and Asia

Year	India	China	USSR	Total
1984	45	45	2268	2358
1985	45	45	2268	2358
1986	68	45	2268	2381
1987	68	45	2268	2381
1988	68	113	2268	2449
1989	91	136	2268	2495
1990	91	136	2268	2495
Totals	476	565	15 876	16 917

1992, Clary 1992). Consequently, in 1992 an international meeting of interested parties took place in Washington, D.C., at which both the atmospheric science of methyl bromide and the technical and economical aspects of phasing out the fumigant were reviewed and reported (Watson *et al* 1992). Also in 1992, in Copenhagen, the *Fourth Meeting of the Parties to the Montreal Protocol on Substances that Deplete the Ozone Layer* (an international agreement that controls the production and consumption of ozone depleting substances) decided to place methyl bromide on the list of ozone-depleting substances. It was also agreed that, from January 1995, the production levels of methyl bromide should be no greater than those of 1991, but with exemptions for the amounts used for quarantine and pre-shipment purposes and for developing countries. The United Nations Environment Programme (UNEP) was requested to set up a Methyl Bromide Technical Options Committee (MBTOC), to investigate current uses of the fumigant and the availability and cost-effectiveness of substitutes for pest control purposes. The report of the MBTOC (UNEP, 1995) was submitted to assist the decisions made by the Parties to the Protocol, at their seventh meeting in November 1995, on what further controls on methyl bromide might be necessary (see Chapter 3).

There is considerable concern in the USA about air pollutants and, under the Clean Air Act, all substances with an ozone-depleting potential (ODP) of > 0.2 (baseline reference ODP of CFC $11 = 1.0$) must be withdrawn within 7 years. In the USA therefore, independently of decisions to be taken under the Montreal Protocol Agreement, and on the basis of the latest internationally accepted ODP value of 0.6, methyl bromide is scheduled to be phased out by the year 2001. Europe has also gone ahead of the Montreal Protocol in agreeing in October 1994 at 25% reduction of methyl bromide supply from 1st January 1998, based on the 1991 production figure, in the EU.

The future for methyl bromide looks uncertain. A great deal of claim and counter-claim is being broadcast worldwide, and a great deal of political, scientific and emotional energy is being directed at this low use, high value commodity chemical. The following chapters, written by some of the major players in the international debate are designed to shed some light on the critical issues.

REFERENCES

Alexeev GV and Kilgore WW, 1983, *Residue Reviews*, **88**, 101.
Baumann H, Heumann KG and Fresenius Z, *Anal. Chem.*, **327**, 186.
Blackburn EGH and Phillips H, 1941, *Biochem. J.*, **35** (5&6), 627.
Blackburn S, Consden R and Phillips H, 1944, *Biochem. J.*, **38** (1), 25.

Bond EJ, 1956, *Canadian Journal of Zoology*, **34**, 405.
Bond EJ, 1984, Food Agric. Org., UN Plants Prod. Prot. Paper, **54**, 71.
Bond JA, Dutcher JS, Medinsky MA, Henderson RF and Birnbaum LS, 1985, *Toxicol. Appl. Pharmacol.*, **78**, 259.
Bridges RG, 1955, *J. Sci. Food Agric.*, **5**, 261.
Chemistry and Industry (19 July 1993) p 528.
Clary P, 1992, *Global Pest. Campaign*, **2** (2), 13.
Clegg KM and Lewis SE, 1953, *J. Sci. Food Agric.*, **11**, 548.
Daft JL, 1992, *J. Assoc. Off. Anal. Chem.*, **75** (4), 701.
Daft JL, 1993, *J. Assoc. Off. Anal. Chem.*, **76** (5), 1083.
Davenport CJ, Ali SF, Miller J, Lipe GW, Morgan KT and Bonnefoi MS, 1992, *Toxicol. Appl. Pharmacol.*, **112**, 120.
Dawson GW, Jennings AL, Drozdowski D and Rider E, 1977, *J. Hazard. Mater.*, **1** (4), 303.
de Francolini J, 1935a, *Revue Path. Veg. Ent. Agric. Fr.*, **22**, 1–8.
de Francolini J, 1935b, *Revue Path. Veg. Ent. Agric. Fr.*, **22**, 9–12.
Djalali-Behzad G, Hussain S, Osterman-Golkar S and Segerbaeck D, 1981, *Mutat. Res.*, **84** (1), 1.
Drew RT, Haber SB and Tice RR, 1984, *Toxicologist*, **4**, 1.
Eustis SL, Haber SB, Drew RT and Yang RSH, 1988, *Appl. Toxicol.*, **11**, 594.
Fairall RJ and Scudamore KA, 1980, *Analyst*, **105**, 251.
Friends of the Earth, 1992, in: *Into the Sunlight, Exposing Methyl Bromide's Threat to the Ozone Layer*, edited by D Straub, (Washington, DC: Friends of the Earth).
Garry VF, Nelson RL, Griffith J and Hoskins M, 1990, *Teratogenis, Carcinog. Mutag.*, **10**, 21–29.
Haber SB, Drew RT, Eustis S and Yang RSH, 1988, *The Toxicologist*, **5**, 130 (Abstract No 518).
Hatch GG, Mamay PD, Ayer ML, Casto BC and Nesnow S, 1983, *Cancer Res.*, **43**, 1945–1950.
Hayes WJ, 1963, Clinical Handbook on Economic Poisons, Washington D.C., U.S. Public Health Service, Publication No. 476.
Heseltine HK, Pearson JD and Wainman HE, 1958, *Chem. Ind. Lond.*, 1287.
Hill EG and Border BSJ, 1953, *Milling*, **121**, pp 488, 490, 492, 494–5.
Honma T, Sudo A, Miyagawa M and Sato M, 1982, *Neurobehav. Toxicol. Teratol.*, **4**, 521.
Honma T, Sudo A, Miyagawa M, Sato M and Hasegawa H, 1983, *Toxicol. Lett.*, **15**, 317.
Ikawa N, Araki A, Nozaki K and Matsushima T, 1986, *Mutat. Res.*, 164–269.
Kaneda M, Hatakenaka N, Teramoto S and Maita K, 1993, *Fd. Chem. Toxic.*, **31** (8), 533.
Kato N, Morinobu S and Ishizu S, 1986, *Indust. Health*, **24**, 87.
Kempton RJ, and Maw GA, 1973, *Ann. Appl. Biol.*, **74**, 91.
Kohn S, 1962, *Mushroom Growers Assoc. Bull.*, **152**, 329.
Kramers PGN, Bissumbhor B and Mout HCA, 1985a, *Mutat. Res.*, **155**, 41–47.
Kramers PGN, Bissumbhor B and Mout HCA, 1985b, *Short-term Bioassays in the Analysis of Complex Environmental Mixtures IV*, (New York: Plenum Press) 65–73.
Lewis SE, 1948, *Nature*, **161**, 692.
Loveday PM and Winteringham FPW, 1951, *Pest. Infest. Res.*, **29**.
McGregor, 1981, Report number 32. Cincinnati, Ohio, National Institute of Occupational Safety and Health, (PB83-130211).
Medinsky MA, Bond JA, Dutcher JS and Birnbaum LS, 1984, *Toxicology*, **32**, 187.

Moriya M, Ohta T, Watanabe K, Miyazawa T, Kata K and Shirasu Y, 1983, *Mutat. Res.*, **116**, 185–216.

Morrissey RE, Schwetz BA, Lamb JCIV, Ross MD, Teague JL and Morris RW, 1988, *Fundam. Appl. Toxicol.*, **11**, 343.

Nishimura M, Umeda M, Ishizu S and Sato M, 1980, *J. Toxicol. Sci.*, **5** (4), 321

Norman KNT, 1991, *Pestic. Sci.*, **33**, 23.

NTP, 1992, NTP TR 385. NIH Publication No 91-2840, CAS0No 74-83-9.

Ong JM, Stewart J, Wen Y and Whong W, 1987, *Environ Mutagen*, **9**, 171–176.

Phillips GL and Bulger JW, 1953, *US Dep Agric, Bur. Ent. plt. Quarant*, E-851.

Powell DF, 1975, *Ann. Biol.*, **81**, 425.

Scudamore KA, 1988, *Analytical Methods for Pesticides and Plant Growth Regulators*, **16**, 207.

Shepard HH and Buzicky AW, 1939, *J. Economic Entomol.*, **32**, 854.

Simmon VF, Kauhauen K and Taudiff RG, 1971, in D Scott, BA Bridges and FH Sobels, eds. *Progress in Genetic Toxicology*. (Amsterdam, Elsevier: North-Holland biomedical Press) pp 249–258.

SORG (U.K. Stratospheric Ozone Review Group), 1990, *Stratospheric Ozone 1990*, U.K. Department of Environment and Meteorological Office.

SORG (U.K. Stratospheric Ozone Review Group), 1991, *Stratospheric Ozone 1991*, U.K. Department of Environment and Meteorological Office.

Starratt AN and Bond EJ, 1981, *Pestic Biochem Physiol*, **15** (3), 275–281.

Thomas DA and Morgan KT, 1988, *CIIT Activities*, **8**, 1/3–7.

Toennies G and Kolb JJ, 1945, *J. Amer. Chem. Soc.*, **67** (5), 849.

Tucker JD, Xu J, Stewart J and Ong T, 1985, *Environ Mutagen* 7: 48 (Abstract).

UNEP, 1994, Report of Methyl Bromide Technical Options Committee for the 1995 assessment by the UNEP Montreal Protocol on substances that deplete the ozone layer (1995).

Watson RT, Albritton DL, Andersen SO and Lee-Bapty S, 1992, *Methyl Bromide: Its Atmospheric Science, Technology and Economics*. Montreal Protocol Assessment Supplement, UNEP, Nairobi, Kenya.

Winteringham FPW, Bridges PM and Hellyer GC, 1995, *Biochem. J.*, **59** (1), 13.

Winteringham FPW, Loveday PM and Harrison A, 1951, *Nature*, **167**, 106.

Winteringham FPW, Hellyer GC and McKay MA, 1958, *Biochem. J.*, **69** (4), 640.

Winteringham FPW and Barnes JM, 1955, *Physiol. Rev.*, **35**, 701.

Winteringham FPW and Harrison A, 1946, *J.C.I.S.*, **65**, 140.

Winteringham FPW, 1955, *J. Sci. Food Agric.*, **5**, 269.

Winteringham FPW, 1977, *Ecotoxicology and Environmental Safety*, **1** (3), 407.

WHO, 1990, *Scientific Assessment of Ozone Depletion: 1989*. World Meteorological Organization Report No. 20, WMO, Geneva.

WMO, 1991, *Scientific Assessment of Ozone Depletion: 1991*. World Meteorological Organization Report No. 25, WMO, Geneva.

Zwart A, 1988, CIVO Report-No. V88-127/27, Zeist, Netherlands, TNO-CIVO Institute, 17 pp.

2

Methyl Bromide in the Atmosphere

JAMES H. BUTLER* and JOSÉ M. RODRIGUEZ[†]
*Climate Monitoring and Diagnostics Laboratory, National Oceanic and Atmospheric Administration, Boulder, Colorado, USA

[†]Atmospheric and Environmental Research, Cambridge, Massachusetts, USA

C.H. Bell, N. Price and B. Chakrabarti: The Methyl Bromide Issue
© 1996 John Wiley & Sons Ltd

1.1   INTRODUCTION

Methyl Bromide (CH$_3$Br) is an atmospheric trace gas of both natural and anthropogenic origin. Presently in the atmosphere at a mole fraction, or volume-mixing ratio, of around 10 parts per trillion (ppt = 10^{-12}), its known sources include oceanic emissions, biomass burning, agricultural application, leaded gasoline, combustion, and structural fumigation. Because of its low mixing ratio and short atmospheric lifetime, little attention had been paid to this gas in the atmosphere, until recently, as scientists began to recognize and understand the chemistry of stratospheric ozone (O$_3$) depletion. In 1974, Molina and Rowland first reported that anthropogenic, chlorinated hydrocarbons could upset the balance of ozone in the stratosphere. This was of particular interest because stratospheric ozone shields the planet from harmful ultraviolet rays. The ecological and health implications of this finding prompted a number of research programs aimed at documenting the distribution and growth of chlorinated organics in the atmosphere. Efforts at understanding the physics and chemistry responsible for maintaining the balance of ozone in the stratosphere also were begun. Shortly after Molina and Rowland's report, Wofsy, McElroy and Yung, (1975) and Yung *et al* (1980) showed that bromine could enhance considerably the reactions that destroy ozone. As data were accumulated and theories were tested, it became increasingly clear that anthropogenic halogens could indeed lower the amount of ozone in the stratosphere. In 1985, Farman, Gardiner and Shanklin first reported unusually low levels of stratospheric ozone over Antarctica in the austral spring. Earlier models suggesting ozone depletion, based upon what was understood of atmospheric chemistry at the time, did not predict such a large drop. Although several theories proposing naturally occurring processes initially were advanced, experimental results from the National Ozone Expedition (NOZE-I) and the Airborne Antarctic Ozone Experiment (AAOE) established the existence of ClO and BrO radicals in sufficient amount to catalyze the observed O_3 destruction. This evidence indicated that about 20–30% of the observed O_3 depletion over Antarctica could be attributed to a synergistic reaction involving both ClO and BrO (DeZafra *et al*, 1989, Anderson *et al*, 1989).

Just before the Farman *et al* (1985) report, an international meeting was initiated in Vienna to study what should be done about possible ozone depletion by man-made compounds. The Farman *et al* (1985) manuscript and subsequent similar findings by other investigators underscored the necessity for some action to be taken. In a reasonably short time, representatives from both developed and developing nations agreed in 1987 to what is now referred to as the Montreal Protocol (UNEP, 1987). This document called for the ultimate elimination of persistent anthropogenic halocarbons, allowing for a phasing out of the compounds as replacements were developed. It was aimed at chlorofluorocarbons (CFC's), which are used as refrigerants, propellants, and solvents, and the bromine-containing, long-lived halons (H1211, H1301, H2402), which are used primarily in fire suppression. More importantly, it called for continuous updates as more data were obtained and as the physics and chemistry of these processes became better understood. At a second meeting in London in 1990, restrictions on the CFC's and halons were tightened and additional compounds were added to the list, two of which, methyl chloroform (CH_3CCl_3) and carbon tetrachloride (CCl_4), were not CFC's or halons. Finally, as chlorine and bromine became more strongly implicated in the depletion of stratospheric ozone, as the enhancement of reaction rates through heterogeneous chemistry was better understood, and as global ozone depletion became more apparent (e.g., Krzyscin 1994, Hood *et al*, 1993, WMO 1994), the Montreal Protocol was further revised with the Copenhagen Amendments of 1992 (UNEP, 1992). These amendments dramatically accelerated the phasing out of CFC's, CH_3CCl_3, and CCl_4 (by 1996) and halons (by 1994), provided a schedule for the phasing out and replacement of hydrochlorofluorocarbons (HCFC's), and targeted a number of additional anthropogenic halogens, including CH_3Br, for ultimate cessation of production and use.

CH_3Br is of concern and included in the agreements because it is the primary carrier of bromine to the stratosphere (Salawitch, Wofsy and

McElroy, 1988, Schauffler *et al* 1993) and because reaction-rate enhancements due to bromine could make it 20–100 times more effective than chlorine in removing ozone from the stratosphere (Albritton and Watson 1992). Thus, 10 ppt of bromine would be the equivalent of 200–1000 ppt of chlorine. (The present-day amount of organic chlorine in the atmosphere is about 3800 ppt, of which only 600 ppt occurs naturally.) Normally, a short atmospheric lifetime, such as that for CH_3Br, reduces the significance of the compound in depleting stratospheric ozone by maintaining a low mixing ratio in the atmosphere. This is reflected in the calculation of its ozone depletion potential, or ODP (Wuebbles 1983, Solomon *et al* 1992, Solomon and Albritton 1992), which has become the fundamental basis for determining whether a compound should be slated for removal. However, because CH_3Br contains bromine, its ODP remains high, even with its short lifetime. In contrast to the CFC's with their long atmospheric lifetimes, the short lifetime of CH_3Br, along with the high efficiency of Br for removal of stratospheric O_3, has made it particularly attractive for elimination. For example, CFC-12, which is around 500 ppt in the atmosphere and represents 1000 ppt of chlorine, has an atmospheric lifetime of around 140 y (Elkins *et al* 1993). Even with complete and immediate cessation of emissions, half of the CFC-12 would still reside in the atmosphere in 100 y; one-fourth, or 250 ppt of chlorine, would still be around in 200 y. By contrast, it has been suggested that an immediate cessation of all anthropogenic CH_3Br emissions might reduce the equivalent chlorine loading from CH_3Br by 30% in 2–4 y, thus providing an almost instantaneous response. The remaining contributions from CH_3Br would be natural.

Problems arise, however, in determining exactly how the atmosphere will respond to a reduction in anthropogenic use. Large uncertainties remain in the budget of atmospheric CH_3Br. The relative contributions of anthropogenic and natural emissions to the total atmospheric budget of CH_3Br are still in question, the distribution of natural sources and sinks is not well understood, recent accumulation in the atmosphere, if any, is not well documented, and the response of nature to reductions or increases in anthropogenic emissions is poorly quantified. The questions presently standing before scientists and policy-makers are, 'What level of information is necessary before regulation is appropriate, with what level of uncertainty should decisions be made regarding any control of this substance, and, finally, how do the benefits of regulation weigh against the costs?' The purpose of this chapter is not to answer these questions, but rather to provide some insight into our current understanding of the physical and chemical processes determining the impact of CH_3Br on stratospheric O_3. We will examine closely the uncertainties associated with these processes, as well as the uncertainties in the overall budget of atmospheric CH_3Br. Our objective is to identify and separate what is known, what is fairly well understood, what is intuitively believed, and what is simply not known in this picture and, in so doing, to lend some quantitative understanding to this issue.

2.1 OVERVIEW OF ATMOSPHERIC CHEMISTRY

2.1.1 COMPOSITION OF THE ATMOSPHERE

The atmosphere is a dynamic system in a constant state of physical and chemical turnover. Although, for convenience, the atmosphere is typically thought of as having various zones and subzones (e.g., troposphere, stratosphere, etc., Figure 2.1) based upon certain physical or chemical boundaries or properties, these are not static regions, nor are their properties tantamount to some permanently assigned values. For example, the amount of O_3 in the lower stratosphere has been reasonably constant over time, but this does not represent a fixed reservoir of O_3 waiting to be depleted. Rather, the concentration of O_3 in any one part of the stratosphere is the result of a balance between production by photolysis of O_2, mainly in the tropics, transport to mid- and high latitudes, and removal through chemical catalysis and photochemistry. The addition or removal of compounds or elements involved in these reactions can alter the mixing ratio of O_3. In the case of Cl or Br, additions would lower the steady-state amounts of O_3. Alterations in the physical conditions, e.g. temperature or transport rates, can also change the balance. A similar argument can be made for virtually every gas in the atmosphere; each is present at some amount in some location, owing to a number of interacting physical, chemical geological, and biological processes.

Although nitrogen (N_2) and oxygen (O_2) presently constitute about 99% of the gas in the earth's atmosphere (dry), with the inert gas argon (Ar) accounting for almost all of the remaining 1%, most of the chemistry of the atmosphere involves the more reactive trace components which make up the remainder (Table 2.1). Gases that are present at parts per million (ppm) through-ppt and even lower levels are instrumental in driving reactions and properties of the earth's atmosphere. To be sure, however, it is O_2 that maintains an overall oxidative environment and provides a source for O_3 molecules and, less directly, the OH radical. Also, although N_2 and O_2 do not combine readily in nature without some external source of energy such as that provided by lightning, internal combustion engines, or microbial organisms, certain oxides of nitrogen do play important roles in maintaining the earth's radiation balance and in regulating the abundance of stratospheric ozone. The potential role of carbon dioxide in maintaining the earth's radiation balance is broadly recognized, as are those of methane

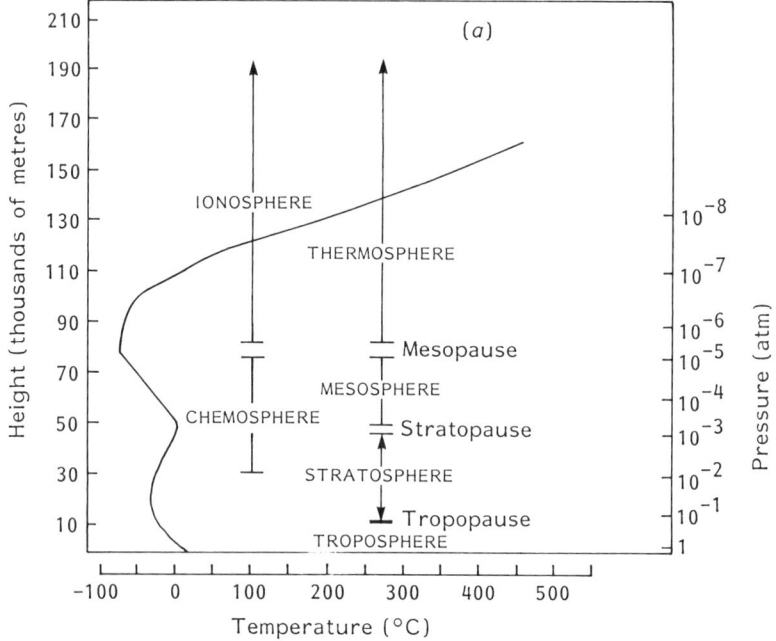

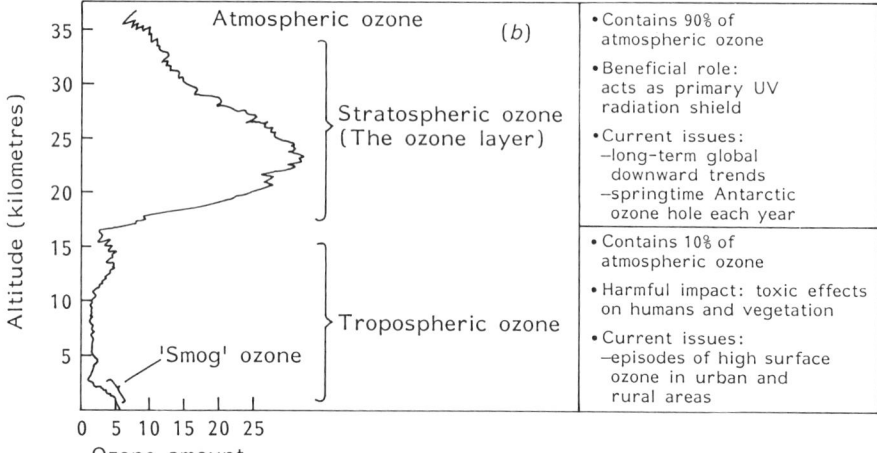

Figure 2.1. (*a*) Vertical structure and temperature of the atmosphere. Demarcations for the various layers are approximate and vary in height with latitude and geography (modified from Cole, 1970). (*b*) Typical vertical profile of ozone through the troposphere and stratosphere (Source: WMO, 1995)

Table 2.1. Properties of some atmospheric gases

Name	Formula	Atmospheric mole fraction	Atmospheric lifetime	Principal sources	Principal Sinks
Nitrogen	N_2	0.78	10^7 y	Geological/primordial;	Biological nitrogen fixation; escape
Oxygen	O_2	0.21	10^4 y	Biological denitrification Photosynthesis	Respiration; oxidation of minerals
Water	H_2O	^{a}0–0.02	3 d	Evaporation; transpiration	Precipitation
Argon	Ar	0.01		Radioactive decay; primordial	Escape
Carbon dioxide	CO_2	3.5×10^4	^{b}22 y	Respiration; fossil fuel burning	Photosynthesis; oxidation of $CaCO_3$; burial of organic matter
Hydrogen	H_2	5×10^{-5}	4 y	Oxidation of hydrocarbons; oceans; soils	Tropospheric oxidation; escape
Helium	He	5×10^{-6}	10^8 y	Radiogenic; primordial	Escape
Ozone	O_3	10^{-6}		Photochemical	Free-radical chemistry; photochemistry
Methane	CH_4	1.7×10^{-6}	10 y	Biogenic; fossil fuel; fires	Tropospheric oxidation
Nitrous oxide	N_2O	3.1×10^{-8}	150 y	Biogenic from oceans, soils; combustion	Stratospheric photochemistry
Carbon monoxide	CO	$2.5\text{–}25 \times 10^{-8}$	2 mol	Oxidation of CH_4; fires; fossil fuels	Tropospheric oxidation
Methyl chloride	CH_3Cl	6×10^{-10}	1–2 y	Ocean; fires; algae	Tropospheric oxidation
CFC-12	CCl_2F_2	5.2×13^{-10}	140 y	Anthropogenic refrigerants, foams	Stratospheric photochemistry
CFC-11	CCl_3F	2.7×10^{-10}	55 y	Anthropogenic refrigerants, foams	Stratospheric photochemistry
Methyl chloroform	CH_3CCl_3	1.2×10^{-10}	6 y	Anthropogenic; solvent	Tropospheric, oceanic, stratospheric removal

continued overleaf

Table 2.1. (*continued*)

Name	Formula	Atmospheric mole fraction	Atmospheric lifetime	Principal sources	Principal Sinks
Sulfur dioxide	SO_2	1.1×10^{-10}	< 2 d	Fossil fuels; H_2S oxidation	Deposition, oxidation
Carbon tetrachloride	CCl_4	1×10^{-10}	50 y	Anthropogenic; solvent, chemical feedstock	Stratospheric photochemistry, oceanic removal
CFC-113	CCl_2FCClF_2	9×10^{-11}	90 y	Anthropogenic; solvent	Stratospheric photochemistry
Methyl bromide	CH_3Br	1×10^{-11}	1 y	Oceanic, fires, fumigation, auto emissions	Tropospheric, oceanic, soil removal
Dibromomethane	CH_2Br_2	2×10^{-12}	< 1 y	Oceanic; industrial	Tropospheric oxidation
Bromoform	CH_2Br_2	$1-2 \times 10^{-12}$	< 1 mo	Oceanic	Trophospheric oxidation
Methyl iodide	CH_3I	$1-2 \times 10^{-12}$	1 wk	Oceanic	Trophospheric oxidation
Radon	Rn	10^{-18}	5 d	Radiogenic	Radioactive decay

[a]Mixing ratios for all gases except water are for a dry atmosphere.
[b]Lifetime for CO_2 is based mainly upon photosynthesis and respiration. It does not represent the time required to remove fossil-fuel CO_2 from the atmosphere. This 'geologic' lifetime of CO_2 is driven by sedimentation and burial and is on the order of 1000's of years.

(CH_4), nitrous oxide (N_2O), and certain CFC's and HCFC's. Hydrocarbons are involved in production of tropospheric ozone, a principle component of photochemical smog. Water is involved in the production of aerosols, the OH radical, O_3, and in numerous other reactions; the transition of water between liquid and gaseous states profoundly affects the physics of the atmosphere.

2.1.2 RADIATION AND THE EARTH'S ENERGY BALANCE

As with chemistry, the temperature of the earth's atmosphere is also a steady-state phenomenon. The total amount of energy arriving at and leaving the earth must be in balance or else the temperature of the earth will rise or fall to accommodate the imbalance. The solar spectrum comprises a wide range of wavelengths, only a small portion of which constitutes what we call visible light. Much of the visible light coming from the sun passes through the atmosphere with little interaction from its component gases. Some of this energy is reflected directly back to space and some light is absorbed directly by component gases at different altitudes. However, much of the radiation is absorbed by the earth and re-radiated in the lower energy, infrared portion of the spectrum. Gases in the atmosphere, mainly water and CO_2, capture some of this long-wave back-radiation, which, in turn causes them to heat; heat is then passed through molecular collision to other gases in the atmosphere. The temperature of the atmosphere, then, is governed overall by the solar input of radiation and the proportion of infrared-absorbing gases present.

Man has been able to alter the radiation balance of the earth in a number of ways. The burning of fossil fuels has caused an increase in the amount of CO_2 in the atmosphere, as has deforestation. Warming of the surface ocean, should that occur, could also release more CO_2 to the atmosphere and atmospheric warming itself could increase rates of respiration, which, in turn would release more CO_2 to the atmosphere. Conversely, plant growth is also enhanced somewhat in a CO_2-rich environment (e.g., Oechel et al 1994, Harmon, Ferrell and Franklin 1990). The feedbacks and implications of the CO_2 cycle and its relation to potential global warming have been the focus of considerable study in the past few decades. We will not delve into this here except to note that the increase in atmospheric CO_2 accounts for about half of the radiative forcing in the earth's atmosphere. The remainder is contributed by CH_4, N_2O, halocarbons, and stratospheric and tropospheric ozone. Gases such as N_2O, CH_4, and certain halocarbons absorb radiation in the $7-14$ μm window where CO_2 and H_2O are less effective. Unlike CO_2, radiative forcing by these gases is linear with increasing amounts in the atmosphere, because absorption by these molecules in this window does not approach saturation. Also, many of them are much more efficient absorbers of radiation in the earth's atmosphere on a per-molecule basis. Thus, increases in the levels of certain trace gases, present at mixing ratios of a few hundred to over a million times less than that of CO_2, can collectively rival the radiative forcing due to increases in CO_2.

Methyl bromide is not a significant greenhouse gas (Anastasi *et al* 1994). Compared to the CFC's, N_2O, and some of the HCFC's that absorb strongly in the 7–14 μm window, CH_3Br has a short lifetime and its absorption in that band is weak. It may contribute indirectly to global warming through ozone depletion, but this is believed to be small and perhaps insignificant. Ozone absorbs both ultraviolet (short wavelength, higher energy) and infrared (long wavelength, lower energy) radiation. The amount of O_3 in the troposphere varies over both space and time, owing to its high reactivity and relatively short atmospheric lifetime, but it nevertheless serves as a 'greenhouse' gas with a small contribution to radiative forcing. However, O_3 in the stratosphere also provides a mechanism for offsetting the radiative forcing due to CFC's, halons, and other reactive halogens. The amount of ozone in the stratosphere has decreased with increasing CFC's in the atmosphere. Thus, some of the warming that might result from infrared absorption by CFC's is offset in part by the lowered amounts of ozone. Unfortunately, quantifying this effect is complex, as it depends strongly upon the vertical profile of O_3 through the atmosphere and upon the distribution of O_3 deficits.

2.1.3   OZONE IN THE STRATOSPHERE

About 90% of atmospheric O_3 is located in the stratosphere, with the bulk of it lying between 19 and 23 km (Figure 2.1(b)). Temperature rises with increasing height through the stratosphere, mainly as a result of the absorption of radiation by O_3. The amount of O_3 in the stratosphere and its vertical distribution are held in balance by the splitting of O_2 by ultraviolet radiation, the combining of free O with O_2 to form O_3, and a number of catalytic reactions that return O_3 to O_2 (Figure 2.2). The ultimate distribution of stratospheric O_3 depends upon the distribution of solar radiation, the relative reaction rates and their temperature dependencies, the concentrations of reactants, and, at the base of all this, the temperature distribution and horizontal and vertical transport rates within the stratosphere.

CFC's, other organic halogens, N_2O, and even water do not react directly with O_3, but rather provide a supply of atoms for reactions that deplete O_3. Only the inorganic, reactive species derived from photolysis of these organic or otherwise stable compounds react with O_3, but the reactive forms rely upon a steady flux of the more stable gases from the troposphere. Before the somewhat refractory, anthropogenic halogens were emitted to the

Inorganic Bromine Cycling

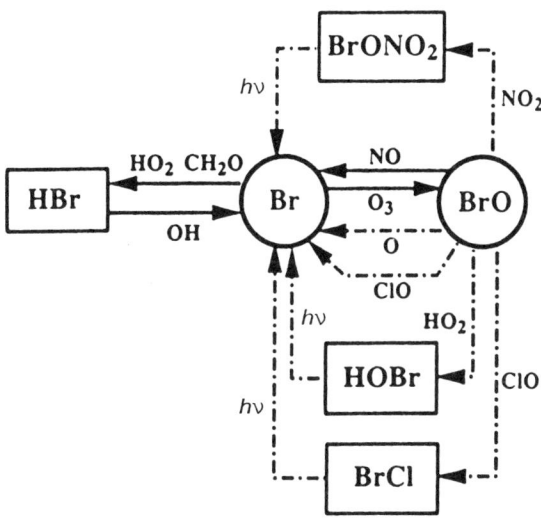

Figure 2.2. Cycle of gas-phase bromine in the stratosphere

atmosphere, reactions with oxides of nitrogen and with HO_x predominated in the destruction of O_3 (McElroy, Wofsy and Yung 1977, Wennberg *et al* 1994). NO and NO_2 (NO_x) react with O_3 in an efficient catalytic cycle, as does HO_x, but, because of their very short lifetimes in the troposphere, are not delivered in significant quantities to the stratosphere. N_2O, a compound that is produced mainly by microbial processes in the soils and oceans, is considerably more stable in the troposphere, with an atmospheric lifetime of 100–150 y. Water, which is very abundant in the lower troposphere, decreases substantially with altitude. Even so, it remains a major component of stratospheric air. The only significant mechanisms for the removal of atmospheric N_2O is photolysis by UV radiation in the stratosphere, followed by conversion to NO, which initiates the cycle of O_3 destruction. The same is true for HO_x, which is derived from water in the stratosphere. (Historically, it was believed that oxides of nitrogen predominated in pre-industrial O_3 removal from the stratosphere, but recent evidence (Wennberg *et al* 1994) suggests that the catalytic cycle involving O_3 with OH and HO_2 accounts for 30–50% of the total photochemical loss in the lower stratosphere at low latitudes.)

A similar cycle also can be envisaged for Cl and ClO, which are delivered to the stratosphere by organic halogens. Before CFC's were emitted to the atmosphere, the amount of organic chlorine in the troposphere was around 600 ppt, predominantly as CH_3Cl, which is produced mainly in the oceans and, to a lesser extent, during biomass burning. Today, the total amount of organic chlorine in the troposphere amounts to 3800 ppt, most of which is present as CFC's. Reactions involving chlorine now destroy about as much ozone as those involving oxides of nitrogen. In addition, anthropogenic activities apparently have elevated atmospheric N_2O levels by about 5–10% over the past century, thus making more NO_x available in the stratosphere.

The role of bromine in destroying O_3 in the atmosphere was first elucidated by Wofsy, McElroy and Yung (1975). Yung *et al* (1980) and later McElroy *et al* (1986) pointed out the importance of coupled reactions involving both bromine and chlorine. (See section 7.1, this chapter.) Yung *et al* (1980) suggested that reactions involving bromine could enhance the depletion of O_3 beyond that caused by chlorine reactions alone by 10–20% and that the reaction involving BrO and ClO together was more important than reactions involving only bromine. In an extensive investigation of the Antarctic polar vortex, Anderson *et al* (1989) concluded that the catalytic cycle that included reaction of BrO with ClO could account for 20% of the depletion of O_3 in this region. Reaction rates listed in Anderson *et al* (1989) indicate that the cycle involving these two reactants was, on a per atom basis for Br and in today's polar atmosphere, 35–65 times more efficient at removing O_3 than reactions involving just chlorine. Thus, it is clear from these investigations that inorganic bromine has a large effect upon the depletion of stratospheric ozone, particularly at high latitudes, even though it is present in much smaller amounts than inorganic chlorine. Part of this large effect results from a larger partitioning of the inorganic bromine into reactive radicals, and part of it has to do with the kinetics of BrO + ClO.

2.1.4 THE CONCEPT OF OZONE DEPLETION POTENTIAL (ODP)

Policymakers have found it useful to adopt a single index to classify substances with respect to their effectiveness in catalyzing stratospheric ozone removal. A single, time-independent index, the so-called steady-state ozone depletion potential (ODP), has been introduced to quantify the steady-state depletion of ozone per unit mass emission of a given trace species, relative to the same steady-state ozone reduction per unit mass emission of CCl_3F (CFC-11). These steady-state depletions are those predicted values that the atmosphere would exhibit if the adopted emissions and background atmospheric composition, temperature and circulation were maintained constant indefinitely.

The concept of ozone depletion potential (ODP) has been discussed extensively in the literature (Wuebbles 1983, WMO 1990, 1992, Solomon *et al* 1992, Solomon and Albritton 1992). Usually, these values are calculated by multi-dimensional models which incorporate the latest understanding of stratospheric chemistry and dynamics. We must keep in mind, however, the limitations inherent in both the definition of ODP and the methods used to derive it. First, the steady-state ODP as defined, represents a global average. It does not contain information as to the impact of a specific substance on a given geographical region or season (for example, Antarctica in the spring). Second, the steady-state ODP is independent of the actual emission of a compound, the relative contributions of anthropogenic vs. natural sources of the compound, and the life-cycle of this substance. A large ozone-depletion potential for a given substance does not necessarily mean that it is actually causing a large ozone depletion, since this also will depend on the actual emissions of the compound in the atmosphere and how they compare to natural sources. This consideration is particularly relevant for methyl bromide, and should be kept in mind when one considers the combined effect of actual emissions (e.g., Ko, Sze and Prather 1994). Third, the steady-state ODP is time-independent. Time-dependent calculations show that the relative short-term impact of certain compounds on ozone can be much greater than their steady-state reductions (Solomon *et al* 1992). However, short-term reductions of ozone are usually very small in absolute values, and time-dependent ODPs have not been used extensively in regulatory policy. Fourth, and most important, the ODP calculated by a given model is only as good as the model is in describing the atmosphere. Ozone-depletion potentials have been calculated with two-dimensional (altitude–latitude) models of the stratosphere (WMO 1990). These models have been shown to give an adequate description of existing observations (WMO 1990). At the same time, they have not been successful in completely accounting for ozone trends observed during the last decades (WMO 1989), nor have they reproduced some of the barriers to transport of species observed at the subtropics or at the polar vortices. Atmospheric observations of brominated and chlorinated compounds can be incorporated into estimating a semi-empirical ODP, at least at the locations where measurements exist (Solomon *et al* 1992). However, extrapolation of these values to the global atmosphere is fraught with uncertainties.

The above considerations indicate the need not only to evaluate ozone depletion potentials, but also to estimate their overall uncertainties due to uncertainties in the different processes determining this index. In this regard, it is useful to apply an approximate expression for ODP, motivated by its definition. For CH_3Br, this would be

$$ODP_{CH_3Br} = \left[\frac{1}{3}\frac{MW_{CFC-11}}{MW_{CH_3Br}}\frac{\tau_{CH_3Br}}{\tau_{CFC11}}\beta\right]\left[\left\langle\frac{F_{CH3Br}(z)}{F_{CFC11}(z)}\alpha\right\rangle\right] \tag{1}$$

$$ODP_{CH3Br} = [BLP][BEF] \tag{2}$$

where MW_{CH_3Br} and MW_{CFC-11} denote the molecular weight of methyl bromide and CFC-11, $F_{CH_3Br}(z)/F_{CFC-11}(z)$ represent the bromine release from CH_3Br relative to that of CFC-11 in the stratosphere, α denotes the efficiency of the released bromine in catalytic removal of ozone, relative to chlorine; β is the decrease in the mixing ratio of CH_3Br at the tropical tropopause, relative to the mixing ratio at the earth's surface; and $\langle \rangle$ denotes spatial averaging of the quantity with the appropriate weighting. The ratio of $1/3$ is the ratio of the number of halogen atoms in CH_3Br to that in CFC-11.

The term in the first bracket represents the amount of bromine delivered to the stratosphere by methyl bromide relative to CFC-11, per unit mass (e.g., kg) emission. This is the so-called Bromine Loading Potential (BLP). The second term, the Bromine Efficiency Factor (BEF), denotes the amount of stratospheric ozone removed per unit mass of methyl bromide delivered to the stratosphere, relative to CFC-11. Bromine-loading potentials depend upon tropospheric removal processes, whereas the BEFs are determined by the chemistry of methyl bromide and inorganic bromine in the stratosphere.

The time constants τ_{CFC-11} and τ_{CH_3Br} relate the change in steady-state atmospheric burden M_g to a change in emission F_s

$$\Delta M_g = \frac{\Delta F_s}{\tau_g} \tag{3}$$

These time constants, or lifetimes, can be obtained by considering all removal processes for the species in question, both within the atmosphere and at the surface. Alternatively, they can be estimated from atmospheric growth rates and burdens, and source strength estimate of the gases in question.

2.1.5   BUDGETS AND LIFETIMES OF ATMOSPHERIC GASES

The amount of a gas in the atmosphere is a direct function of the magnitude of its sources and sinks. The atmospheric lifetime of a given gas can be estimated from computations involving its sources, its mass in the atmosphere, and its atmospheric growth rate, or from computations involving its sinks and its mass in the atmosphere. In theory, the two approaches should yield the same result. In general, we have available box models, one dimensional (1-D) models, 2-D models, and 3-D models to perform these calculations. Ultimately, however, data are necessary to verify the model predictions and to confirm the processes invoked, if the models are to be useful learning tools. The most simple case is the one-box model with limited inward and outward fluxes. CFC's provide a good example of this approach (Figure 2.3(a)). In this example, there is one inward flux (anthropogenic emissions) and one outward flux (photochemical removal in

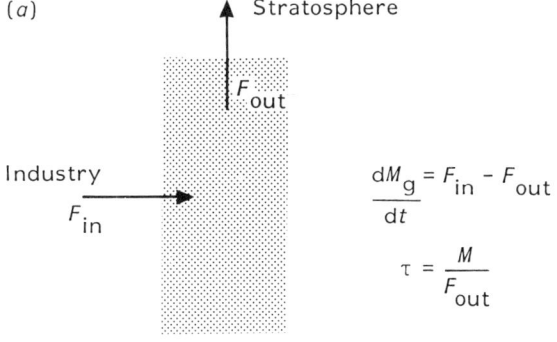

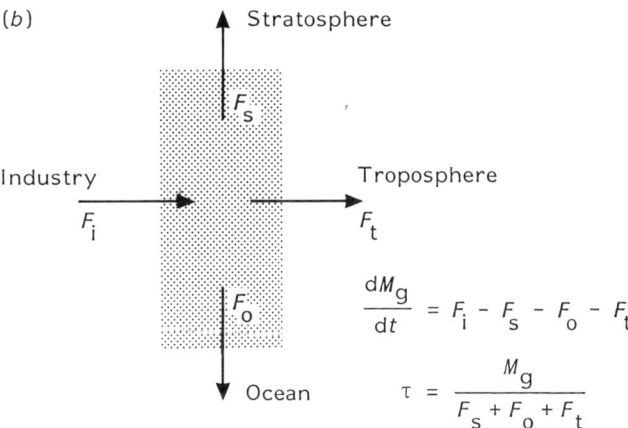

Figure 2.3. (a) Simple, one-box model for long-lived atmospheric gas (e.g., chlorofluorcarbons) where the only significant sink is loss in the stratosphere. The atmospheric lifetime is estimated from the mass of the gas in the atmosphere (M) and the outward flux of the gas (F_{out}). (b) Same model, but for an atmospheric gas with sinks in the stratosphere, troposphere, and ocean. The atmospheric lifetime of the gas is estimated from its mass in the atmosphere and the sum of the inward or outward fluxes

the stratosphere). The resulting differential equation supporting this is straightforward:

$$\frac{dM_g}{dt} = F_{in} - F_{out} \qquad (4)$$

where M_g is the mass of gas in the atmosphere, and F_{in} and F_{out} are the inward and outward fluxes of the gas. (Flux, as defined here can represent physical or chemical removal of the gas from the atmosphere.) For compounds like the CFC's, which are entirely anthropogenic and are removed by way of simple photochemistry, the assumptions of zero- and first-order kinetics for the inward and outward fluxes are not unreasonable, yielding,

$$\frac{dM_g}{dt} = F_{in} - kM_g = F_{in} - \frac{1}{\tau}M_g \tag{5}$$

where k is a pseudo-first-order rate constant for the removal of the CFC from the atmosphere. By definition, k is the inverse lifetime of the CFC in the atmosphere $(1/\tau)$. There are essentially four variables or parameters in this equation—the atmospheric growth rate of the gas (dM_g/dt), its emissions (F_{in}), the rate constant (k) or inverse-lifetime (τ), and the amount of the gas in the atmosphere (M_g). If three of these are known or can be estimated within reason, then it is possible to determine the fourth. For CFC's and halons the growth rates, amounts, and emission rate are known within reason, so it is possible to calculate atmospheric lifetimes from these variables. However, because the atmosphere in truth is not a simple box and because there are uncertainties in the variables, particularly emission estimates, some uncertainty must be assigned to the calculated value. It is possible to place further constraints upon the system, for example the observed interhemispheric difference, by invoking a two-box model (Section 6.1.1). Although this can put more certainty into the calculations, it requires some understanding of the distribution of emissions and losses between the hemispheres, which, depending upon the gas, have their own uncertainties.

In many cases, the atmosphere cannot be considered an entity independent of other influences and reservoirs. For example, methyl chloroform (CH_3CCl_3) is destroyed in the troposphere by reaction with OH and by photolysis in the stratosphere (Talukdar *et al* 1992). It also hydrolyzes in seawater, which, in turn, affects the atmospheric burden and growth rate of the gas (Gerkens and Franklin 1989, Jeffers *et al* 1989, Butler *et al* 1991). In this instance, a simple model can still apply, but with additional sinks incorporated (Figure 2.3(b)). The differential equation describing this system then becomes

$$\frac{dM_g}{dt} = F_{in} - F_s - F_t - F_o \tag{6}$$

where F_s, F_t, and F_o represent losses to reaction in the stratosphere, troposphere and ocean. Because CH_3CCl_3 losses in the stratosphere and troposphere can be treated as first-order processes and because there are no sources of CH_3CCl_3 in the ocean, Eq. (6) can be rewritten as

$$\frac{dM_g}{dt} = F_{in} - (k_s + k_t)M_g + A_o z \frac{K_{gw}}{H_g}(P_{gw} - P_{ga}) \qquad (7)$$

where A_o is the area of the ocean, z is the depth of the surface mixed layer where hydrolysis is significant, K_{gw} is a coefficient governing the transfer of the gas across the air–sea interface (see Section 5.1.2.2), H_g is a solubility term for the gas, and p_{gw} and p_{ga} are the partial pressure of the gas in water and air. Note that, because the gas is destroyed in the ocean, p_{gw} will be less than p_{ga}, hence the last term in Eq. (7) will be negative.*

In general, similar cases can be made from other boundaries to the atmosphere, such as soils and plant canopies, so long as there is some evidence for these as sources or sinks of the gas in question. We shall see later in this chapter that CH_3Br requires even further treatment because, for it, the ocean acts as both a source and a sink, these processes occurring simultaneously in any parcel of water.

3.1 DISTRIBUTION AND GROWTH OF CH_3Br IN THE ATMOSPHERE

Junge (1957) first reported that about half of the chlorine in air might be in gaseous form. Later, Duce Winchester and Van Nahl (1965), in analyzing 200 samples of rainwater, aerosol, and air from Hawaii by neutron activation analysis, confirmed Junge's findings and further suggested that half of the bromine and an ever greater fraction of the iodine in the atmosphere were gaseous. In 1975, Lovelock, reporting on air and water measurements from the English coast, suggested that CH_3Br might be emitted in large amounts from coastal waters. He also suggested that CH_3Br might be distributed ubiquitously in the atmosphere. This was followed by the work of Singh (1977) and Singh et al (1977, 1979, 1983, a, b), who reported a wide range of values for CH_3Br in the atmosphere. CH_3Br was reported as around 5 ppt in the remote atmosphere, 16–23 ppt in the coastal marine atmosphere, and over 100 ppt in areas that were influenced by urban, coastal, and agricultural activities. Data from Cicerone, Heidt and Pollock (1988), collected monthly at five remote locations over three years, suggested a global mean of 10.4 ppt for 1985–1987. Although the authors observed inter-annual cycles in the atmospheric mixing ratio, no discernible

---

*To be sure, Eq. (7) is somewhat oversimplified for making a global estimate. K_{gw} and H_g both vary with sea surface temperature and z varies with latitude. In application, either these values should be weighted in this equation according to their distributions or the equation should be solved in finite increments corresponding to the spatial distributions of these properties. We will not delve into that here, but simply wish to make the point that properties of the ocean must be considered in the atmospheric budget of this gas.

trend over time was evident. The longest data set on atmospheric CH_3Br reported to date is that of Khalil, Rasmussen and Gunawardena (1993), comprising nine years of data (1983–1992) collected from five remote locations spanning both hemispheres, five additional years of data (1978–1983) from one location in the mid-NH, and an additional two years of data from the South Pole. These data indicate that CH_3Br over the past decade has increased at 0.15 ± 0.1 (1 s.d.) ppt y^{-1} and that the mean global mixing ratio of CH_3Br in the remote troposphere has ranged from 9 to 10.5 ppt over the past decade. The most recent global atmospheric data, collected in both hemispheres during a research cruise in early 1994, suggest a global mean of 9.8 ± 0.6 ppt (Lobert *et al* 1995).

There is still some question as to how much the atmospheric burden of CH_3Br is increasing, or even if it is increasing at all. The growth rate obtained from Khalil *et al* (1993) is small and has a sizable uncertainty. Unfortunately, no other data are available for the same period to confirm these findings. The only other extended data set, that from Cicerone, Heidt and Pollack (1988), covered only three years. As such, it is not sufficiently long as to allow the measurement or even detection of such a small growth rate, especially in light of the large inter-annual variations that were noted at these sites. Unfortunately, it is not possible to deduce a reliable growth rate by comparing results from different studies over time, because no mechanism was in place for intercalibration of measurements. The only assumption that can be made is that data within each study are likely to be internally consistent.

From data collected on separate research cruises, Singh *et al* (1983a) and Penkett *et al* (1985) were the first to report higher amounts of CH_3Br in the northern hemisphere (NH) than in the southern hemisphere (SH), with an inter-hemispheric ratio (IHR = NH/SH) of CH_3Br of 1.4–1.5. Data from Cicerone *et al* (1988), collected from fixed, remote locations, indicated a much smaller difference, with a mean IHR of 1.10 ± 0.05 for 1985–1987, but those of Khalil *et al* (1993), also obtained at fixed, remote sites, show a mean IHR of 1.38 ± 0.06 for over a decade. Lobert *et al* (1995) reported a one-time-only IHR of 1.31 ± 0.08 ppt for early 1994. An IHR different than 1.0 implies an imbalance among sources and sinks in the two hemispheres. These reported values, although differing somewhat, all imply a greater source strength in the NH, a stronger net sink in the SH, or some combination of the two. From a two-box model similar to that used to deduce relative emissions of N_2O to the atmosphere (Butler *et al* 1989, Cicerone 1989), Lobert *et al* (1995) suggested that the sum of sources and sinks in the NH should be about twice that in the SH in order to maintain the observed IHR.

In summary, from our present-day understanding of the distribution and growth of CH_3Br in the atmosphere, its mean mixing ratio in the atmosphere is around 10 ppt and the most probable inter-hemispheric ratio is 1.3, although the IHR could vary seasonally or inter-annually. At any one location, the mixing ratio of CH_3Br can be expected to vary by 1–2 ppt over a year or between years. The most probable growth rate for atmospheric CH_3Br over the past decade lies between 0 and 0.4 ppt y^{-1} (0–4% y^{-1}).

4.1 SOURCES OF ATMOSPHERIC CH₃Br

Unlike CFC's and halons, the origins of which are entirely anthropogenic, CH_3Br is emitted into the atmosphere from a number of natural and anthropogenic sources. These sources can be classified as industrial, agricultural, pyrrhic (biomass burning), and oceanic. Although evidence is limited, the possibility of sources other than those listed here cannot be ruled out at this time. For the purpose of this discussion, industrial and agricultural emissions will be merged, as they both involve direct emission of the anthropogenically produced gas. This distinguishes them from biomass-burning emissions, which occur naturally, but which have been enhanced to a considerable degree by man's activities. It also distinguishes them from automobile emissions, which are linked industrially to a different brominated compound.

4.1.1 EMISSION OF INDUSTRIALLY-PRODUCED CH₃Br

4.1.1.1 Direct Production and Subsequent Emission of CH₃Br

Global, industrial production of CH_3Br in 1992 was around 76 Gg, representing an annual average increase of 3.7 Gg y^{-1} from 1984 to 1991 (UNEP 1994; see also Chapter 1). In accordance with the Copenhagen amendments to the Montreal Protocol, industrial production was subsequently frozen at 1991 levels (UNEP, 1992). Of the total production, about 73% is used in soil applications, 13% for fumigation of durables such as grain, nuts, and raw timber, 8% for perishables, 3% as feedstock for chemical synthesis, and 3% for fumigation of structures and vehicles. Emissions of industrially produced CH_3Br, however, are not equivalent to production, because of chemical and biochemical losses during use. Historically, the estimation of losses, particularly those associated with application to soils, has been difficult owing to high variability in loss rates. Rolston and Glauz (1982), using soil measurements and theoretical calculations, indicated that most of the CH_3Br used in soil applications escaped eventually to the atmosphere. However, initial estimates of degradation in soils following fumigation suggested that only 30–60% of the CH_3Br used in soil treatment makes it to the atmosphere (Singh and Kanakidou 1993, Albritton and Watson 1992). Recent, carefully conducted studies of CH_3Br loss in soils following application, show that the amount escaping to the atmosphere depends strongly upon soil pH, organic content, moisture, injection depth and injection method (Yagi *et al* 1993, 1995). These studies, showing relative emissions of 34% and 87% from two markedly different fields, support the earlier estimates of degradation in the soils, but also underscore the high variability in this flux.

The relative amount of CH_3Br emitted to the atmosphere from fumigation of durables, defined as dried agricultural products, is also an uncertain number. However, this has only a small impact on the budget estimate because the total amount of CH_3Br used in this process is much smaller than that for soils. Uncertainty in the atmospheric flux from fumigation of durables only amounts to $4.4\,Ggy^{-1}$, whereas that for soil applications is almost $31\,Ggy^{-1}$. Similarly, the combined uncertainty for atmospheric emissions of CH_3Br associated with treatment of perishables and fumigation of structures and vehicles is also small, amounting to less than $1\,Ggy^{-1}$. Thus, it is clear that most of the uncertainty in emissions of industrially produced CH_3Br derives from soil applications. The observed variability does not result so much from difficulty in obtaining reliable measurements as it does from the variations in soil types and properties and in methods of application. Extrapolation of results from studies on individual fields to a larger scale will remain fraught with potential error, until the relative degradation rates are linked quantitatively to specific soil properties or specific methods of application.

4.1.1.2 Indirect Emission of CH_3Br From Leaded Gasoline

Ethylene dibromide $(C_2H_4Br_2)$ is added along with tetraethyl lead to 'leaded' gasoline. During combustion, much of the $C_2H_4Br_2$ is converted to

Table 2.2. 1992 Emissions of Industrially Produced CH_3Br (adapted from UNEP, 1994)

Use	Production		Emissions	
	Amount Gg	Relative to total (%)	Amount Gg	Relative to production (%)
Chemical feedstock	2.4	3%	n/a	n/a
Soil fumigation	55.6	73%	16.7–47.3	30–85%
Durable disinfestation	9.5	13%	4.8–8.4	51–88%
Perishable disinfestation	6.3	8%	5.4–6.0	85–95%
Structural disinfestation	2.2	3%	2.0–2.1	90–95%
TOTAL	76.0	100%	28.9–63.8	38–84%

CH_3Br. Baumann and Heumann (1989) estimated that 22–44% of the $C_2H_4Br_2$ in gasoline is emitted in organic form, 62–82% of which is CH_3Br. Thus, roughly 14–36% of the $C_2H_4Br_2$ in gasoline is emitted as CH_3Br in the exhaust fumes. One estimate of global emissions from this source suggests that 7.5–22 Gg of CH_3Br are emitted annually (Penkett *et al* 1995). Another estimate, however, indicates that only 0.01–0.03 Gg of CH_3Br are emitted in the United States. Extrapolated globally, this would amount to 0.5–1.5 $Gg\,y^{-1}$ from this source. The poor agreement between these studies is not currently resolved. According to Penkett *et al* 1995, this discrepancy is traceable mainly to estimates of the amounts of $C_2H_4Br_2$ used in the United States.

There is some need to obtain a better understanding of present and past emissions of CH_3Br from internal combustion engines, as it impinges directly upon interpretations of the atmospheric growth rate, or lack of a growth rate, of CH_3Br. Leaded gasoline use and, presumably, emissions of CH_3Br from this source in the United States have declined dramatically in the past two decades (Penkett *et al* 1995). At the same time, use of CH_3Br in agriculture has increased. Such changes in emission rates would tend to compensate one another, but providing accurate, quantitative estimates of emissions is difficult at this time. For example, from reported increases in CH_3Br production, the atmospheric growth rate due solely to anthropogenic emissions should have been on the order of 3–5% y^{-1} during the past 15 years or so. Data from Khalil *et al* (1993) do show a slightly increasing growth rate after 1988 ($0.26 \pm 0.15\,ppt\,y^{-1}$, i.e., about 2.5% y^{-1}), but almost no growth at all from 1978–1988. These discrepancies suggest a more complicated picture for CH_3Br in the atmosphere, one that involves not only variability in emission patterns, but in the cycling of CH_3Br in nature as well.

4.1.2 EMISSION FROM BIOMASS BURNING

The combustion of vegetation has considerable influence on the atmosphere (e.g., Crutzen and Goldammer 1993, Levine 1991). Whether the source material is firewood, trees, brush, or grass, growing in tropical or temperate forests or savannas, or whether it is agricultural waste, biomass burning contributes dramatically to the trace gas composition and chemistry of the atmosphere (Crutzen *et al* 1979, Crutzen and Andreae 1990). It presently is believed that the burning of biomass produces 2–5 petagrams (peta- or $P = 10^{15}$) of carbon from CO_2 annually. Although wildfires have always existed and may even have been more widespread in the past than they are today, recent estimates indicate that most of the biomass burning that occurs today is the direct result of fires set intentionally by man. Crutzen *et al* (1979) estimated that 2–5% of the biomass burned annually resulted from wildfires. The remainder was mainly due to deforestation, shifting agriculture, prescribed fires in temperate forests, burning of firewood and agricultural wastes—all fires set intentionally by man. Most of biomass burning occurs in the tropics and subtropics (e.g., Levine, 1990), making it difficult to estimate the contribution of biomass burning to the inter-hemispheric differences in gas composition. Estimates have included both a slightly dominant NH source and a slightly dominant SH source. The most recent study suggests that 55–60% of biomass burning occurs in the southern hemisphere (Hao and Liu 1994).

Halogens such as chlorine and bromine are not part of plant tissue, but they are present in all plants as dissolved electrolytes. Burning of plants allows the halogens to combine with organic matter, thus releasing organic halogens in the smoke. Fires have been known to emit a variety of gases, e.g., CO_2, CO, hydrocarbons, NO_x, and sulphur compounds, for some time. Crutzen *et al* (1979) first identified CH_3Cl as a component of smoke plumes, estimating a range in the ratio of CH_3Cl emitted to CO_2 emitted ($\Delta CH_3Cl/\Delta CO_2$) of 10^{-5} to 10^{-6}. Watson, Lovelock and Stedman (1980) further noted that CH_3Cl was probably produced during the smoldering of plant tissue, as it would decompose at temperatures above 300 °C. Lobert *et al* (1991) presented data showing the production of CH_3Cl during the smoldering phase, but not so much during the flaming stage, of burn experiments. The authors suggested that about 1.5 Tg of CH_3Cl was emitted annually to the atmosphere from fires alone. This is about three times less than the amount of CH_3Cl destroyed annually by tropospheric OH, which indicates that biomass burning could contribute 30–40% of the atmospheric budget of CH_3Cl.

Although several investigators have pointed out the possibility of CH_3Br emission from biomass burning (e.g., Khalil *et al* 1993, Singh and Kanakidou 1993, Albritton and Watson 1992), it was only recently that data have been made available. Manö and Andreae (1994) reported on data collected from chaparral, savanna, and boreal forest fires, as well as from controlled fires in their laboratory. Their results indicate that $\Delta CH_3Br/\Delta CO_2$ can be expected to fall between 10^{-7} and 10^{-6}, and that the fires emit 100–1000 times less CH_3Br than CH_3Cl. From emission ratios of CH_3Br to both CO_2 and CH_3Cl, the authors estimated that global biomass burning could contribute between 10 and 50 Gg, with a best estimate of 30 Gg of CH_3Br, annually to the atmosphere. The authors suggested that the 30 $Gg\,y^{-1}$ would represent about 30% of the annual budget of atmospheric CH_3Br. Considering uncertainties in both the emission estimate and the total atmospheric budget, it could amount to 8–50% of the annual flux of CH_3Br to the atmosphere. Recently, Andreae *et al* (1995) reported on methyl halide emissions from savanna fires in southern Africa, showing that concentrations of CH_3Cl, CH_3Br, and CH_3I were all enhanced in smoke plumes and that the production of these methyl halides was correlated with the smoldering phase of the fires. CH_3Br emissions from savannah fires were estimated at 7 Gg CH_3Br y^{-1} from correlations with CO, CO_2, and CH_3Cl. The authors suggested that global emissions from all kinds of fires were on the order of 20 Gg y^{-1}. No additional studies of the emission of CH_3Br from fires have been published as of this writing, so the possible range for this flux remains large.

4.1.3 OCEANIC PRODUCTION AND EMISSION

4.1.3.1 Marine Production Mechanisms

The ocean has been considered a source of halomethanes for a least two decades. Lovelock, Maggs and Wade (1973) first showed that CH_3I was probably produced in seawater; later Lovelock (1975), from water measurements alone, implicated CH_3Br as well. Although numerous investigations have aimed at determining marine sources of a variety of volatile organic halogens, only a couple of studies have addressed the production of CH_3Br. Sturges et al (1993) reported on CH_3Br fluxes from a crack in the Antarctic ice, but none of their laboratory incubations included measurements of CH_3Br. Wever et al (1991), in evaluating organic bromine emissions from marine macroalgae, suggested that CH_3Br might be produced extracellularly following excretion of hypobromous acid (HOBr) from macroalgae, but this mechanism has not been confirmed. A number of possible mechanisms exist. Most of these involve biological processes, either in direct production of CH_3Br or in the production of a precursor that is converted to CH_3Br through abiotic means. Associations have been suggested for macroalgae, phytoplankton, and bacteria.

Most work on marine production of organic halogens to date has focused on $CHBr_3$, $CHBr_2$, $CHBr_2Cl$, $CHBrCl_2$, CH_3I, CH_2I_2, and CH_2ClI. These studies indicate a clear association of halogen production with the presence or senescence of macroalgae. Gschwend, McFarlane and Newman (1985) studied the emission of $CHBr_3$, $CHBr_2$, and $CHBr_2Cl$ from incubated samples of marine macroalgae (seaweed). Their results showed a high rate of emissions from brown algae (150–12 500 ng g^{-1} — dry wt) and green algae (0–14 000 ng g^{-1}), and much smaller amounts coming from red algae (0–2100 ng g^{-1}). CH_2Br_2 and $CHBr_2Cl$ were also produced in lesser, but significant, amounts by brown and green seaweeds. The authors were not able to determine if microbes or abiotic reactions were involved, but suggested by reason and inference that the compounds were produced within and exuded by the algae. Work by Class and Ballschmitter (1988) and later others (e.g., Klick 1992, Moore and Tokarczyk 1993) showed high associations of elevated halomethanes in air or water in near-shore or coastal waters. Additional incubations of macroalgae confirmed that an entire suite of halogenated methanes is produced by these seaweeds (Klick 1993).

Macroalgae, however, are not the only marine plants implicated in production of organic halogens. Oram and Penkett (1994) suggested that elevated CH_3I in the air of eastern England most likely came from phytoplankton blooms in the open Atlantic ocean. Moore and Tokarczyk (1993) showed that CH_2ClI in seawater probably came from phytoplankton blooms in the open ocean, but their data also suggested a probable coastal origin for CH_3I. Klick and Abrahamsson (1992) also noted an open water source for CH_2ClI. Sturges, Cota and Buckley (1992) and Sturges *et al* (1993) found that both $CHBr_3$ and CH_3Br were produced in large amounts by ice algae (mainly diatoms) and that a range of halomethanes could be found in the cells of these diatoms. Tokarczyk and Moore (1994) found significant production of volatile organohalogens in phytoplankton cultures. Manley and Dastoor (1988) found that CH_3I was produced during the degradation of seaweeds, thus implicating micro-organisms in its production. According to Moore *et al* (1995a) the question of production of halomethanes by non-photosynthetic microbes in association with phytoplankton is still unresolved.

Only a few laboratory studies implicating the production of CH_3Br by particular organisms have been conducted to date. Sturges *et al* (1993) showed production of CH_3Br in a suspension of the ice-diatom *Nitschia stellata* and Manley and Dastoor (1988), showed that CH_3Br is a direct product of kelp (*Macrocyctis pyrifera*) metabolism. Recent work of Moore *et al* (1995b) suggests that a number of diatom species common to the marine environment produce CH_3Br, but not in sufficient quantities to explain the observed distributions and fluxes. Production rates by the organisms tested were about 100 times lower than the production rates inferred by Lobert *et al* (1995) for the open ocean. Zafiriou (1975) did suggest the possibility of CH_3Br production from the abiotic bromination of CH_3I, but the reaction would be much slower in seawater compared to probable biological production. Hu and Moore *et al* (1995) showed that reaction of Br^- with dimethylsulphinopropionate (DMSP), a compound produced by marine phytoplankton, was probably not a significant production mechanism for CH_3Br in seawater. Thus, to date, the processes producing CH_3Br in seawater in quantities sufficient to explain the observed distributions have yet to be explained or identified.

4.1.3.2 Oceanic Source Estimates

At this point in our discussions of oceanic sources and sinks, it is important to introduce various terms specific to the ocean-atmosphere system. These terms, regarding production, emission, uptake and loss of CH_3Br, are defined in Table 2.3 and illustrated in Figure 2.4. The approach given here is necessary because CH_3Br is simultaneously produced and consumed in the ocean. Thus, all CH_3Br produced in the ocean will not escape to the atmosphere. Similarly, some CH_3Br entering the ocean will be re-emitted to the atmosphere and some will be destroyed by *in situ* reactions.

Ever since Lovelock's (1975) study of the coastal waters of southern England, the ocean has been considered a potentially large source of atmospheric CH_3Br. Further evidence for the ocean as a large source was given by Singh, Salas and Styles (1983a) and Khalil *et al* (1993). Although both studies indicated that the ocean was highly supersaturated in CH_3Br everywhere, their results differed significantly. Singh, Salas and Styles (1983a) reported a mean saturation anomaly* of 180–250% for CH_3Br from an expedition in the east Pacific Ocean, ranging from 40 °N to 35 °S; Khalil *et al* (1993) reported a mean saturation anomaly of 40–80% for data from two cruises transversing the Pacific Ocean from about 40 °N to 40 °S. The study area of Singh *et al* (1983a) was located close to the Americas, whereas data of Khalil *et al* (1993) consisted mostly of samples from the open ocean. It is possible that the differences in saturation resulted from this difference in sampling geography. However, there were sampling and analytical differences as well. Singh *et al*'s (1983a) samples were stored briefly in the aqueous phase and analyzed as soon as possible aboard ship. Khalil *et al*'s (1993) samples were collected in the gas phase from both air and water and stored in metal flasks for analysis weeks to months later on shore. Saturation anomalies could be calculated directly from Khalil *et al*'s (1993) data as they essentially were measurements of partial pressure, but results of Singh *et al* (1983a) required solubility data to obtain saturation anomalies.

In an effort to resolve the apparent discrepancy between the earlier studies, Lobert *et al* (1995) conducted another investigation of the saturation of CH_3Br in ocean waters. They selected a cruise track in the east Pacific Ocean that covered portions of the oceanic central gyres, current divergences in the open ocean, and coastal waters, all in both hemispheres. Lobert *et al* (1995) directly measured the gas phase mole fraction (roughly equivalent to partial pressure) of CH_3Br in the air and in equilibrated surface water aboard ship. This avoided problems with sample storage and obviated the need for solubility data in obtaining partial pressure differences across the air–sea interface. Finally, to preclude possible interferences from co-eluting compounds, they selected an automated, gas chromatograph-mass

---

*The term 'saturation anomaly' refers to the percent departure of the dissolved gas from equilibrium with the atmosphere. Mathematically, this is defined as:

$$\Delta\% = \frac{100(p_{gw} - p_{ga})}{p_{ga}}$$

Table 2.3. Flux terms and their derivations

Term	Formula*	Definition	Equivalent
Invasion	$\dfrac{K_w A p_{ga}}{H_g}$	The rate of CH$_3$Br going into the ocean	Uptake + Return flux
Evasion	$\dfrac{K_w A p_{gw}}{H_g}$	The rate of CH$_3$Br leaving the ocean	Emission + Return flux
Return flux	$(1-R)\dfrac{K_w A p_{ga}}{H_g}$	The rate of CH$_3$Br going into the ocean and returning to the atmosphere without being destroyed in the water	Invasion − Uptake Evasion − Emission
Uptake	$\dfrac{M_{tr} p_{ga}}{0.95\,\tau_o}$	The rate of CH$_3$Br going into the ocean and not returning to the atmosphere	Invasion − Return flux
Production	P	The rate of CH$_3$Br produced by biological or chemical processes in the ocean	
Loss	$k_d M_o$	The rate of destruction in the ocean	Production − Emission + Uptake
Emission	$(1-R)P$	The rate of aquatically produced CH$_3$Br emitted to the atmosphere	Evasion − Return flux
Net flux	$\dfrac{K_w}{H_g}(p_{gw}-p_{ga})$	The net rate of CH$_3$Br leaving the ocean	Production − Loss Evasion − Invasion Emission − Uptake

*Formulas from Butler (1994) except uptake (Yvon and Butler, 1996) and net flux (e.g., Kanwisher, 1963; Lobert *et al* 1995).

K_w Air–sea transfer velocity
A Surface area of ocean
p_{ga} Partial pressure of CH$_3$Br in the troposphere, roughly equal to the mole fraction
p_{gw} Partial pressure of CH$_3$Br in the surface water
H_g Solubility of CH$_3$Br
τ_o Partial lifetime of atmospheric CH$_3$Br with respect to oceanic loss
M_a Amount of CH$_3$Br in the troposphere (moles)
k_d Aquatic degradation constant for CH$_3$Br
M_o Amount of CH$_3$Br in the surface ocean (moles)
M_{tr} Amount of CH$_3$Br in the troposphere (moles)
R Relative amount of CH$_3$Br destroyed in solution.

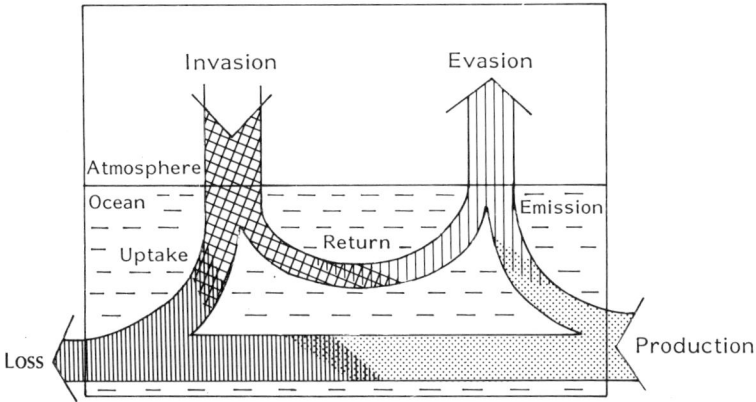

Figure 2.4. Fluxes of CH_3Br between the ocean and atmosphere. Terms are defined in Table 2.3. These distinctions are necessary to obtain the lifetime of a gas that is simultaneously produced and consumed in the ocean, or to determine the response of the atmosphere to an independent perturbation of the atmospheric burden

spectrometer (GCMS) over an electron capture gas chromatograph (ECGC) for the separation and detection of CH_3Br. Results from this expedition suggest that most of the ocean is undersaturated in atmospheric CH_3Br, not supersaturated, but that the saturation of CH_3Br varied dramatically with oceanic regime. Coastal waters, representing a small part of the ocean, were supersaturated up to almost 100%, but the open ocean was undersaturated almost everywhere, in some cases by as much as 50% (Figure 2.5). A second expedition, using the same approach, yielded similar results for a transect through the Atlantic Ocean (NOAA/CMDL unpublished data).

The Lobert *et al* (1995) data from the E. Pacific suggest that the ocean is actually a net sink (i.e., emissions from the ocean are smaller than uptake by the ocean) for atmospheric CH_3Br. The authors estimated that the net flux into the ocean was on the order of $13\,\mathrm{Gg\,y^{-1}}$. Given that reactions of dissolved CH_3Br with Cl^- and H_2O are on the order of days, depending mainly upon the sea surface temperature (SST) (Elliot and Rowland 1993, 1994 – Section 5.1.2.1), total removal of CH_3Br within the ocean was estimated at $164\,\mathrm{Gg\,y^{-1}}$. Production in the ocean, therefore, could be calculated as the difference between the net flux, which is based mainly upon concentration and the measured degree of saturation, and the total loss, which is based mainly upon concentration and sea-surface temperature (e.g., Table 2.3). Mean aquatic production was thus estimated at 151 $\mathrm{Gg\,y^{-1}}$, ranging from $0.37\,\mathrm{mg}$ ($3.9\,\mu\mathrm{mol}$) $CH_3Br\,m^{-2}\,y^{-1}$ in the open ocean to $0.76\,\mathrm{mg}$ ($8.0\,\mu\mathrm{mol}$) $CH_3Br\,m^{-2}\,y^{-1}$ in coastal waters. Of the methyl bromide produced in the ocean, 60–75% of it is destroyed *in situ*, with the

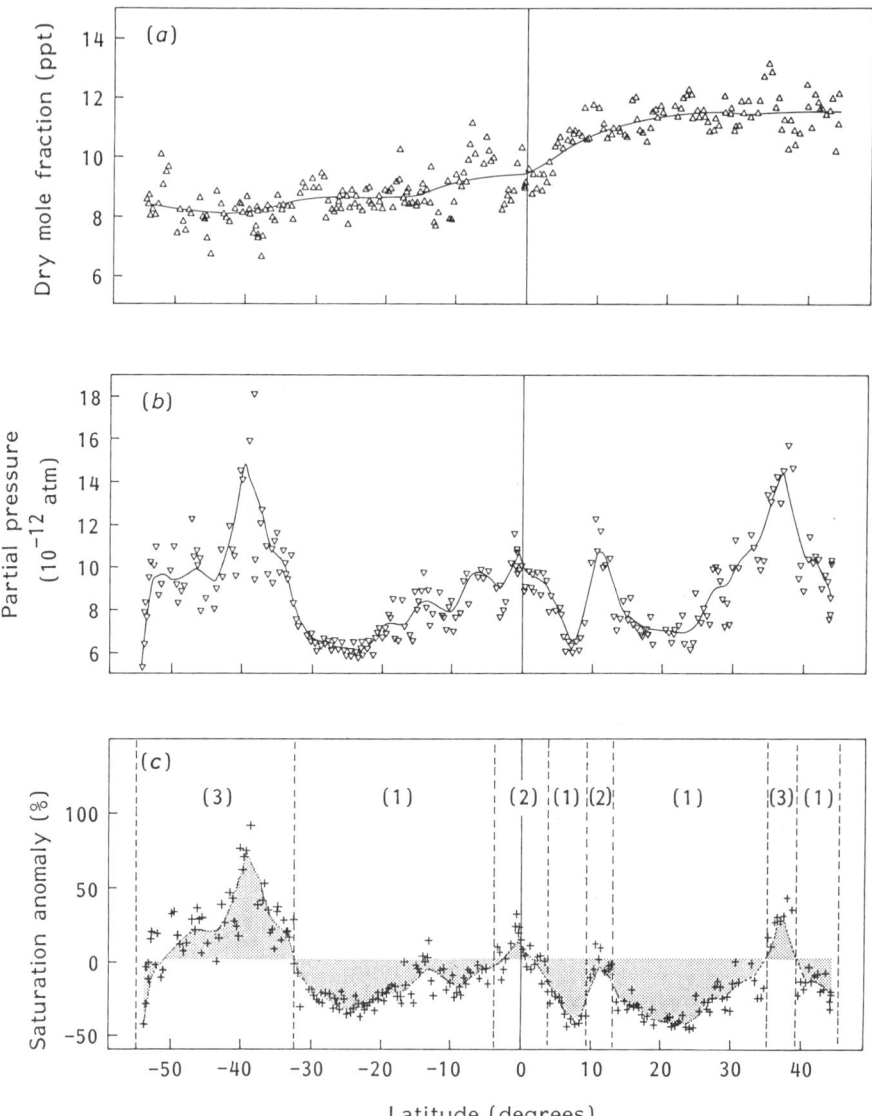

Figure 2.5. Methyl bromide in the air and surface water of the east Pacific Ocean (adapted from Lobert *et al* 1995): (*a*) dry mole fraction of atmospheric CH$_3$Br; (*b*) partial pressure of CH$_3$Br in the surface water; and (c) the saturation anomaly. Dashed vertical lines indicate oceanic regimes: Data from region (1) represent most of the open ocean, which is about 80–90% of the global coverage; region (2) represents areas of upwelling in the open ocean and region (3) represents coastal and coastally influenced areas. Regions (2) and (3) together are about 10–20% of the global ocean coverage

remainder being emitted to the atmosphere. Data from the Lobert *et al* (1995) study, extrapolated globally, suggested that the gross flux of CH_3Br to the atmosphere is 39 Gg y^{-1}.

It is possible that the results of Lobert *et al* (1995) are not inconsistent with those of Singh *et al* (1983a), as the Singh *et al* study was conducted much nearer the coast of the Americas and they might never have been outside of coastally influenced waters. Thus, extrapolation of Singh *et al*'s data to a global scale might not have been valid. Also, the study of Lobert *et al*, although designed to cover most oceanic regimes, did not touch upon coastal waters in the tropics, where production could be higher than that in the temperate regions where coastal waters were sampled. It also missed sub-polar waters where aquatic degration would be low. If aquatic production of CH_3Br is high in polar and subpolar waters, emissions would approximate to production, quite possibly making the ocean a net source of atmospheric CH_3Br. The seawater data of Khalil *et al* (1993) do not agree in any way with those of Lobert *et al* (1995), suggesting that sampling or analytical artifacts may have affected the results of the earlier study (e.g., Montzka *et al* 1995).

4.1.4 OTHER NATURAL SOURCES

No natural sources of atmospheric CH_3Br other than biomass burning and production in the marine environment have yet been identified. Given the role of biology in producing halomethanes in the marine environment, it is not unreasonable to expect that there may be additional sources of CH_3Br in soils and perhaps plants. We do not know whether these contribute significantly to the atmospheric budget of CH_3Br. The molecule apparently is cycled rapidly in nature, and natural soil sinks recently have been identified (Shorter *et al* 1995). These sinks may consume what CH_3Br might be produced by soil micro-organisms before it reaches the atmosphere.

5.1 SINKS FOR ATMOSPHERIC CH₃Br

Once in the atmosphere, CH_3Br is reasonably short-lived, with a lifetime on the order of one year. It differs from the fully halogenated, anthropogenic CFC's in that it is destroyed by OH in the troposphere, which appears to be the main pathway for its removal. Dissolution and reaction in seawater is also significant, as may be biological removal in soil. Lesser pathways of removal include photolysis in the stratosphere and hydrolysis in continental surface waters. Dissolution and hydrolysis in water droplets in the atmosphere is a very small contribution.

5.1.1 REACTIONS WITHIN THE ATMOSPHERE

Gas-phase degradation of methyl bromide in the troposphere occurs through reaction with the radicals Cl, NO_3 and OH. Heterogeneous removal by cloud hydrolysis and rainout is also a mechanism for removal of CH_3Br from the atmosphere. However, simple consideration of the concentrations and rates involved (MBGC 1992, Penkett *et al* 1995), indicate that the rates of removal by Cl, NO_3, and by cloud and rain water are negligible. This leaves reaction with the hydroxyl radical (OH) as the major *in situ* removal process for CH_3Br in the bulk troposphere.

The time constant against removal by the reaction

$$CH_3Br + OH \rightarrow CH_2Br + H_2O \qquad (8)$$

is determined by both laboratory measurements of the reaction rate constant, and the average concentrations of OH in the troposphere. Recent studies (Mellouki *et al* 1992, Zhang *et al* 1992, Poulet *et al* 1992) are in good agreement, and yield rates for reaction (1) to within $\pm 10\%$ uncertainty at tropospheric temperatures (DeMore *et al* 1994).

Concentrations of OH in the troposphere are determined by a balance between production initiated by the reactions:

$$O_3 + h\nu \rightarrow O_2 + O(^1D) \qquad (9)$$

$$O(^1D) + H_2O \rightarrow 2OH \qquad (10)$$

and removal, primarily through oxidation of CO,

$$CO + OH \rightarrow H + CO_2 \qquad (11)$$

The H atom produced in (6) is rapidly converted to the radical HO_2 through three-body reaction with atmospheric O_2. Hydroxyl can be reformed from HO_2 through the reaction:

$$HO_2 + NO \rightarrow OH + NO_2 \qquad (12)$$

Measurements of OH in different tropospheric environments are just becoming available, but they still exhibit large uncertainties. Even if the experimental error could be reduced, it would be extremely difficult to derive a global average OH field from measurements at different locations, since the variability of the reactants in Eqs. (9–12) implies a large temporal and spatial variability in OH.

Global, yearly averaged OH fields have been deduced indirectly from analysis of observed concentrations, seasonal variation, and trends of methyl chloroform (CH_3CCl_3). This compound is emitted solely by anthropogenic activity, and its historical emission rates and distribution are reasonably well quantified. Methyl chloroform is removed in the atmosphere mostly through reaction with OH. Its atmospheric mole fraction, trend, and geographic distribution have been measured on a daily basis by five stations in the ALE/GAGE network for the past 25 years (e.g., DeMore *et al* 1994, Prinn *et al* 1995) and at seven stations in the NOAA/CMDL network for the past five years. Statistical fits of these data and the emissions (Prinn *et al* 1995) yield a total atmospheric lifetime of less than 5 years for this gas. Finally, consideration of removal processes for CH_3CCl_3 in the stratosphere and oceans allows determination of the lifetime for removal of CH_3CCl_3 by OH, from which a global average OH can be inferred by incorporating the measured reaction rate of OH with CH_3CCl_3.

Alternatively, lifetimes of CH_3Br and CH_3CCl_3 are also calculated by multi-dimensional models that include a complete description of tropospheric chemistry (e.g., Reeves and Penkett, 1993, Spivakovsky *et al* 1990). However, since the input conditions for NO_x emissions, CO, CH_4, and ozone production have a reasonable degree of uncertainty, results of these models do not always reproduce the lifetime deduced for methyl chloroform. Thus, the standard practice for models is to calculate lifetimes, and rescale the methyl bromide OH removal lifetime by the ratio of the 'observed' to calculated methyl chloroform lifetime.

Determination of the atmospheric lifetime of CH_3CCl_3 from ALE/GAGE data has improved over the past decades, as the database for methyl chloroform increases, as analysis tools improve, and as calibration discrepancies are resolved. The most recent analysis and recalibration of these data (Prinn *et al* 1995) yields a methyl chloroform lifetime of 4.9 ± 0.2 years for removal by OH in the lower troposphere, corresponding to a density-weighted OH tropospheric concentration of $(9.7 \pm 0.4) \times 10^5 \, cm^{-3}$. The ALE/GAGE data now agree more closely with those of NOAA/CMDL (e.g., Butler *et al* 1991). This yields a lifetime for CH_3Br of about 1.8 years for removal by OH. This lifetime is about 10% lower than the value quoted in the latest WMO scientific assessment (Penkett *et al* 1995), since the latest analysis of ALE/GAGE data took place after this report was put together. The implications of this reassessment for the ODP of CH_3Br are explored in Section 7.1.1.

Methyl bromide can also be transported to the stratosphere, where it is removed by reaction with OH and photolysis. The calculated stratospheric lifetime of methyl bromide is 35 years (Prather 1995), similar to the 50–55 years estimated for CFC-11 (Kaye, Penkett and Ormond 1994, Elkins *et al* 1993). This is confirmed by the similarity in methyl bromide and CFC-11 profiles measured by aircraft and balloon instruments (Schauffler *et al* 1993, Lal *et al* 1994).

5.1.2 DISSOLUTION AND REACTION IN SEAWATER

The reaction of CH_3Br with Cl^- and, to a lesser extent, H_2O in seawater is reasonably fast, with a mean aquatic degradation rate near 15% d^{-1} (Yvon and Butler 1996). For these reactions to be of significance to CH_3Br in the atmosphere, the gas must first pass from the atmosphere into the ocean. As it turns out, it is a combination of the reaction and transport rates that ultimately determines the lifetime of atmospheric CH_3Br with respect to the ocean. The picture is seemingly complicated by the production of CH_3Br in seawater, but is fairly straightforward if oceanic production and loss are treated separately in the budget (Figure 2.4, Table 2.3).

5.1.2.1 Aquatic Degradation of Methyl Bromide

It was clear from the work of Swain and Scott (1953) that CH_3Br in seawater would be likely to be subject to nucleophillic attack by dissolved Cl^- and water. Extrapolating on their findings for nucleophillicity, Zafiriou (1975) pointed out that the reaction of CH_3Br with Cl^- could proceed about 9.5 times faster than that with H_2O. Mabey and Mill (1978) showed that neutral hydrolysis of CH_3Br proceeded at 2% d^{-1} at 20 °C and as high as 8% d^{-1} at 30 °C. Fells and Moelwyn-Hughes (1959) suggested that basic hydrolysis (i.e., nucleophillic attack by OH^-) would probably be small in natural waters, a point that was consistent with the nucelophillic constants given in Swain and Scott (1953). Gentile, Ferraris and Crespi (1989) later claimed that reaction with OH^- should account for losses of 10% d^{-1} in surface water at a pH of 8, but it is likely that their observed rates were driven predominantly by reaction with phosphate buffers used in the experiment. Data from Elliott (1984) and Elliot and Rowland (1994) show that the apparent neutral hydrolysis rate of CH_3Br is on the order of 0–10% d^{-1} over the full range of seawater temperatures (0–30 °C). The reaction with Cl^- clearly proceeds faster in seawater, ranging from 0–30% d^{-1} over the full range of seawater temperatures (Elliot and Rowland 1993, Zafiriou 1975, Swain and Scott 1953). Recent work by King, Pilinis and Saltzman (1995), although in agreement with Elliott (1984) and Elliot and Rowland (1993) at the very low temperatures, shows that the total degradation rate for CH_3Br in seawater ranges from 8–42% d^{-1} for a temperature range of 20–30 °C, and that the reaction with Cl^- is around 6.5 times that of the reaction with H_2O.

Other processes may also contribute to the aquatic degradation of CH_3Br. Castro and Belser (1981) and Gentile *et al* (1989) found that ultraviolet (UV) light can enhance substantially the hydrolysis of dissolved CH_3Br. Although both studies found that hydrolysis could be increased by a factor of about 7 under direct, high-intensity, UV irradiation, the extent to which this occurs in nature has not been quantified. It is likely to be much less, however, because the amount of UV light penetrating into the ocean is small. Similar arguments can be made for the biological degradation of dissolved CH_3Br. Organic halogens in general are degraded by micro-organisms under the appropriate conditions (Oremland, Miller and Strohmaier 1994, Ghosal *et al* 1985) and there is specific evidence that the ammonium oxidase enzyme common to nitrifying and methanogenic bacteria oxidizes CH_3Br (Rasche *et al.* 1990a, b, Keener and Arp 1993). Unfortunately, the subject of microbial degradation of CH_3Br in aquatic environments has not received much attention, so it is not possible at this time to make quantitative, global estimates. However, recent findings show a 30% higher degradation rate in unfiltered seawater from a single, coastal location than in filtered, presumably abiotic, seawater (Eric Saltzman, personal communication). This indicates that biological effects may be significant in the degradation of CH_3Br in seawater.

5.1.2.2 Air−Sea Exchange−A Critical Step in Aquatic Removal of CH_3Br from the Atmosphere

Although consumption of CH_3Br in oceanic surface waters proceeds rather quickly over much of the globe, the effect of this aquatic degradation on CH_3Br in the atmosphere is tempered by the rate of gas exchange between the air and sea. Considerable attention has been given over the years to this topic, yet there is still some uncertainty regarding the actual transfer rate. We will not go into all of the detail and controversy surrounding air−sea exchange in this chapter, only to note that ventilation of the surface layer is on the order of a few weeks and that any figure given for the intrinsic rate of exchange will have an uncertainty on the order of $\pm 50\%$.

The mechanisms driving air−sea exchange have been elucidated by Kanwisher (1963), Liss (1973), Broecker and Peng (1974), and Liss and Slater (1974). This concept suggests that there is a thin layer at the interface of air and liquid that restricts the flow of gas between the two phases; its thickness would depend heavily upon wind speed. Originally conceived as a 'laminar layer', this film could be thought of as existing not only at the immediate interface, but also around bubbles in the water or spray in the air. Many earlier studies addressed the thickness of this film, where molecular diffusion is believed to predominate, in assessing air−sea exchange rates (e.g., Broecker and Peng 1982). However, as sound as this approach might appear conceptually, measurement of the actual film thickness is impractical. Also, mechanisms other than molecular diffusion are probably involved to some degree in this process. Most studies today simply use an air−sea transfer velocity that incorporates wind speed, molecular diffusion, water viscosity, and other adjustments into one term. The rate of gas exchange for most gases, then, is defined as the air−sea transfer velocity times the difference in partial pressure of the gas across the air−sea interface,

$$F_g = \frac{K_{gw}}{H_g}(p_{gw} - p_{ga}) \qquad (13)$$

where F_g is the flux of the gas out of the water $(mol\,m^{-2}\,d^{-1})$, K_{gw} is the air–sea transfer velocity for the gas $(m\,d^{-1})$, H_g is the solubility of the gas $(m^3\,atm\,mol^{-1})$, and p_{gw} and p_{ga} are the partial pressures of the gas in the water and air (atm).

Different approaches have yielded different values for the air–sea exchange velocity, but almost all rely heavily upon wind speed. Values are scaled to global $^{14}CO_2$ or ^{222}Rn in seawater (e.g., Craig 1957, Broecker et al 1980, Broecker 1965, Peng et al 1979), or to specific tracers such as He and SF_6 (e.g., Wanninkhof et al 1985, Watson et al 1991, Upstill-Goddard et al 1990). Differences in mean values derived from these various approaches fall within about $\pm50\%$, although specific circumstances can lead either to more or to less agreement, depending upon the physical parameters involved (Liss and Merlivat 1986, Wanninkhof 1992, Erickson 1993, Smethie et al 1985).

From data on naturally occurring $^{14}CO_2$ and ^{222}Rn in the ocean, the global mean transfer velocity for CO_2 appears to be around $2.5–5\,m\,d^{-1}$. This value, of course, will be somewhat different for each gas because of the dependence of the exchange velocity upon molecular diffusivity (e.g., Liss et al 1993). Despite the uncertainties and differences among various approaches, it is still possible to estimate a range of air–sea exchange velocities for any given gas. From mean oceanic surface layer thickness of 75 m (Li et al 1984), the mean ventilation time for the surface or 'mixed' layer would be around 15–30 d.

The large uncertainty in the air–sea transfer can be a problem in modelling atmospheric gases, particularly those that are produced or destroyed in seawater. There is still considerable need to reduce this uncertainty, not only for understanding how the ocean affects atmospheric CH_3Br, but also how it regulates the distribution and composition of numerous trace gases in the atmosphere. This problem should not be considered insurmountable. As will be seen, as long as the uncertainty is addressed, it is possible to derive reasonable and believable limits for the budget of an atmospheric gas that interacts with the ocean.

5.1.2.3 Loss of Atmospheric CH₃Br to the Ocean

So far we have seen that the average degradation rate of CH_3Br in ocean surface waters is about $10\%\,d^{-1}$, thus representing a lifetime of about 10 d or so for dissolved CH_3Br, and that the surface layer of the ocean, where most of this degradation takes place, is ventilated with a time constant of 15–30 d. Although air–sea transfer would thus appear to be the rate-limiting step over most of the ocean, it is of the same order as aquatic degradation.

Consequently, both processes must be considered simultaneously to estimate the rate of consumption of atmospheric CH_3Br by the ocean. By coupling the degradation of CH_3Br in the ocean with CH_3Br in the atmosphere, it is possible to derive an expression for the lifetime of atmospheric CH_3Br with respect to the ocean (Butler 1994). Expanding that expression to incorporate independent terms (e.g., Lobert *et al* 1995, Yvon and Butler 1996), yields

$$k_o = \frac{1}{\tau_o} = \frac{AK_{gw}}{H_g M} \frac{[zk_d + (k_z D_z)^{0.5}]}{[zk_d + K_{gw} + (k_z D_z)^{0.5}]} \tag{14}$$

where k_o is the oceanic loss rate of CH_3Br in the atmosphere (y^{-1}), τ_o is the lifetime of CH_3Br with respect to the ocean (y), A is the area of the ocean (m^2), M is the total mass of the atmosphere, z is the thickness of the mixed layer (m), k_d is the aquatic degradation rate constant for CH_3Br in the ocean surface waters (y^{-1}), k_z is the mean degradation rate constant for CH_3Br in the thermocline below the mixed layer (y^{-1}), D_z is the eddy diffusion rate for downward mixing of oceanic surface waters $(m^2\,y^{-1})$, and K_w and H_g are as previously defined. The term for downward removal, $(k_z D_z)^{0.5}$, represents degradation during downward mixing. Overall, this is a small contribution, because the water temperature below the mixed layer is much lower, thus slowing the reaction rates. However, in colder waters of higher latitudes, it can be the locally dominant process in removing CH_3Br from the surface waters. From the weighted, globally averaged estimates and uncertainties given in Butler (1994), the most probable value for τ_o would be 3.7 y, with a possible range of 1.3–14 y. Lobert *et al* (1995), extrapolating data from a research cruise through both hemispheres in the east Pacific Ocean, calculated a τ_o of 3.0 (2.9–3.6) y for CH_3Br. The ranges of Lobert *et al* (1995), however, represent only the mean values for the three oceanic regions defined in that study, not the full range of possible partial lifetimes.

In an effort to define more fully the uncertainty associated with oceanic loss, Yvon and Butler (1996) computed global distributions of aquatic degradation, air–sea exchange, and the overall, intrinsic loss of atmospheric CH_3Br to the ocean from a 40 y data set of ocean and air temperature, wind speed, and other variables critical to the calculation of the partial atmospheric lifetime (Combined Ocean and Atmosphere Data Set, e.g., Wright 1988). From this tightly gridded set of data covering 85–98% of the world's oceans on a $2° \times 2°$ grid and using various algorithms for air–sea exchange, solubility, diffusivity, and degradation, they determined that the most probable lifetime of atmospheric CH_3Br with respect to the ocean was 2.7 (2.4–6.5) y. This uncertainty in partial oceanic lifetime is much narrower than earlier estimates and induces a much smaller uncertainty in the total atmospheric lifetime of atmospheric CH_3Br.

5.1.3 LOSSES TO SOILS AND OTHER POTENTIAL SINKS FOR ATMOSPHERIC CH₃Br

CH_3Br is a biologically active compound and is involved in a number of biochemical reactions. There is clear evidence that it is destroyed by the ubiquitous ammonium oxidizing micro-organisms, including methane oxidizers (Rasche *et al* 1990a,b, Keener and Arp 1993). These organisms inhabit virtually all environments, including soils, oceans, sediments, etc. Oremland *et al* (1994) studied both cultures of methanotrophs in the laboratory and *in situ* fluxes from lake-bed sediments. Their results showed a lifetime of CH_3Br on the order of hours in cultures and in isolated headspace above the sediments.

Arvieu (1983) found a wide variation in the destruction rate of CH_3Br within soils. Reactions with inorganic substrates such as clays or carbonates were deemed insignificant. Although reaction with soil organic matter is much more probable, the authors found no clear relationship between CH_3Br degradation rate and soil organic matter content. Observed degradation rates ranged from 24.6% d^{-1} in composted soil (45% organic carbon) to as low as 1.4% d^{-1} in carbon poor soil (0.23% organic carbon). Nevertheless, CH_3Br degradation was only 3.7% d^{-1} in peat (48.6% organic carbon) yet as high as 8–10% d^{-1} in soils having only 1–2% organic carbon.

In contrast to the work of Arvieu (1983), Shorter *et al* (1995) found that CH_3Br was destroyed with a residence time on the order of hours in many soils. The main reason given for these higher rates was that Shorter *et al* (1995) had worked with much lower mixing ratios of CH_3Br (ppb levels), so that it was not toxic to soil organisms. Degradation rates were highest in moist soils of high organic content. Uptake by most soils was rapid and irreversible and apparently driven by aerobic prokaryots. Given such rapid destruction rates, even those on the order of days, it seems possible that terrestrial soils might be analogous to the ocean, whereby the removal of CH_3Br from the atmosphere would be limited in part by the rate of exchange of CH_3Br between the soils and the atmosphere. However, Shorter *et al* (1995) suggested that for temperate forests, woodlands, or shrublands, loss fluxes were not limited by diffusion. Soil diffusion was limiting in temperate, agricultural, sandy boreal, and tropical soils. Extrapolating from these data, the authors suggested that annual, atmospheric losses totalled 42 ± 32 Gg y^{-1}. Given an atmospheric burden of 143 Gg for CH_3Br this would represent a partial atmospheric lifetime of 3.4 (1.9–14.3) y for losses to soils.

Montzka, Myers and Harris (1994) found that CH_3Br in the atmosphere within the plant canopy of an Alabama forest was slightly lower at night, when the canopy was better isolated, than during the day, when turbulence opened up the canopy layer. In night-time samples, air containing higher levels of α-pinene always contained lower levels of CH_3Br, suggesting that

CH_3Br was depleted in air that had been in close contact with the forest canopy, soils and leaf litter. Although the data were collected over only one month during the summer, the diurnal pattern persisted throughout the study and was significant at the 95% confidence level. The mean atmospheric loss from 6:00 pm to 6:00 am was about 0.04 ± 0.02 ppt h^{-1}. Assuming a canopy layer thickness of 10 to 50 m, Montzka et al (1994) roughly estimated a mean night-time depositional velocity of $0.001–0.005$ cm s^{-1}. The authors did not tender a value for daytime loss, but suggested that it could be higher if plants were involved. The reported night-time deposition velocity, extrapolated over all land and vegetative surfaces, would represent 2–10% of the CH_3Br loss due to reaction with tropospheric OH, e.g., a partial atmospheric lifetime of 17–84 y. These results must be interpreted with caution as they represent one season at one location and it is not known if the night-time losses are reversible or not. However, they are evidence of the cycling of atmospheric CH_3Br in nature, that it potentially is turned over on the order of hours to days within terrestrial systems, and that it probably is not a biochemically inert compound such as the CFC's.

Although information is available linking soils and biological systems to the rapid destruction of CH_3Br, the key question is whether or not this has a measurable effect upon CH_3Br in the atmosphere. The Shorter et al (1995) and Montzka et al (1994) studies and the in situ measurements of Oremland et al (1994) show that it might be significant, but extrapolation to a global scale remains a tenuous undertaking. Variability among plant and soil communities is one barrier, but variability in the ventilation rates of soils and sediments is also limiting. Finally, more studies that address the impact of the soils or natural systems upon CH_3Br in the atmosphere are needed before the significance of these other sinks can be addressed adequately and the uncertainties of global estimates can be reduced.

6.1 BUDGET OF ATMOSPHERIC CH₃Br

6.1.1 BOX MODELS OF ATMOSPHERIC CH₃Br

From the foregoing material, we know that atmospheric CH_3Br cannot be budgeted independent of processes occurring outside of the atmosphere, most notably those occurring in the ocean, but perhaps those in soils as well. For simplicity, this section will first address the coupled ocean–atmosphere, as interaction with the ocean is reasonably well defined. Then soils and other terrestrial influences will be assessed as to their potential for affecting the budget of atmospheric CH_3Br. The objective here is to evaluate quantitatively the uncertainties in this budget with the ultimate goal of identifying future research needs.

A balanced budget simply requires that the growth rate of the substance in question equal the sum of all sources and sinks. Mathematically, this can be expressed in its most simple form as an expansion of Eq. (4).

$$\frac{dM_g}{dt} = \sum F_{in} - \sum F_{out}. \tag{15}$$

where dM_g/dt is the growth rate of the gas in the atmosphere, M_g being the amount of the gas in the atmosphere, and F_{in} and F_{out} represent inward fluxes (sources) and outward fluxes (sinks) of the gas. If the outward fluxes can be approximated as pseudo-first-order losses, then this equation becomes

$$\frac{dM_g}{dt} = \sum F_{in} - \left(\frac{1}{\tau}\right) M x_g \tag{16}$$

where τ is the total lifetime of the gas in the atmosphere, M is the mass of the atmosphere (moles), and x_g is the mole fraction of the gas in the atmosphere. Constraints on such a simple budget include the measured growth rate, the atmospheric lifetime, and the atmospheric mole fraction of the gas in question. Further constraint is placed upon the model if the sink term is divided into atmospheric (i.e., *in situ* reactions), oceanic, and soil contributions, yielding

$$\frac{dM_g}{dt} = \sum F_i - \left(\frac{1}{\tau_a} + \frac{1}{\tau_o} + \frac{1}{\tau_s}\right) M x_g \tag{17}$$

where τ_a is the lifetime due to *in situ* atmospheric reactions (i.e. removal by OH in the troposphere with a time constant of 1.8 y and photolysis in the stratosphere with a time constant of 35 y), τ_o is the atmospheric lifetime resulting from oceanic losses, and τ_s is the atmospheric lifetime due to natural consumption by soils. Two additional constraints to the budget, the IHD of CH_3Br and the overall interhemispheric exchange time for the atmosphere, can be included if we expand the model to two boxes, northern and southern hemispheres. This would be expressed as

$$\frac{dM_{gN}}{dt} = \sum F_{inN} - \left(\frac{1}{\tau_a} + \frac{1}{\tau_o} + \frac{1}{\tau_s}\right) \frac{M}{2} x_{gN} - \frac{M\Delta x_{gNS}}{2\tau_e} \tag{18}$$

$$\frac{dM_{gS}}{dt} = \sum F_{inS} - \left(\frac{1}{\tau_a} + \frac{1}{\tau_o} + \frac{1}{\tau_s}\right) \frac{M}{2} x_{gS} - \frac{M\Delta x_{gNS}}{2\tau_e} \tag{19}$$

where the added subscrips N and S refer to northern and southern hemispheres, τ_e is the interhemispheric exchange time, and Δx_{NS} is the difference between the northern and southern hemispheric mole fractions of atmospheric CH_3Br. (The use of $(M/2)$ presumes equal atmospheric masses in the two hemispheres. For simplicity we also show the equation with equal partial lifetimes in the two hemispheres).

Finally, to test constraints of various oceanic variables upon the atmospheric budget of CH_3Br, we need to couple the atmosphere to the ocean. This can be expressed as

$$\frac{dM_{ga}}{dt} = \sum F_{in} - \left(\frac{1}{\tau_a} + \frac{1}{\tau_s}\right)Mx_{ga} + \sum \left[\frac{K_{gw}A_o}{H_g}(p_{gw} - p_{ga})\right] \quad (20)$$

$$\frac{dM_{gw}}{dt} = \sum P_o - \sum \left[\frac{A_o z k_d}{H_g}p_{gw}\right] + \sum \left[\frac{K_{gw}A_o}{H_g}(p_{gw} - p_{ga})\right] \quad (21)$$

where the P_o refers to production in the ocean, the added subscript w refers to water, and all other terms are as previously defined.

Various iterations and combinations of these equations, or equations similar to these, have been used to evaluate interhemispheric differences (e.g., Reeves and Penkett 1993, Khalil *et al* 1993) and the partial lifetime with respect to oceanic losses (e.g. Yvon and Butler, 1996). Uncertainties are typically much smaller for models that incorporate many small boxes, because they can account for co-variation of non-linear terms. Simple one- or two-box models can yield mean values for lifetimes or fluxes that do not differ significantly from those calculated with more complex models, but uncertainties are almost always larger with the simpler schemes.

6.1.2 UNCERTAINTIES IN EMISSIONS CALCULATED FROM SINK TERMS

The most direct way to estimate emissions is that of adding up the various components. However, because reasonably good, global information is now available for reaction with tropospheric OH, the atmospheric mixing ratio, and oceanic properties, the calculation of emissions from these and other sink terms is more precise than that from emissions alone. To make possible a comparison of results from adding sources and adding sinks, all model results are expressed as emissions of $Gg\,y^{-1}$. This means that, whereas source estimates can stand alone, sink estimates must be added to growth to be comparable. From a simple one-box model (e.g., Eq. (17)), total CH_3Br emissions should be around $185\,Gg\,y^{-1}$ to balance the losses and atmospheric growth, with a possible low of $91\,Gg\,y^{-1}$ and a possible high value of $271\,Gg\,y^{-1}$ (Table 2.4). This range seems extraordinarily large, but it represents the extremes. A better approximation of the probable range for global emissions can be expressed as the root-mean-square (rms) of the individual uncertainties listed in Table 2.4, which yields a range of

Table 2.4. Uncertainty in emissions calculated from sink terms

Variable	*Best estimate	High value	Low value	†High emission estimate	†Low emission estimate	‡Relative induced uncertainty
§X (ppt)	10	12	9	222	167	30%
dX/dt (ppt/y)	0.15	0.3	−0.1	188	182	3%
τ_a (y)	1.7	1.9	1.5	197	178	11%
τ_s (y)	3.4	15	2.1	212	152	32%
τ_o (y)	2.7	6.5	2.4	192	154	21%
All variables combined (Gg/y)	185	Full range		271	91	97%
		Probable (rms) range		232	135	52%

*Best estimate of tropospheric mixing ratio is 1 ppt lower than Penkett *et al* (1995); dX/dt is from Khalil *et al* 1993, τ_a is modified from Penkett *et al* (1995) according to data in Prinn *et al* (1995); τ_s is from Shorter *et al* (1995); τ_o is from Yvon and Butler (1996). High and low values represent full possible ranges and, in the total, probable (rms) ranges for the best estimates.

†High and low estimates are computed by holding all terms at their best estimated values, except for the variables in question, for which the high and low values are used.

‡Relative induced uncertainty in each case is the percentage uncertainty in emissions relative to total emissions for all variables combined (e.g., 185 Gg/y). This also is expressed as the range of possible values, except for the total which includes full and probable ranges.

§The value given here for X is the tropospheric mixing ratio, which is roughly equivalent to the partial pressure at the air–sea interface. The average for the atmosphere is a factor of 1.16 lower than this (Lal *et al* 1994, Yvon and Butler 1996).

$135–232\ Gg\,y^{-1}$, or 55% of the best estimate. Of these emissions, results of a two-box model indicate that 66% should emanate from the northern hemisphere, with a probable range of 60%–74% necessary to sustain the observed interhemispheric difference in atmospheric burden (Table 2.5). This places particularly tight constraints upon the distribution of sources of atmospheric CH_3Br. No matter what sources or sinks are involved, they must collectively sustain this rather robust observation while simultaneously maintaining a balanced budget.

It is most informative, however, to see how changes in the input variables individually affect total emissions required to balance the budget (Tables 2.4–2.6). For example, uncertainty in the interhemispheric exchange time of 1.1 y, which encompasses a range of 36% of its mean value (i.e. ±0.2 y), induces a possible range of only 4% in the uncertainty of the final emission estimates (Table 2.5). In other words, it can be considered insignificant at this point and not in immediate need of further study in resolving this budget. Note also that uncertainty in the growth rate also has only a small effect upon the budget outcome (Table 2.4). Differences in the interhemispheric ratio can induce proportional uncertainty in relative emission estimates, as can changes in estimates of the *in situ* atmospheric lifetime (Table 2.5). Uncertainty in the atmospheric mixing ratio also has a direct effect upon emission estimates, yielding a range in the order of 30% of the mean (Table 2.4). Clearly, this number can be improved dramatically with continued laboratory intercalibrations and the results will have a strong impact upon the budget of atmospheric CH_3Br.

Another significant uncertainty is the lifetime with respect to oceanic removal, with an induced range of 21% of the mean. Although the error is large, this is a composite term and the product of a number of independent uncertainties (Table 2.6). It is only a little larger than the uncertainty in the reaction with OH, however. A test of the different oceanic variables shows that uncertainties associated with loss through the thermocline, solubility, and the degradation rate are insignificant. Most of the uncertainty induced by oceanic loss (8%) is driven by uncertainty in the air-sea exchange coefficient. This is an important term in the atmospheric budget of many ozone-depleting and greenhouse gases, not the least of which are CO_2, N_2O, dimethyl sulphide (DMS), and an array of organic halogens, and can be expected to receive considerable attention in coming years.

The largest uncertainty in the indirect emission estimates (36%) is induced by the soil sink, for which quantitative information was suggested only recently (Shorter *et al* 1995). This term has a large inherent variability, but also very little data in support of it. Clearly, it is a significant term representing a subject in need of additional study. At this time there is little data available for quantitatively assessing individual sources of uncertainty in the soil–atmosphere system.

Table 2.5. Relative hemispheric emissions from interhemispheric ratio

Variable	*Best estimate	High value	Low value	High relative NH emissions	Low relative NH emissions
IH ratio	1.3	1.4	1.2	70%	61%
$\tau_{equator}$ (y)	1.1	1.3	0.9	68%	64%
τ (y)	0.8	1.4	0.6	73%	64%
All variables combined		Full range		82%	59%
(% − NH)	66%	Probable (rms) range		74%	60%

Sources for best estimates are listed with Table 4.

Table 2.6. Contribution of oceanic variables to budget uncertainties

Ocean variable	Lifetime (sinks)	Emissions (sources)
Air–sea transfer velocity	11%	15%
Mixed layer depth	5%	10%
Solubility	3%	3%
Aquatic loss rate	2%	4%
Saturation/anomaly	–	10%
Atmospheric mixing ratio	–	10%
Total induced uncertainty (c.f. Table 2.4 and 2.7)	21%	52%

The recently introduced soil sink actually offsets by about $40\,\mathrm{Gg\,y^{-1}}$ what was believed to be a reasonably balanced budget, one which could be obtained if we considered atmospheric and oceanic sinks alone. This does not mean that the soil sink can be simply dismissed, but there are some constraints imposed by what we do know about the budget and the distribution of sources and sinks. As noted earlier, the IHR of 1.3 ± 0.1 for atmospheric CH_3Br requires that the difference between sources and sinks in the NH exceed that for the SH by a about factor of two (Table 2.5). This, upon inspection, is difficult to reconcile with what is currently understood. We know that anthropogenic emissions of industrially-produced CH_3Br emanate almost exclusively from the NH; emissions from the combustion of leaded gasoline also are likely to predominate in the north. Biomass burning is roughly even. According to Lobert *et al* (1995), we can assume that the oceanic net flux could account for only about 8% of the interhemispheric difference, representing a factor of only 0.05 in the IHR. Finally, the relative interhemispheric strength of the tropospheric OH sink for various hydrocarbons has been debated for some time, with arguments presented for stronger sinks in both the northern and southern hemispheres (e.g., Chameides and Tan 1981, Spivakovsky *et al* 1990, Brenninkmeijer *et al* 1992). A stronger southern hemispheric OH sink combined with a stronger relative oceanic sink in the SH together might be able to account for the effect of an evenly distributed soil sink upon the IHR. However, unless addition sources were invoked, it would still leave a less than satisfactory picture for the budget as a whole. More plausible solutions would involve adjustments to our estimates of the magnitude and distribution of a number of sources and sinks, including the soil sink, in the budget. Such adjustments can come only from further study in these areas.

It is possible to demonstrate the effect of the overall uncertainty in loss to soils, and other sinks for that matter, upon the total atmospheric lifetime, τ. This approach is similar to that used in generating emission estimates for

Table 2.4. The equation for the atmospheric lifetime of CH_3Br as a function of *in situ* losses, losses to the ocean, and losses to soils is

$$\frac{1}{\tau} = \frac{1}{\tau_a} + \frac{1}{\tau_o} + \frac{1}{\tau_s} \qquad (22)$$

where τ_s is the partial lifetime for CH_3Br with respect to soil losses and the other terms are as previously defined. Using the most recent estimates and their uncertainties for τ_a (Prinn *et al* 1995), τ_o (Yvon and Butler 1995), and τ_s (Shorter *et al* 1995), we calculate a lifetime of 0.8 (0.6–1.4) y for atmospheric CH_3Br. About 60% (0.48 y) of the total uncertainty in the lifetime comes from soils, 23% (0.18 y) comes from the oceans, and 17% (0.14 y) comes from reactions in the atmosphere.

6.1.3 UNCERTAINTIES IN DIRECT EMISSION ESTIMATES

All of the significant direct emission estimates are highly uncertain (Table 2.7). Uncertainties induced by oceanic emissions are largest (70%) followed by biomass burning (26%) and soil fumigation (16%). As with the sinks, the contribution from the ocean is a function of a number of variables. Air–sea exchange again provides the largest individual uncertainty (Table 2.6). However, uncertainties in the ocean term induced by the atmospheric mixing ratio, saturation anomaly, and mixed layer depth are also significant.

It is clear that the fraction of CH_3Br that is produced in or enters the ocean and is removed by aquatic processes is around two-thirds* (Butler 1994, Yvon and Butler 1995). Applying this approach to their data set alone, Lobert *et al* (1995) calculated gross oceanic emissions of 27–78 $Gg\,y^{-1}$, with a best estimate of 39 $Gg\,y^{-1}$. Although the Lobert *et al* (1995) range for oceanic emissions is similar to the rms range in Table 2.7, their best estimate is lower mainly because the mean R for that data set was slightly smaller than the globally calculated value. A more likely global value would fall near 60 Gg/g. Best estimates also vary for contributions from other sources. From Manö and Andreae (1994), biomass burning can contribute 10–50 $Gg\,y^{-1}$, with a best estimate of 30 $Gg\,y^{-1}$. In a different study, Andreae *et al* (1995) later suggested that a better value for biomass emissions would be 20 $Gg\,y^{-1}$, although the range was still high. Emissions associated with

---

*CH_3Br entering oceanic surface waters is either removed by aquatic processes or emitted to the atmosphere. The proportion removed by aquatic processes is given by where R is the fraction removed and other terms are as previously defined. The term, K_w/z, is the rate constant governing evasion to the atmosphere.

$$R = \frac{k_d}{\dfrac{K_w}{z} + k_d}$$

Table 2.7. Uncertainties in direct methyl bromide emission estimates

Source		*Best estimate	High value	Low value	Range (%)
Oceans	Full range	60	100	26	52%
	Probable (rms) range		73	46	18%
Fumigation–soils		32	47.3	16.7	16%
Fumigation–durables		6.6	8.4	4.8	2%
Fumigation–perishables		5.7	6	5.4	0%
Fumigation–structures		2	2	2	0%
Gasoline		15	22	0.5	14%
Biomass burning		20	50	10	26%
Total direct	Full range	141	236	65	121%
emissions (Gg/y)	Probable (rms) range	141	194	100	65%

*The estimate for oceanic production is calculated from data in Lobert et al (1995) and Yvon and Butler (1995). Fumigation values are adjusted from UNEP (1994) and biomass burning estimates are taken directly from Manö and Andreae (1994) and Andreae et al (1995). Gasoline emissions are from Penkett et al (1995). Highs and lows for oceanic emissions and total direct emissions represent full possible ranges and probable (rms) ranges.

fumigation of durables, perishables, and structures are small and more or less fixed (Tables 2.2 and 2.7). The conversion of ethylene dibromide to CH_3Br during combustion of leaded gasoline is highly uncertain and could range from small to moderate amounts. Some of the larger uncertainties are those with respect to agricultural use, specifically pretreatment of soils before planting.

6.1.4 BUDGET SUMMARY

The sum of the estimates for direct emissions is considerably less than that for emissions derived from sinks alone. Both estimates fall within the quoted uncertainties, although the best estimates of 141 Gg/y derived from sources (Table 2.7) is near the bottom of the probable range derived from sink terms (Table 2.4) suggesting that our best estimates of sources may be too small or that the best estimates for sinks are too large. Also total uncertainty associated with the sink calculations is substantially smaller than that obtained from direct emission calculations. However, it should be remembered that the numerical approach shown in Tables 2.4 and 2.7 yields only approximations. The relative induced uncertainties given in the tables are for comparative purposes, to help identify areas for needed research. In some cases these, too, are not without some caveat. The uncertainty in the mean mixed layer depth, for example, is given as ± 25 m, which is the value cited by Li *et al* (1984), and it may be a little large. Uncertainty in the degradation rate appears to have only a small effect upon lifetime and emission estimates. Although higher degradation requires higher productivity to sustain the observed saturation anomalies, the concomitant increase in emissions simply compensates for the increased uptake, because R approaches 1.0 as k_d approaches infinity. An average increase of the mean degradation rate from 10% d^{-1} to infinity would decrease the partial atmospheric lifetime due to oceanic losses by only 30% and would decrease the total atmospheric lifetime by about 10–15%. The degradation rates used in generating Tables 2.4 through 2.7 are those due to nucleophillic displacement reactions. Biological and even photochemical effects could be much larger, but we have little or no information at this time regarding either of these effects in nature.

At this time a tighter budget is obtained by evaluating the sink terms for CH_3Br (Table 2.4). Although the best estimates of the two approaches are over 40 Gg y^{-1} apart, they both fall within the probable ranges denoted by the uncertainties. From consideration of both sources and sinks, the CH_3Br emissions should fall between 135 Gg y^{-1} (the probable low value in the sink budget) and 194 Gg y^{-1} (the probable high value in the source budget). The biggest uncertainty in the entire picture is the soil sink. If the soil sink is as large as indicated by Shorter *et al* (1995), then either northern hemisphere sources amounting to $40\,\mathrm{Gg\,y^{-1}}$ must be found, other identified sinks must fall near their lower estimates and must predominate in the SH, or some combination of both sources and sinks must accommodate this effect.

7.1 STRATOSPHERIC BROMINE CHEMISTRY

7.1.1 CYCLING OF INORGANIC BROMINE

Once CH_3Br is delivered to the stratosphere, it joins other brominated compounds to contribute to a pool of inorganic bromine. Methyl bromide is believed to be the largest single contributor to bromine in the stratosphere followed by the halons and CH_2Br_2 (Schauffler *et al* 1993). Depending upon the mechanism of delivery, however, it is possible that shorter-lived brominated compounds are getting into the stratosphere as well. Regardless of their source or mechanism of delivery, these organic source gases, upon entering the stratosphere, are degraded by photolysis, reaction with OH, and reaction with $O(^1D)$, yielding various forms of inorganic brominated compounds.

Our current understanding of the chemistry controlling the relative partitioning of bromine compounds in the stratosphere is summarized in Figure 2.2. The relative concentrations of the various forms of bromine are controlled by temperature, pressure, and the background concentrations of important reactants such as HO_2, OH, O_3 and NO_2. Concentrations of BrO are of particular importance, since they determine the rate-limiting step for the different bromine-related catalytic cycles in the lower stratosphere. In this regard, BrO plays a role analogous to ClO in the chlorine family.

There is, however, an important difference between the chemistries of inorganic chlorine and bromine. Most of the inorganic chlorine emitted in the lower stratosphere is rapidly transformed into the relatively long-lived HCl form (lifetime of weeks), which does not participate in ozone removal. Calculations indicate that HCl should account for about 80–90% of inorganic chlorine in the lower stratosphere, although aircraft measurements (Webster *et al* 1993) indicate a somewhat lower fraction. The remaining inorganic chlorine resides mostly in the $ClONO_2$ reservoir, which has a lifetime of a few hours. As a consequence, concentrations of ClO under normal stratospheric conditions are only a few percent of the total inorganic chlorine.

By contrast, the HBr reservoir is very short lived (lifetime of hours), and probably constitutes only a small fraction of the total inorganic bromine. In addition, $BrONO_2$ is more unstable than chlorine nitrate. As a result, a substantial fraction of the inorganic bromine (30–50%) is in the form of BrO, making inorganic bromine more efficient in chemically removing

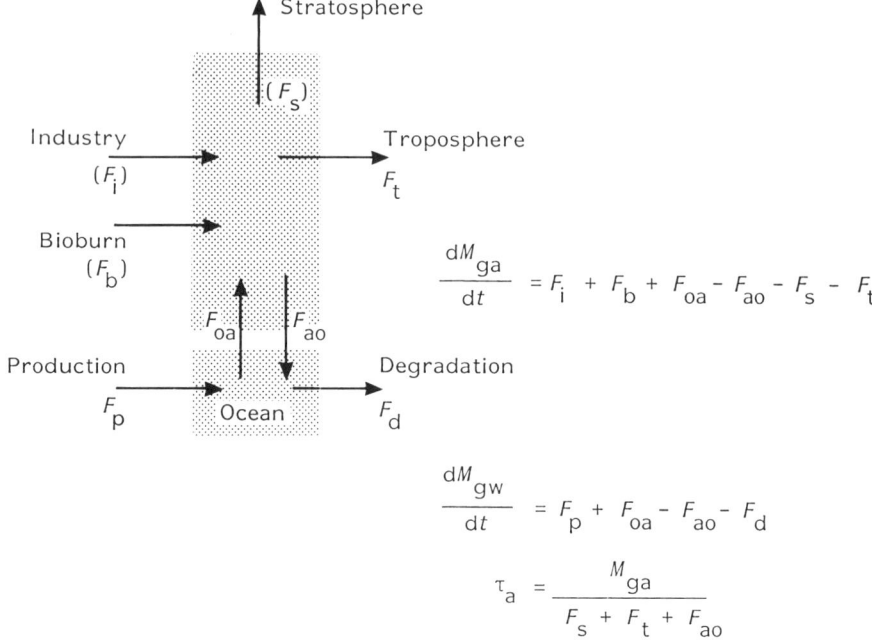

Figure 2.6. Coupled model of the ocean–atmosphere system for CH_3Br. Note that because the computation of atmospheric lifetime can include only the outward fluxes, uptake by the oceanic must be separated from emissions. This is significantly different than the model for a gas that is only consumed in the ocean (e.g., CH_3CCl_3, Figure 2.3(b))

ozone. Model calculations suggest that, on average, a molecule of inorganic bromine produced in the lower stratosphere is about 50 times more efficient in removing ozone than a molecule of inorganic chlorine (Penkett *et al* 1995). This efficiency is included in the so-called 'bromine efficiency factor' (BEF) in the calculation of methyl bromide (Eq. (1)).

The distribution of bromine species calculated by the AER two-dimensional model (Prather and Remsberg, 1992) is illustrated in Figure 2.7(a). The only significant changes in the bromine chemistry since the 1994 WMO Scientific Assessment are (a) measurements of rapid heterogeneous reaction of $BrONO_2$ with H_2O in sulfate aerosols, with a reaction efficiency of about 0.4 (Hanson and Ravishankara 1995) and (b) slower photolysis of the reservoir HOBr from the new absorption cross-sections measured by Orlando (1995). The net effect is an increase the BrO concentrations (and ODP of methyl bromide) by about 5% relative to the numbers cited in the WMO Scientific Assessment.

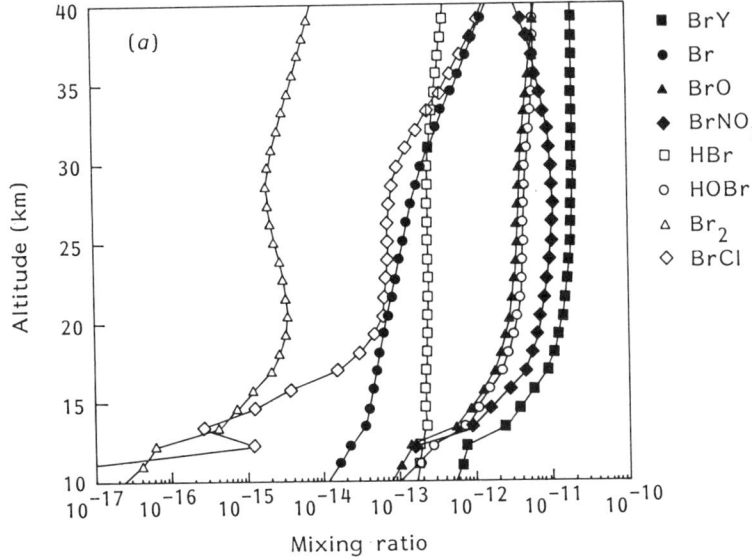

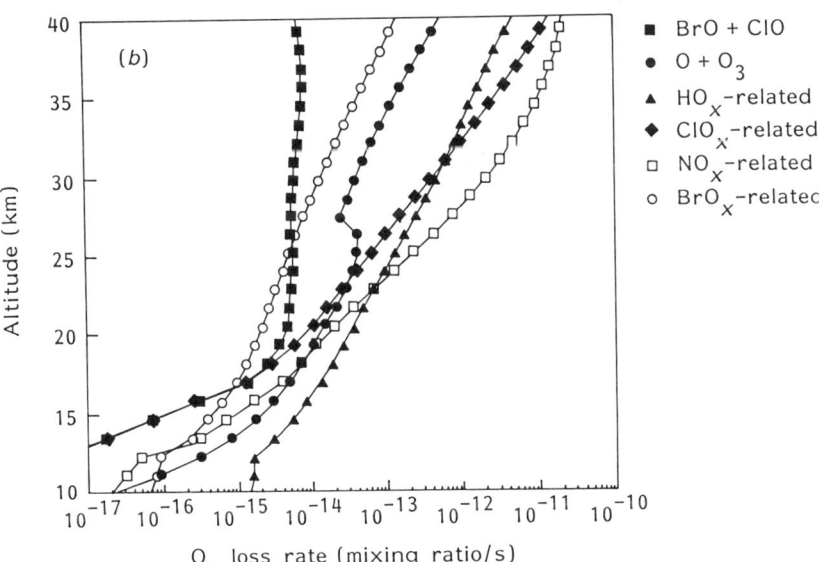

Figure 2.7. Results of the AER two-dimensional model for: (*a*) mixing ratios of inorganic bromine species; and (*b*) loss rates of odd oxygen ($O_3 + O$) through the stratosphere. Calculations were made for 38 °N during March

Ozone removal by bromine-related cycles occurs mostly in the lower stratosphere (altitudes less than 22 km):

Cycle I.

$$Br + O_3 \rightarrow BrO + O_2 \tag{Ia}$$

$$Cl + O_3 \rightarrow ClO + O_2 \tag{Ib}$$

$$*BrO + ClO \rightarrow Br + Cl + O_2 \tag{Ic}$$

$$2O_3 \rightarrow 3O_2 \tag{Id}$$

Cycle II:

$$Br + O_3 \rightarrow BrO + O_2 \tag{IIa}$$

$$*BrO + HO_2 \rightarrow HOBr + O_2 \tag{IIb}$$

$$HOBr + h\nu \rightarrow OH + Br \tag{IIc}$$

$$OH + O_3 \rightarrow HO_2 + O_2 \tag{IId}$$

$$2O_3 \rightarrow 3O_2 \tag{IIe}$$

The above reactions marked with asterisks denote the so-called 'rate-limiting steps', i.e., the reaction in the cycle that occurs at the slowest rate, thus determining the actual rate for ozone removal.

Measurements of the rates of all reactions involved in cycles I and II now exist, most of them to an accuracy of 10–15%. An important change in our assessment of the ODP of methyl bromide occurred with the remeasurement of reaction (IIb) (Poulet *et al* 1992, Bridier, Veyret and Lesclaux 1993, Maguin *et al* 1995), which yielded rates a factor of 6 faster than previous measurements at room temperature, thus increasing the bromine efficiency in the lower stratosphere. The present uncertainty in the above reaction rate is about a factor of 3 at stratospheric temperatures (DeMore *et al* 1994).

Our current understanding of stratospheric chemistry allows us to calculate the rate of catalytic ozone removal at different altitudes, latitudes, and seasons. An example of the calculated rates is shown in Figure 2.7b, for March, mid-latitude conditions (see also Avallone *et al* 1993b). These calculations indicate that bromine-related mechanisms account for a substantial portion of the catalytic ozone removal below 20 km. These calculated ozone losses are one of the inputs that multi-dimensional models use to estimate the ozone depletion potential of methyl bromide.

It is crucial that the photochemical mechanisms proposed in Figure 2.2, and used to derive Figure 2.7, be tested against atmospheric observations. A large number of measurements of NO_x, ClO_x and HO_x species by aircraft campaigns, balloons, and satellites exists for such testing. In particular, aircraft data from the various NASA expeditions (AAOE, AASE-I, AASE-II, SPADE, ASHOE/MAESA) have yielded a complete set of measurements which allows 'fixing' the concentrations of long-lived compounds, thus eliminating the uncertainties due to dynamical motions not reproduced in multi-dimensional models. These data indicate that the understanding of chlorine, nitrogen, and hydroxyl chemistry in the stratosphere is correct to a first approximation, (e.g., Salawitch *et al* 1993).

Unfortunately, bromine chemistry has not been tested to the same degree of accuracy. The only *in situ* measurements of BrO available are from the aircraft campaigns mentioned above. However, the large signal-to-noise ratio in the measurements require averaging over long spatial domains of the data. The latest reanalysis of the aircraft data from the AASE-II mission indicates that mean values of BrO mixing ratios in the lower stratosphere are around 5.4 pptv, with a standard deviation of 3.1 pptv (Avallone *et al* 1995). The authors note that these values are about 30% lower than those calculated from models, although they still fall within the 2-σ uncertainty. Nevertheless, this apparent discrepancy suggests that our understanding of stratospheric bromine chemistry may not be complete. It should be emphasized, however, that the above discrepancy does not necessarily mean that the ODP of methyl bromide should be reduced by 30%, since whatever chemical mechanisms could be responsible for this discrepancy could also affect the chlorine efficiency. Since ODPs are always calculated relative to the chlorine effect, we cannot estimate the impact of a given discrepancy in measurements unless we include the appropriate chemistry in a model.

Another potential uncertainty in stratospheric chemistry is the possibility of the following pathway existing for reaction (IIb):

$$BrO + HO_2 \rightarrow HBr + O_3 \qquad \text{(IIb′)}$$

The above branch would produce, not destroy ozone. Original estimates indicate that branching ratios as small as 10% could significantly reduce the calculated ODP of methyl bromide.

There are two ways to determine the potential importance of (IIb′): direct measurement of its rate, and indirect constraints placed by concentrations of HBr in the stratosphere. No direct measurements of (IIb′) exist, but thermodynamic considerations (Mellouki *et al* 1992) indicate that this branching should be much less than 1%, which would have a negligible impact on the calculated ODP. The maximum impact of (IIb′) is estimated to occur in the lower stratosphere, where HBr measurements are lacking. Attempts to measure HBr by far-infrared emission techniques yield upper limits of about 1 pptv at 30 km (Traub *et al* 1994), which are consistent with branching ratios of less than 5%. This issue has thus not been completely resolved. The 1994 WMO Scientific Assessment adopted a conservative upper bound of 2% for the branching ratio of (IIb′).

7.1.2   VALUES AND UNCERTAINTIES FOR THE ODP OF METHYL BROMIDE

We use Eq. (1) (Section 2.1.4) to estimate the steady-state ODP of methyl bromide and to assess its uncertainty due to uncertainties in sinks and stratospheric chemistry. As already mentioned previously, the uncertainties associated at present with the sinks are better constrained than those associated with the sources, allowing a better analysis of error.

The total lifetime for removal of methyl bromide is obtained from Eq. (22) (Section 6.1.2). Adopting the value of τ_{CH_3Br} given in Table 2.4, and letting $\tau_{CFC-11} = 50$ y (Kaye, Penkett and Ormond 1994) we calculate a bromide loading potential of 0.008 from the expression in the first brackets of Eq. (1) A Bromine Efficiency Factor (BEF) of 54 is calculated by the AER two-dimensional model, adopting heterogeneous chemistry on background aerosols, including reaction of $ClNO_3$, the kinetic recommendations of DeMore *et al* 1994, and the latest absorption cross sections of HOBr. Based upon these current data, the best estimate for the ODP of CH_3Br is

$$ODP_{(CH_3Br)} = 0.43$$

This estimate is smaller than that cited in the latest WMO scientific assessment (Penkett *et al.* 1995). The primary reasons for the difference are (a) a re-evaluation of the lifetime for OH removal of CH_3Br due to the latest analysis from ALE/GAGE data Prinn *et al* 1995); (b) the shorter lifetime for ocean removal (2.7 versus 3.7 years) estimated recently by Yvon and Butler (1995) and (c) the inclusion of a natural soil sink.

The algorithm given by Eq. (1) provides a useful framework to estimate the uncertainties in the calculated ODP of methyl bromide due to uncertainties in the different input parameters. Uncertainties in the input parameters, and their impact on the calculated ODP are listed in Table 2.8. Uncertainties in the bromine loading potentials are directly calculated from Eq. (1), while the AER 2-D model has been used to calculate the bromine loading potentials.

The largest uncertainties in ODP are due to the following:

- Uncertainties in the lifetime of CH_3Br. The impact of uncertainties in this parameter can be illustrated by comparing the values estimated for the WMO scientific assessment (Penkett *et al* 1995) versus those using the best available information at present. In particular, the ocean removal lifetime has been better constrained, the uncertainty in OH removal has been reduced, and a soil sink with a large uncertainty suggested.
- Uncertainties in the kinetics of $BrO + HO_2$. Atmospheric measurements of HBr indicate that the branching of the $BrO + HO_2$ reaction to the HBr channel is probably much less than 5%. Measurements of HBr

below 30 km would further constrain this parameter. There are at present no measurements of either the rate or branching of the above reaction at stratospheric temperatures. The uncertainty in the rate of $BrO + HO_2$ produces uncertainties in the Bromine Efficiency Factor of about 50%.

The 'semi-empirical' ODPs discussed by Solomon *et al* (1992) provide a valuable constraint to the model-based results, particularly if we are interested in the ODP for a particular region of the atmosphere. Larger uncertainties are introduced when steady-state ODPs are derived from the semi-empirical approach, since the necessary observations are usually not available for a global coverage. This is particularly true for methyl bromide, where the coincident measurements of BrO, ClO and HO_2 are sparse, particularly at mid-latitudes, and where the existing measurements have uncertainties of 25% for ClO, 35% for BrO, and 50% for HO_2. Based upon this analysis, we can say that, although no single process brings the estimated ODP of CH_3Br below 0.2, some of the individual uncertainties bring the calculated values within 30–50% of that value (Table 2.8).

8.1 RESEARCH NEEDS

Much of the recent attention given atmospheric CH_3Br derives from its potential to deplete stratospheric ozone. From this perspective, the distribution, transport, and chemistry of CH_3Br is of interest to scientists who are trying to understand the causes and effects of ozone depletion. Nevertheless, information is also needed by policy-makers, who are trying to decide how best to deal with anthropogenic causes of lowered stratospheric ozone. The potential for contributions by chlorine, bromine, ice, and aerosol particles, as well as transport mechanisms, is recognized at this time. Beyond identification, however, these effects must be quantified and this is where research regarding the role of CH_3Br is most needed. Research should focus upon answering those questions which can help us understand both the role of bromine and methyl bromide in stratospheric chemistry and the relative effect of anthropogenic activities upon the atmospheric budget of CH_3Br.

Uncertainties in the atmospheric budget ultimately translate to uncertainties in estimates of the atmospheric lifetime, which, in turn, show up in estimates of the ODP for CH_3Br. Also, our ability to assess the relative contribution of anthropogenic emissions of CH_3Br to bromine in the stratosphere depends in good part upon how well we understand the budget. For these reasons, this chapter has focused upon budget issues and the uncertainties underlying various estimates. Improving these estimates will require research in a number of areas (Table 2.9). Of these, obtaining a better estimate of the atmospheric mole fraction can more tightly constrain

Table 2.8. Steady-state ODP uncertainty

Parameter	Value-range[a]	Induced range in τ (CH_3Br) (yrs)	Induced range in BLP[b]	Induced range in BEF[b]	Induced range in ODP[b]
τ_{atm}	1.8 y ($\pm 16\%$)*	0.76–0.83	0.0076–0.0083	54	0.41–0.45
τ_{ocean}	2.7 y (2.4–6.5 y)[†]	0.77–0.97	0.0077–0.0097	54	0.42–0.52
τ_{CFC-11}	50 y ($\pm 10\%$)[‡]	0.80	0.0072–0.0088	54	0.39–0.48
τ_{soil}	3.4 (1.9–14.3 y)[§]	0.68–0.98	0.0068–0.0098	54	0.36–0.53
F_{CH_3Br}/F_{CFC-11}	1.08 ($\pm 15\%$)[¶]	0.80	0.0080	46–62	0.37–0.50
$k_{(BrO+HO_2)}$	6.3 (2.2–18) $\times 10^{-11}$ $cm^3 s^{-1}$**	0.80	0.008	36–56	0.29–0.45
Branching of $BrO + HO_2 \rightarrow HBr + O_3$	0 ($<2\%$)[††]	0.80	0.008	33–54	0.26–0.43

[a]The first number denotes the current best estimate for each term. Numbers in parentheses denote the best current estimate for the uncertainty.
[b]The ranges quoted in this column are calculated by assuming the best estimates in column 2 and varying the single parameter in the corresponding row according to the uncertainty estimates in column 2.
*Mellouki *et al* (1992) Prinn *et al* (1995)
[†]Yvon and Butler (1996)
[‡]Kaye *et al* (1994)
[§]Shorter *et al* (1995) a conservative estimate of a factor of 2 has been adopted
[¶]Pollock *et al* (1992)
**DeMore *et al* (1994) evaluated at 215K
[††]M.K.W. Ko, private communication.

Table 2.9. Research needs for resolving the budget of atmospheric CH_3Br

Variable	Variability in emission estimate induced by uncertainty	Ease of improvement	Expected improvement
Atmospheric mixing ratio-calibration	Uncertainty of 3 ppt yields 30% range in emission estimates	Straightforward, but care and inter-calibration necessary.	Uncertainty down to ±0.5 ppt; induced emission range down to 9%
Natural soil sink	32% range in emission estimate	Difficult, high variability in soil type and deposition rates.	?
Biomass burning	26% range in emission estimate	Difficult to quantify. Variability in source distribution and smoke composition is high.	?
Agricultural emissions	16% range in emission estimate	Difficult to quantify. Variability among fields and techniques very high.	?
Oceanic source	52% range in emission estimate	Source distribution highly variable; dependent on saturation anomaly, but also air–sea exchange coefficient.	Emission range down to 30%
Air–sea transfer velocity	11–15% range in emission estimate	Not simple, involves dependence upon a number of physical variables.	?
Kinetics of aquatic loss and solubility	7% range in emission estimate	Straightforward, but care and inter-calibration necessary.	Induced emission range down to 5%
Interhemispheric gradient	13% range in estimate of NH emissions	Straightforward, intercalibrated measurement program.	Induced emission range down to 5%

the budget in a relatively short time. Other large uncertainties are not so easily dealt with. Resolving contributions from the soil sink, biomass burning, and agricultural uses will require considerable effort and ingenuity. The ocean is a big player in summations of both sources and sinks of atmospheric CH_3Br, yet we know little of the organisms responsible for CH_3Br production and are unable to quantify their contribution to global oceanic emissions. Uncertainties in the air–sea exchange velocity affect calculations involving both source terms and sink terms. Finally, monitoring of CH_3Br in the atmosphere has been limited in the past. Today, it is essential that various global monitoring programs include CH_3Br in their suites of potential ozone-depleting compounds, so that trends and correlations with physical phenomena can be better established.

Uncertainties regarding stratospheric bromine chemistry also can be tightened with further research. Measurements of total inorganic bromine in the stratosphere and total organic bromine in the troposphere are useful in determining contributions by various source gases. However, speciation of the inorganic bromine and the rates of various reactions are also important. Atmospheric models could further reduce the existing uncertainties in methyl bromide lifetime and ODP by trying to incorporate the whole suite of information we now have, both on the troposphere and stratosphere. These models should incorporate, for example, constraints placed by measured interhemispheric gradients, distribution of land types, and expected ocean uptake as calculated from existing climatologies of sea surface temperature and wind velocities. Such approaches can place additional constraints upon the budget and the distribution of sources.

A final question concerns the constraints placed by laboratory kinetics and stratospheric measurements upon estimates of the bromine efficiency factor in the stratosphere. Uncertainties in the absolute rate and branching of the $BrO + HO_2$ reaction induce some of the largest uncertainties in the ODP of methyl bromide (Table 2.8). Because the impact of such changes is non-linear, they must be assessed by stratospheric models incorporating complete chemistry. These models should also verify their adopted chemical mechanisms by comparing the calculated partitioning of bromine and chlorine species with those measured or derived from *in situ* aircraft and balloon measurements.

9.1 SUMMARY

New information regarding the emission, transport, and reaction of atmospheric CH_3Br has been rapidly forthcoming over the past few years, often requiring change in the selected paradigms describing the behaviour of this compound. Significant changes have already occurred since publication of the 1994 WMO Scientific Assessment. The atmospheric burden and sink have been estimated more accurately, uncertainties in the estimate of the oceanic sink have been substantially reduced, and a potentially significant terrestrial sink for atmospheric methyl bromide has been identified. Although these represent substantial improvements in our understanding of atmospheric CH_3Br, some aspects of the budget of atmospheric CH_3Br remain uncertain, as do estimates of the relative contribution of bromine in methyl bromide to the total destruction of stratospheric ozone. Reducing these uncertainties is at the heart of continued efforts by scientists from a variety of disciplines. The goal of this research is to provide a more solid understanding of the cycling of this compound in nature and to determine more accurately the contribution of anthropogenic CH_3Br to the depletion of stratospheric ozone.

REFERENCES

Albritton DL and Watson RT, 1992, Methyl bromide and the ozone layer: A summary of current understanding, *Methyl Bromide: Its Atmospheric Science, Technology, and Economics, Montreal Protocol Assessment Supplement*, edited by RT Watson, DL Albritton, SO Anderson and S Lee-Bapty, Nairobi, Kenya: United Nations Environment Programme.

Anastasi C, Heathfield AE, Knight GP and Nicolaisen F, 1994, Integrated absorption coefficients of $CHClF_2$ (HCFC-22) and CH_3Br in the atmospheric infrared window region. *Spectrochimica Acta*, **50**A (10), 1791–8.

Anderson JG, Bruhne WH, Lloyd SA, Toohey DW, Sander SP, Starr WL, Loewenstein M and Podolske JR, 1989, Kinetics of O_3 destruction by ClO and BrO within the Antarctic vortex: an analysis based on in situ ER-2 data. *Journal of Geophysical Research*, **94** (D9), 11 480–520.

Andreae MO, Atlas E, Harris GW, Helas G, de Kock A, Koppmann R, Mano S, Pollock WH, Rudolph J, Scharffe D, Schebeske G and Welling M, 1995, (submitted). Methyl halide emissions from savanna fires in southern Africa. *Journal of Geophysical Research*.

Aneja VP, 1984, Environmental Impact of Natural Emissions, Transactions, APCA Specialty Conference (Pittsburg, PA: Air Pollution Control Association).

Arvieu JC, 1983, Some physico-chemical aspects of methyl bromide behavior in soil. *Acta Horticulturea*, **152**, 267–74.

Avallone LM, Toohey DW, Brune WH, Salawitch RJ, Dessler AE and Anderson JG, 1993a, Balloon-borne in situ measurements of ClO and ozone: Implications for heterogeneous chemistry and mid-latitude ozone loss. *Geophysical Research Letters*, **20**, 1795–8.

Avallone LM, Toohey DW, Proffitt MH, Margitan JJ, Chan KR and Anderson JG, 1993b, In situ measurements of ClO at midlatitudes: Is there an effect from Mt. Pinatubo? *Geophysical Research Letters*, **20**, 2519–22.

Avallone LM, Toohey DW, Schauffler SM and Pollock WH, 1995. In situ measurements of BrO during AASE II, *Geophysical Research Letters*, **22** (7), 831–4.

Baumann H and Heumann KG, 1989, Analysis of organobromine compounds and HBr in motor car exhaust gases with a GC/microwave plasma system, *Fresenius Zeitschrift Analytische Chemie*, **333**, 186–92.

Brenninkmeijer CAM, Manning MR, Lowe DC, Wallace G, Sparks RJ and Volz-Thomas A, 1992, Interhemispheric asymmetry in OH abundance inferred from measurements of atmospheric [14]CO, *Nature*, **356**, 50–2.

Bridier I, Veyret V and Lesclaux R, 1993, Flash photolysis kinetic study of reactions of the BrO radical with BrO and H_2O, *Chem. Phys. Lett.*, **201**, 563–8.

Broecker WS, 1965, An application of natural radon to problems in ocean circulation, in *Symposium on Diffusion in Oceans and Fresh Water*, pp. 116–145. Palisades, New York: Lamont-Doherty Geological Observatory, Columbia University.

Broecker WS and Peng T-H, 1974, Gas exchange rates between air and sea. *Tellus*, **26**, 21–35.

Broecker WS and Peng T-H, 1982, *Tracers in the Sea*, Palisades, New York: Lamont-Doherty Geological Observatory, Columbia University.

Broecker WS, Peng T-H and Takahashi T, 1980, A strategy for the use of bomb-produced radiocarbon as a tracer of fossil fuel CO_2 into the deep sea source regions, *Earth and Planetary Science Letter*, **49**, 463–8.

Butler JH, 1994, The potential role of the ocean in regulating atmospheric CH_3Br, *Geophysical Research Letters*, **21** (3), 185–8.

Butler JH, Elkins JW, Thompson TM and Egan KB, 1989, Tropospheric and dissolved N_2O of the west Pacific and east Indian Oceans during the El Niño-Southern Oscillation event of 1987, *Journal of Geophysical Research*, **94** (D12), 14 865–77.

Butler JH, Elkins JW, Thompson TM, Hall BD, Swanson TH and Koropalov V, 1991, Oceanic consumption of CH_3CCl_3: Implications for Tropospheric OH, *Journal of Geophysical Research*, **96** (D12), 22 347–55.

Castro CE and Belser NO, 1981, Photohydrolysis of methyl bromide and chloropicrin, *Journal of Agricultural Food Chemistry*, **29**, 1005–8.

Chameides WL and Tan A, 1981, The two-dimensional diagnostic model for tropospheric OH: An uncertainty analysis, *Journal of Geophysical Research*, **86** (C6), 5209–23.

Cicerone R, 1989, Analysis of sources and sinks of atmospheric nitrous oxide (N_2O), *Journal of Geophysical Research*, **94**, 18 265–271.

Cicerone RJ, Heidt LE and Pollock WH, 1988, Measurements of atmospheric methyl bromide and bromoform, *Journal of Geophysical Research*, **93**, 3745–9.

Class T and Ballschmiter K, 1988, Chemistry of organic traces in air VIII: Sources and distribution of bromo- and bromochloromethanes in marine air and surface water of the Atlantic Ocean, *Journal of Atmospheric Chemistry*, **6**, 35–46.

Cole FW, 1970, *Introduction to Meteorology*, (New York: Wiley).

Craig H, 1957, The natural distribution of radiocarbon and the exchange time of carbon dioxide between the atmosphere and sea, *Tellus*, **9**, 1–17.

Crutzen PJ and Andreae MO, 1990, Biomass burning in the tropics: Impact on atmospheric chemistry and biogeochemical cycles, *Science*, **250**, 1669–77.

Crutzen PJ and Goldammer JG, 1993, *Fire in the Environment. The Ecological, Atmospheric, and Climatic Importance of Vegetation Fires* (New York: Wiley).

Crutzen PJ, Heidt LE, Krasnec JP, Pollock WH and Seiler W, 1979, Biomass burning as a source of atmospheric trace gases: CO, H_2, H_2O, NO, CH_3Cl and COS, *Nature*, **282**, 253–6.

DeMore WB, Sander SP, Golden DM, Hampson RF, Kurylo MJ, Howard CJ, Ravishankara AR, Kolb CE and Molina MJ, 1994, Chemical kinetics and photochemical data for use in stratospheric modeling: Evaluation No. 11, Jet Propulsion Laboratory, Pasadena, CA.

DeZafra RL, Jaramillo M, Barrett J, Emmons LK, Solomon PM and Parrish A,

1989, New observations of a large concentration of ClO in the springtime lower stratosphere over Antarctica and its implications for ozone-depleting chemistry, *Journal of Geophysical Research*, **94** (D9), 11 423–8.

Duce RA, Winchester JW and VanNahl TW, 1965, Iodine, bromine, and chlorine in the Hawaiian marine atmosphere, *Journal of Geophysical Research*, **70** (8), 1775–99.

Elkins JW, Thompson TM, Swanson TH, Butler JH, Hall BD, Cummings SO, Fisher DA and Raffo AG, 1993, Decrease in the growth rates of atmospheric chloro-fluorocarbons 11 and 12, *Nature*, **364**, 780–3.

Elliot S and Rowland FS, 1993, Nucleophilic substitution rates and solubilities for methyl halides in seawater, *Geophysical Research Letters*, **20**, 1043–6.

Elliot S and Rowland FS, 1994, Methyl halide hydrolysis rates in natural waters, *EOS* **75** (44), 110.

Elliott SM, 1984, The chemistry of some atmospheric gases in the ocean, PhD Thesis, University of California, Irvine.

Erickson III DJ, 1993, A stability dependent theory for air–sea gas exchange, *Journal of Geophysical Research*, **98**, 8471–88.

Farman JC, Gardiner BG and Shanklin JD, 1985, Large losses of total ozone in Antarctica reveal seasonal ClO_x/NO_x interaction, *Nature*, **315**, 207–10.

Fells I and Moelwyn-Hughes EA, 1959, The kinetics of the hydrolysis of chlorinated methanes, *Journal of the Chemical Society*, **1959**, 398–409.

Gentile IA, Ferraris L and Crespi S, 1989, The degradation of methyl bromide in some natural freshwaters: Influence of temperature, pH, and light, *Pesticide Science*, **25**, 261–72.

Gerkens RR and Franklin JA, 1989, The rate of degradation of 1,1,1,–trichloro-ethane in water by hydrolysis and dehydrochlorination, *Chemosphere*, **19**, 1929–37.

Ghosal D, You I-S, Chatterjee DK and Chakrabarty AM, 1985, Microbial degradation of halogenated compounds, *Science*, **228**, 135–42.

Graedel TE and Crutzen PJ, 1993, *Atmospheric Change: An Earth System Perspective*, Vol. xiii (New York: W.H. Freeman and Company).

Gschwend PM, MacFarlane JK and Newman KA, 1985, Volatile halogenated organic compounds released to seawater from temperate marine macroalgae, *Science*, **227**, 1033–5.

Hanson DR and Ravishankara AR, 1995, Heterogeneous chemistry of bromine species in sulfuric acid under stratospheric conditions, *Geophysical Research Letters*, **22** (4), 385–8.

Hao WM and Liu M-H, 1994, Spatial and temporal distribution of tropical biomass burning, *Global Biogeochemical Cycles*, **8** (4), 495–503.

Harmon ME, Ferrell WK and Franklin JF, 1990, Effects on carbon storage of conversion of old-growth forests to young forests, *Science*, **247**, 699–702.

Hood LL, McPeters RD, McCormick JP, Flynn LE, Hollandsworth SM and Gleason JF, 1993, Altitude dependence of stratospheric ozone trends based on Nimbus 7 SBUV data, *Geophysical Research Letters*, **20**, 2667–70.

Hu Z and Moore RM, 1995, (submitted) Kinetics of methyl halide production by reaction of DMSP with halide ion, *Marine Chemistry*.

Jeffers PM, Ward LM, Woytowitch LM and Wolfe NL, 1989, Homogenous hydrolysis rate constants for selected chlorinated methanes, ethanes, ethenes, and propanes, *Environmental Science Technology*, **23**, 965–9.

Junge CE, 1957, Chemical analysis of aerosol particles and of gas traces on the island of Hawaii, *Tellus*, **9**, 528–37.

Kanwisher J, 1963, On the exchange of gases between the atmosphere and the sea,

Deep-Sea Research, **10**, 195–207.

Kaye JA, Penkett SA and Ormond FM, 1994, Report on concentrations, lifetimes and trends of CFCs, halons and related species, National Aeronautics and Space Administration, NASA Reference Publication, 1339.

Keener WK and Arp DJ, 1993, Kinetic studies of ammonia mono-oxygenase inhibition in nitrosomonas-europea by hydrocarbons and halogenated hydrocarbons in an optimized whole-cell assay, *Applied Environmental Microbiology*, **59** (8), 2501–10.

Khalil MAK, Rasmussen RA and Gunawardena R, 1993, Atmospheric methyl bromide: trends and global mass balance, *Journal of Geophysical Research*, **98**, 2887–96.

King DB, Pilinis C and Saltzman ES, 1995, Measurement of the total degradation rate of methyl bromide in seawater, *EOS, Transactions of the American Geophysical Union*, **76** (17), S162.

Klick S, 1992, Seasonal variations of biogenic and anthropogenic halocarbons in seawater from a coastal site, *Limnology Oceanography*, **37** (7), 1579–85.

Klick S, 1993, The release of volatile halocarbons to seawater by untreated and heavy metal exposed samples of the brown seaweed Fucus Vesiculosus, *Marine Chemistry*, **42**, 211–21.

Klick S and Abrahamsson K, 1992, Biogenic volatile iodated hydrocarbons in the ocean, *Journal of Geophysical Research*, **97**, 12 683–7.

Ko MKW, Sze N-D and Prather MJ, 1994, Protection of the ozone layer: Adequacy of existing controls, *Nature*, **367**, 545–8.

Krzyscin JW, 1994, Total ozone changes in the Northern Hemisphere mid-latitudinal belt (30–60 °N) derived from Dobson spectrophotometer measurements, 1964–1988, *Journal Atmospheric and Terrestrial Physics*, **56**, 1051–6.

Lal S, Borchers R, Fabian P, Patra PK and Subbaraya BH, 1994, Vertical distribution of methyl bromide over Hydarabad, India, *Tellus*, **46B**, 373–7.

Levine JS, 1990, Global biomass burning: Atmospheric, climatic and biospheric implications, *EOS*, **71**, 1075–7.

Levine JS, 1991, *Global Biomass Burning*, (Cambridge: MIT Press).

Levitus S, 1982, Climatological Atlas of the World Ocean, *NOAA Professional Paper*, **13**, 173.

Li YH, Peng TH, Broecker WS and Ostlund HG, 1984, The average vertical mixing coefficient for the oceanic thermocline, *Tellus*, **36B**, 212–7.

Liss PS, 1973, Processes of gas exchange across an air-water interface, *Deep-Sea Research*, **20**, 221–38.

Liss PS and Merlivat L, 1986, Air-sea gas exchange rates: Introduction and synthesis, in *The Role of Air-Sea Exchange in Geochemical Cycling*, edited by P Buat-Menard (Norwell, MA: D. Reidel).

Liss PS and Slater PG, 1974, Flux of gases across the air-sea interface, *Nature*, **274**, 181–4.

Liss PS, Watson AJ, Liddicoat MI, Malin G, Nightingale PD, Turner SM and Upstill-Goddard RC, 1993, Trace gases and air-sea exchange, *Philosophical Transactions of the Royal Society of London–Series A, Physical Sciences and Engineering*, **343**, 531–41.

Lobert JM, Butler JH, Montzka SA, Geller LS, Myers RC and Elkins JW, 1995, A net sink for atmospheric CH_3Br in the East Pacific Ocean, *Science*, **267**, 1002–5.

Lobert JM, Scharffe DH, Hao W-M, Kuhlbusch TA, Seuwen R, Warneck P and Crutzen P, 1991, Experimental evaluation of biomass burning emissions: nitrogen and carbon containing compounds, in *Global Biomass Burning. Atmospheric, Climatic, and Biospheric Implications*, edited by JS Levine (Cambridge, MA: MIT

Press).

Lovelock JE, 1975, Natural halocarbons in the air and in the sea, *Nature*, **256**, 193–4.

Lovelock JE, Maggs RJ and Wade RJ, 1973, Halogenated hydrocarbons in and over the Atlantic, *Nature*, **241**, 194–6.

Mabey W and Mill T, 1978, Critical review of hydrolysis of organic compounds in water under environmental conditions, *Journal of Physical and Chemical Reference Data*, **7**, 383–92.

Maguin F, Laverdet G, LeBras G and Poulet G, 1995, (in press), Discharge-flow mass spectrometric study of the reaction $BrO + HO_2$.

Manley SL and Dastoor MN, 1988, Methyl iodide (CH_3I) production by kelp and associated microbes, *Marine Biology*, **98**, 477–82.

Manö S and Andreae MO, 1994, Emission of methyl bromide from biomass burning, *Science*, **263**, 1255–8.

Methyl Bromide Global Coalition (MBGC), 1992, Proceedings of The Methyl Bromide Science Workshop Atmospheric Environmental Research, Inc., Cambridge, MA.

McElroy MB, Salawitch RJ, Wofsy SC and Logan JA, 1986, Reductions of Antarctic ozone due to synergistic interactions of chlorine and bromine, *Nature*, **321**, 759–62.

McElroy MB, Wofsy SC and Yung YL, 1977, The nitrogen cycle: perturbations due to man and their impact on atmospheric N_2O and O_3, *Philosophical Transactions Royal Society, London*, **277**, 159–81.

Mellouki A, Talukdar RK, Schmoltner A, Gierczak T, Mills MJ, Solomon S and Ravishankara AR, 1992, Atmospheric lifetimes and ozone depletion potentials of methyl bromide (CH_3Br) and dibromomethane (CH_2Br_2), *Geophysical Research Letters*, **19**, 2059–62.

Molina MJ and Rowland FS, 1974, Stratospheric sink for chlorofluoromethanes: Chlorine atom catalyzed destruction of ozone, *Nature*, **249**, 810–4.

Montzka SA, Butler JH, Elkins JW, Yvon S, Clarke A, Lobert J and Locke L, 1995, Difficulties associated with measuring atmospheric levels of methyl bromide and other methyl halides, *EOS, Transactions of the American Geophysical Union*, **76** (17), S160–1.

Montzka SA, Myers RC and Harris J, 1994, Methyl bromide at a rural site in the southeastern U.S., *EOS*, **75** (44), 110.

Moore RM and Tokarczyk R, 1993, Volatile biogenic halocarbons in the northwest Atlantic, *Global Biogeochemical Cycles*, **7**, 195–210.

Moore RM, Tokarczyk R, Tait VK, Poulin M and Geen C, 1995a, Marine phytoplankton as a natural source of volatile organohalogens, in *Naturally-Produced Organohalogens*, edited by A Grimvall and EWB deLeer (Dordrecht: Kluwer Academic).

Moore RM, Tokarczyk R, Webb ME and Wever R, 1995b, Marine phytoplankton as a natural source of halocarbons, *EOS*, **76** (17), S161.

Oechel WC, Cowles S, Grulke N, Hastings SJ, Lawrence B, Prudhomme T, Riechers G, Strain B, Tissue D and Vourlitis G, 1994, Transient nature of CO_2 fertilization in Arctic tundra, *Nature*, **371**, 500–3.

Oram DE and Penkett SA, 1994, Observations in eastern England of elevated methyl iodide concentrations in air of Atlantic origin, *Atmospheric Environment*, **28** (6), 1159–74.

Oremland RS, Miller LG and Strohmaier FE, 1994 (submitted), Degradation of atmospheric halogenated methanes. II: Chemical and bacterial attack of methyl-bromide in anaerobic sediments, *Applied and Environmental Biology*.

Orlando J, 1995, Gas-phase UV/visible absorption spectra of HOBr and Br_2O, *Journal of Physical Chemistry*, **99**, 1143–50.

Peng T-H, Broecker WS, Mathieu GG and Li Y-H, 1979, Radon evasion rates in the Atlantic and Pacific Oceans as determined during the GEOSECS program, *Journal of Geophysical Research*, **84**, 2471–86.

Penkett SA, Butler JH, Kurylo MJ, Reeves JM, Singh H, Toohey D and Weiss R, 1995, Chapter 10—Methyl Bromide, in *Scientific Assessment of Ozone Depletion: 1994*, edited by CA Ennis (WMO: Geneva).

Penkett SA, Jones BMR, Rycroft MJ and Simmons DA, 1985, An interhemispheric comparison of the concentrations of bromine compounds in the atmosphere, *Nature*, **318**, 550–3.

Pollock WH, Heidt LE, Lueb RA, Vedder JF, Mills MJ and Solomon S, 1992, On the age of stratospheric air and ozone depletion potentials in polar regions, *Journal of Geophysical Research*, **97**, 12993–9.

Poulet G, Pirre M, Maguin F, Ramaroson R and LeBras G, 1992, Role of the $BrO + HO_2$ reaction in the stratospheric chemistry of bromine, *Geophysical Research Letters*, **19**, 2305–8.

Prather M and Remsberg E (editors), 1992, The atmospheric effects of stratospheric aircraft. Report of the 1992 Models and Measurements Workshop. National Aeronautics and Space Administration. NASA 1292, 1, page 63.

Prather MJ, 1995, Atmospheric lifetimes of HCFCs and HFCs: Current estimates and uncertainties. Proceedings of the NASA/NOAA/AFEAS Workshop on the Atmospheric Degradation of HCFCs and HFCs, November 17–19, Boulder, CO.

Prinn RG, Weiss RF, Miller BR, Huang J, Alyea FN, Cunnold DM, Fraser PB, Hartley DE and Simmonds PG, 1995, Atmospheric trends and lifetime of trichloroethane and global average hydroxyl radical concentrations based on 1978–1994 ALE/GAGE measurements, *Science*, **269**, 187–92.

Rasche ME, Hicks RE, Hyman MR and Arp DJ, 1990a, Oxidation of monohalogenated ethanes and N chlorinated alkanes by whole cells of nitrosomonas-europea, *Journal of Bacteriology*, **172** (9), 5368–73.

Rasche ME, Hyman MR and Arp DJ, 1990b, Biodegradation of halogenated hydrocarbon fumigants by nitrifying bacteria, *Applied Environmental Microbiology*, **56** (8), 2568–71.

Reeves CE and Penkett SA, 1993, An estimate of the anthropogenic contribution of atmospheric methyl bromide, *Geophysical Research Letters*, **20** (15), 1563–6.

Rolston DE and Glauz RD, 1982, Comparisons of simulated with measured transport and transformation of methyl bromide gas in soils, *Geophysical Research Letters*, **17**, 561–4.

Salawitch RJ, Wofsy SC, Gottlieb EW, Lait LR, Newman PA, Schoeberl MR, Loewenstein M, Podolske JR, Strahan SE, Proffitt MH, Webster CR, May RD, Fahey DW, Baumgardner D, Dye JE, Wilson JC, Kelly KK, Elkins JW, Chan KR and Anderson JG, 1993, Chemical loss of ozone in the Arctic polar vortex in the winter of 1991–1992, *Science*, **261**, 1146–9.

Salawitch RJ, Wofsy SC and McElroy MB, 1988, Chemistry of OClO in the Antarctic stratosphere: implications for bromine, *Planetary Space Science*, **36**, 213–24.

Schauffler SM, Heidt LE, Pollock WH, Gilpin TM, Vedder JF, Solomon S, Lueb RA and Atlas EL, 1993, Measurements of halogenated organic compounds near the tropical tropopause, *Geophysical Research Letters*, **20** (22), 2567–70.

Shorter JH, Kolb CE, Crill PM, Kerwin RA, Talbot RW, Hines ME and Harriss RC, 1995, An effective soil surface sink for atmospheric methyl bromide, *Nature* **377**, 717–719.

Singh HB, 1977, Atmospheric halocarbons: evidence in favor of reduced average hydroxyl radical concentration in the troposphere, *Geophysical Research Letters*, **4**, 101–4.

Singh HB and Kanakidou M, 1993, An investigation of the atmospheric sources and sinks of methyl bromide, *Geophysical Research Letters*, **20** (12), 133–6.

Singh HB, Salas L, Shigeishi H and Crawford A, 1977, Urban-nonurban relationships of halocarbons, SF6, N_2O and other atmospheric trace constituents, *Atmospheric Environment* **11**, 819–28.

Singh HB, Salas LJ, Shigeishi H and Scribner E, 1979, Atmospheric halocarbons, hydrocarbons, and sulfur hexafluoride: Global distributions, sources, and sinks, *Science*, **203**, 899–903.

Singh HB, Salas LJ and Stiles RE, 1983a, Methyl halides in and over the eastern Pacific (40°N–32°S), *Journal of Geophysical Research*, **88**, 3684–90.

Singh HB, Salas LJ and Stiles RE, 1983b, Selected man-made halogenated chemicals in the air and oceanic environment, *Journal of Geophysical Research*, **88**, 3675–83.

Smethie WM, Takahashi TT, Chipman DW and Ledwell JR, 1985, Gas exchange and CO_2 flux in the tropical Atlantic Ocean determined from ^{222}Rn and pCO_2 measurements, *Journal of Geophysical Research*, **97** (C4), 7373–82.

Solomon S and Albritton DL, 1992, Time-dependent ozone depletion potentials for short-term and long-term forecasts, *Nature*, **357**, 33–7.

Solomon S, Mills M, Heidt LE, Pollock WH and Tuck AF, 1992, On the evaluation of ozone depletion potentials, *Journal of Geophysical Research*, **97**, 825–42.

Spivakovsky CM, Yevich R, Logan JA, Wofsy SC and McElroy MB, 1990, Tropospheric OH in a three-dimensional chemical tracer model: An assessment based on observations of CH_3CCl_3, *Journal of Geophysical Research*, **95** (D11), 18 441–471.

Sturges WT, Cota GF and Buckley PT, 1992, Bromoform emission from Arctic ice algae, *Nature*, **358**, 660–2.

Sturges WT, Sullivan CW, Schnell RC, Heidt LE and Pollock WH, 1993, Bromoalkane production by Antarctic ice algae, *Tellus*, **45B**, 120–6.

Swain CG and Scott CB, 1953, Quantitative correlation of relative rates. Comparison of hydroxide ion with other nucelophilic reagents toward alkyl halides, esters, expoxides, and acyl halides, *Journal of the American Chemical Society*, **75**, 141–7.

Talukdar RK, Mellouki A, Schmoltner AM, Watson T and Ravishankara AR, 1992, Kinetics of the OH reaction with 1,1,1-trichloroethane (methyl chloroform) and its atmospheric implications, *Science*, **257**, 227–30.

Tokarczyk R and Moore RM, 1994, Production of volatile organohalogens by phytoplankton cultures, *Geophysical Research Letters*, **21** (4), 285–8.

Traub WA, Jucks KW, Johnson DG and Coffey MT, 1994, Comparison of column abundances from three infrared spectrometers during AASE II, *Geophysical Research Letters*, **21** (23), 2591–8.

United Nations Environment Programme (UNEP), 1987, Montreal Protocol on Substances that Deplete the Ozone Layer, Final Act, Nairobi: UNEP.

United Nations Environment Programme (UNEP), 1992, Report of the Fourth Meeting of the Parties to the Montreal Protocol on Substances that Deplete the Ozone Layer, at Copenhagen.

United Nations Environment Programme (UNEP), 1994, 1994 Report of the Methyl Bromide Technical Options Committee for the 1995: Assessment of the UNEP Montreal Protocol on Substances that Deplete the Ozone Layer, Kenya, UNEP.

Upstill-Goddard RC, Watson AJ, Liss PS and Liddicoat MI, 1990, Gas transfer in lakes measured with SF6, *Tellus*, **42B**, 364–77.

Wanninkhof R, 1992, Relationship between wind speed and gas exchange over the ocean, *Journal of Geophysical Research*, **97** (C5), 7373–82.

Wanninkhof R, Ledwell JR and Broecker WS, 1985, Gas exchange wind speed relationship measured with sulfur hexafluoride on a lake, *Science*, **227**, 1224–6.

Watson AJ, Lovelock JE and Stedman DH, 1980, The problem of atmospheric methyl chloride, in *Proceedings of the NATO Advanced Study Institution on Atmospheric Ozone: Its Variation and Human Influences*, edited by AC Aikin, Federal Aviation Administration, Washington D.C.

Watson AJ, Upstill-Goddard RC and Liss PS, 1991, Air–sea gas exchange in rough and stormy seas measured by a dual tracer technique, *Nature*, **349**, 145–7.

Webster CR, May RD, Toohey DW, Avallone LM, Anderson JG, Newman P, Lait L, Schoeber MR, Elkins JW and Chan KR, 1993, Chlorine chemistry on polar stratospheric clouds in the arctic winter, *Science*, **261**, 1130–4.

Wennberg PO, Cohen RC, Stimpfle RM, Koplow JP, Anderson JG, Salawitch RJ, Fahey DW, Woodbridge EL, Keim ER, Gao RS, Webster CR, May RD, Toohey DW, Avallone LM, Proffitt MH, Loewenstein M, Podolske JR, Chan KR and Wofsy SC, 1994, Removal of stratospheric O_3 by radicals: In situ measurements of OH, HO_2, NO, NO_2, ClO, and BrO, *Science*, **266**, 398–404.

Wever R, Tromp MGM, Krenn BE, Marjani A and Tol MV, 1991, Brominating activity of the seaweed Ascophyllum nodosum: Impact on the biosphere, in *Environmental Science and Technology*, **25**, 446–9.

World Meteorological Organization (WMO), 1989, *Report of the International Ozone Trends Panel 1988*, WMO Report No. 18, Nairobi, Kenya.

World Meteorological Organization (WMO), 1990, *Scientific Assessment of Stratospheric Ozone: 1989*, Vol. 20 (Geneva: World Meteorological Organization).

World Meteorological Organization (WMO), 1992, *Scientific Assessment of Ozone Depletion: 1991*, Vol. 25 (Geneva: World Meteorological Organization).

World Meteorological Organization (WMO), 1995, *Scientific Assessment of Ozone Depletion: 1994*, edited by DL Albritton, RT Watson, PJ Aucamp, Vol. 37 (Geneva: World Meteorological Organization).

Wofsy SC, McElroy MB and Yung YL, 1975, The chemistry of atmospheric bromine, *Geophysical Research Letters*, **2** (6), 215–8.

Wright PB, 1988, An atlas based on the 'COADS' data set: Fields of mean wind, cloudiness and humidity at the surface of the global ocean, Report No. 14, Max-Planck-Institute für Meteorologie, Hamburg, Germany.

Wuebbles DJ, 1983, Chlorocarbon emission scenarios: Potential impact on stratospheric ozone, *Journal of Geophysical Research*, **88**, 1433–43.

Yagi K, Williams J, Wang N-Y and Cicerone RJ, 1993, Agricultural soil fumigation as a source of atmospheric methyl bromide, *Proceedings, National Academy of Science*, **90**, 8420–3.

Yagi K, Williams J, Wang N-Y and Cicerone RJ, 1995, Atmospheric methyl bromide (CH_3Br) from agricultural soil fumigations, *Science*, **267**, 1979–81.

Yung YL, Pinto JP, Watson RT and Sander SP, 1980, Atmospheric bromine and ozone perturbations in the lower stratosphere, *Journal of the Atmospheric Sciences*, **37**, 339–53.

Yvon SA and Butler JH, 1996, Uncertainties in the effect of the ocean on the atmospheric lifetime of CH_3Br, *Geophysical Research Letters* **23**, 53–56.

Zafiriou OC, 1975, Reaction of methyl halides with seawater and marine aerosols, *Journal of Marine Research*, **33** (1), 75–81.

Zhang Z, Saini RD, Kurylo MJ and Huie RE, 1992, A temperature-dependent kinetic study of the reaction of the hydroxyl radical with CH_3Br, *Geophysical Research Letters*, **19**, 2413–6.

3

Methyl Bromide and the Environment

MELANIE MILLER
Environmental Policy Analyst, PO Box 665, Napier, New Zealand

C.H. Bell, N. Price and B. Chakrabarti: The Methyl Bromide Issue
© 1996 John Wiley & Sons Ltd

1.1 OVERVIEW

Methyl bromide is produced by several human activities. The chemical industry manufactured approximately 75 500 tonnes in 1992—about 72 977 tonnes (96%) was sold as a fumigant, and the remainder used in chemical processing (TEAP 1995: 67). On average 64% of the fumigant is estimated to be emitted to the atmosphere (MBTOC 1994: 48). Methyl bromide is also generated as a by-product of leaded fuel combustion, forest and savannah burning, and certain industrial processes (WMO 1994: 10.11). Emissions to the atmosphere from all anthropogenic (human) sources have been esti-mated to be approximately 90 000 tonnes p.a. in 1991/2—with approximately 52% coming from the fumigant, approximately 33% from biomass burning, 13% from leaded petrol and 2% from industrial processes (see section 4.1.4).

The oceans also emit methyl bromide and certain other ozone depleting compounds such as methyl chloride. The natural emissions of chlorine and bromine are believed to be part of a balanced rate of natural ozone break-up and replenishment, established over billions of years (NOAA 1992: 6). Human activities have added to the natural background levels, disrupting the equilibrium. The additional chlorine and bromine breaks up ozone faster than it can be replenished, giving a net reduction in the level of ozone.

A recent international scientific panel examined the hypothesis that if human emissions of methyl bromide ceased, the oceans might emit an additional, equivalent amount to compensate. The panel concluded that ocean 'buffering' is realistically limited to only about 2 or 3%—an insignificant fraction (WMO 1994: 10.14).

Since 1991, four authoritative scientific panels have concluded that anthropogenic methyl bromide contributes notably to ozone layer depletion (WMO 1992, UNEP 1992a, SORG 1993, WMO 1994) (see section 6.1). The 1994 World Meterological Organisation (WMO) Scientific Assessment of Ozone Depletion, for example, stated that 'methyl bromide continues to be viewed as a significant ozone-depleting compound' (WMO 1994: xv), and concluded that a cessation of methyl bromide emissions would have a rapid impact on the extent of stratospheric ozone loss (WMO 1994: 10. 23).

Major uncertainties about methyl bromide's chemical reactions in the stratosphere have been mostly resolved (WMO 1994: 10.23). Some other factors, such as the size of certain sources and sinks, have not yet been clearly quantified. The 1994 WMO Scientific Assessment took account of the remaining uncertainties when calculating estimates, and expressed 'considerable confidence' in its findings (WMO 1994: 10.23). In 1994 there was sufficient scientific information for the Scientific Assessment to conclude that eliminating emissions of anthropogenic methyl bromide would reduce ozone losses significantly in future (WMO 1994: xxiii) (see section 7.1.5).

A UK scientific panel concluded that anthropogenic bromine emissions (from methyl bromide and halons) had raised the abundance of bromine in the atmosphere to more than 20 pptv (parts per trillion by volume) from the natural background level of about 9–14 pptv (SORG 1993: 26). CFCs and other anthropogenic chlorine emissions have raised the atmospheric abundance of chlorine from about 600 pptv to about 3500 pptv (SORG 1993: 22).

Although the abundance of anthropogenic bromine is less than 1% of the chlorine, it has a disproportionate impact. Each 1 pptv of bromine is estimated to destroy as much ozone as about 50 pptv of chlorine, because bromine is highly potent (WMO 1994: xxi) (see section 5.1).

The 1991 WMO Scientific Assessment estimated that if the abundance of methyl bromide in the atmosphere could be reduced by as little as 1.5 pptv, this would achieve a rapid reduction in chlorine/bromine loading, comparable to accelerating the CFC phase-out schedule by up to 3 years (WMO 1992: xvii). The total abundance of methyl bromide in the atmosphere— from natural and anthropogenic sources—is approximately 11 pptv (WMO 1994: 10.4). If anthropogenic sources contributed as little as 14% of the total, their elimination would benefit the ozone layer significantly.

Data in the 1994 WMO report indicated that anthropogenic methyl bromide might currently contribute about 4.5 pptv (2.0–7.5 pptv) of atmospheric bromine.

For policy-makers the most important calculations of the WMO Scientific Assessments are the integrated future chlorine/bromine loading scenarios. These estimate the consequence of different actions, such as continuing or eliminating emissions of ozone depleting compounds by certain dates. The scenarios are based on emissions rather than usage, taking account of the fact that methyl bromide fumigant emissions are significantly lower than sales and usage, and thus provides a better indicator of ozone impact than the Ozone Depletion Potential (ODP). The 1994 WMO Scientific Assessment identified four remaining areas where governments could eliminate emissions of ozone depleting compounds to reduce future ozone losses. Of the four, eliminating agricultural and industrial emissions of methyl bromide would have the single greatest impact (see section 6.1). Eliminating these emissions in the year 2001 is calculated to reduce the integrated future chlorine/bromine loading (which is related to the cumulative future loss of ozone) by 13% over the next 50 years (WMO 1994: xxiii). For comparison, rapidly eliminating remaining sources of other ozone depleting compounds would give reductions in integrated future chlorine/bromine loading of 10% for halons, 5% for HCFCs, and 3% for CFCs. Action on methyl bromide represented 42% of the protective action identified by the Scientific Assessment (see section 6.1).

Bromine has a more significant impact on ozone in the northern hemisphere than in the southern hemisphere. In the Antarctic, bromine may account for about 20–30% of springtime ozone losses due to anthropogenic chlorine and bromine, whereas in the Arctic and at mid-latitudes in the northern hemisphere, bromine could be responsible for about 50% of the winter ozone losses (UNEP 1992a: 11).

The atmospheric concentration of chlorine/bromine is still growing, but the rate has been slowed substantially, due to controls agreed under the Montreal Protocol, an international government agreement to protect the ozone layer. Bromine from halons, for example, increased by about 0.2–0.3 ppt/year in 1992 compared to 0.6–1.1 ppt/year in 1989 (WMO 1994: 2.1). Atmospheric abundance of chlorine grew by about 60 ppt/year in 1992 compared to 110 ppt/year in 1989 (WMO 1994: 2.1). Growth is expected to cease around the turn of the century. So average ozone losses are expected to increase for the remainder of the decade, with gradual recovery during the 21st century (WMO 1994: xiii).

Around 1998, ozone depletion is predicted to reach about 6–7% in summer/autumn and about 12–13% in winter/spring over Europe and North America, which is about 1.5% and 2.5% above 1993/4 levels (WMO 1994: xxiii) (see section 2.1). These ozone losses are expected to be accompanied by 8% and 15% increases in the UV radiation reaching the earth's surface, if other influences such as clouds and pollution remain constant (WMO 1994: xxiii).

The additional UV radiation is likely to cause economic disruption to agriculture, forestry, fisheries and tourist industries in many countries in future. For example, the incidence of diseases such as squamous cell carcinomas and bovine infectious keratoconjunctivitis would be expected to increase in outdoor livestock (Environmental Effects Panel 1994: II–23).

Some plant species and cultivars have mechanisms to acclimate to increased levels of UV-B; while in others photosynthesis, growth, reproduction and disease resistance will be inhibited (WHO 1994). Sensitive agricultural crops include types of maize, soybean, oats, barley, sugar beet and rice, and horticultural crops such as tomato, cucumber, melon, cauliflower and broccoli (Krupa and Kickert 1989, Environmental Effects Panel 1994). About half of the species of conifer seedlings so far studied can be adversely affected by UV-B (DoE 1993: 2). Increased UV from ozone depletion is likely to affect farmed fish, and is predicted to reduce ocean fish stocks by several millions of tonnes (Environmental Effects Panel 1994). The additional UV radiation in future is likely to have a significant impact on human health, suppressing certain immune responses and raising the incidence of infectious diseases, non-melanoma skin cancer and eye cataracts (Environmental Effects Panel 1994: iv) (see section 3.1).

Under the Montreal Protocol, phase-out schedules have been established for a number of chlorinated and brominated compounds. According to the WMO Scientific Assessment, the future bromine loading to the atmosphere will depend on choices made about the human production and emissions of methyl bromide (WMO 1994: xxiii).

Methyl bromide is the only widely-used, significant ozone depleting chemical that has not yet been controlled under the Montreal Protocol. The Protocol's Technology and Economic Assessment Panel (TEAP) believes that placing controls on methyl bromide would be more cost-effective than additional controls on other ozone depleting compounds, such as destroying CFCs in old equipment (TEAP 1995: ES. 4). Methyl bromide was officially listed as an ozone depleting compound in 1992. The quantity of methyl bromide manufactured for soil fumigation and most other purposes in industrialised countries was restricted from January 1995 (limited to 1991 levels) (UNEP 1992b: 38). During the Montreal Protocol meeting in 1995 governments agreed international schedules to phase out most uses of methyl bromide by the year 2010 in industrialised countries (section 7.1.4).

TEAP has reported that users of methyl bromide have been reluctant to commercialise and implement alternatives because they believe it to be more difficult to eliminate than CFCs or halons (TEAP 1995: 21). The Panel has examined the lessons for methyl bromide from the successful CFC and halon phase-outs. TEAP believes that the challenges faced by users of methyl bromide, while substantial, are no more difficult than the problems already overcome by users of the other ozone depleting compounds (TEAP 1995: ES.4).

In 1987 CFC and halon users claimed that no single chemical offered the same advantages, and that it would not be possible to develop acceptable alternatives. Numerous reports from industry, governments and independent institutes predicted few prospects for alternatives (TEAP 1995: 21).

TEAP noted that these initial perspectives failed to appreciate the potential for technical innovation, the power of market forces, the efficiency of public/private partnerships and the leadership of specific companies that pledged early phase-outs (TEAP 1995: 22) (see section 7.1.4).

When the Montreal Protocol agreed international controls for CFCs and halons, very few alternatives had been identified (TEAP 1995: 21). Users of methyl bromide are in a superior position at this stage, because technically feasible alternatives, either currently available or at an advanced stage of development, have been identified for a substantial proportion of uses (MBTOC 1994: 3).

Since 1993 some countries have set national reduction or phase-out schedules for methyl bromide in response to the international scientific assessments of ozone layer depletion. The USA, the world's largest user and manufacturer of methyl bromide, has finalised regulations to phase out production by the year 2001 (Federal Register 1993b). Austria will phase out by the year 2000. Denmark and the other Scandinavian countries will phase out by 1998 (Danish EPA 1994). Indonesia has also issued regulations to phase out use of methyl bromide by 1998 (Menteri Pertanian 1994). The European Union and Canada have adopted regulations to cut consumption by 25% in 1998 (Council of EU 1994; Canada Gazette 1994) (see section 7.1.1–7.1.3).

Several countries have already phased out major uses of methyl bromide because of concerns about the levels of residues detected in food, water or air in the vicinity of fumigation sites (Miller 1994). For example, horticulturalists in the Netherlands, Germany and a region of Italy heavily dependent on methyl bromide, have replaced soil fumigation with alternative methods of pest control (see section 9.1). Such cases demonstrate that the development of other methods, though disruptive and in certain cases very difficult, has been technically feasible for diverse crops, pests, climates and conditions.

This chapter describes trends in ozone losses, predicted impacts of increased UV radiation, compounds contributing to ozone losses, comparisons of chlorine and bromine, and scientific assessments of methyl bromide's role in ozone depletion. Subsequent sections discuss policy approaches, reductions and phase-out schedules planned by various governments, phase-outs already implemented, and local environmental impacts.

2.1. OZONE LAYER DEPLETION

2.1.1 THE OZONE LAYER

The ozone layer is a band in the stratosphere, between 10 km and 50 km (about 6 to 30 miles) above the earth's surface, where about 90% of atmospheric ozone is found (WMO 1994: xxv). There are about three molecules of ozone to every ten million air molecules. Ozone molecules are scattered so thinly in the stratosphere that if they were all collected together they would form a band around the earth no thicker than the skin of an orange (DoE 1993: 2).

Ozone in the stratosphere is essential to many living organisms, preventing most of the sun's damaging UV-B radiation and almost all the dangerous UV-C radiation from reaching the earth (UNEP 1993a: 2). A build-up of ozone at ground level, however, can have adverse effects on human health, trees and crops (Environmental Effects Panel 1994). Ground level ozone is produced by photochemical reactions involving pollutants mainly from motor vehicles, power stations and certain industrial activities (DoE 1991: 3).

Ozone molecules in the ozone layer are primarily created by UV radiation from the sun; the molecules are continually broken up and re-formed naturally. Scientists believe that a delicately balanced rate of ozone break-up and replenishment has probably been established over billions of years (NOAA 1992: 6). Gases from natural sources, such as methyl chloride and methyl bromide from the oceans, are part of this equilibrium.

Solar and seasonal cycles cause ozone levels to fluctuate regularly over time. Between summer and winter, for example, ozone concentrations differ by about 25% at mid-latitudes because the rate at which ozone is created depends upon sunlight (DoE 1993: 2). Global ozone levels decrease by 1–2% from the maximum to the minimum of a typical 11-year solar cycle (WMO 1994: xxx).

2.1.2 MEASUREMENTS OF OZONE LEVELS

Continuous ground-based measurements of ozone have been carried out in the Antarctic since 1957. Significant springtime Antarctic ozone reductions started in the mid-1970s, and ozone levels have continued a downward trend since (Shanklin 1994). Figure 3.1 shows that average springtime Antarctic ozone was typically around 300 Dobson units during the 1950s and 1960s. In the 1980s it dropped to around 200, and by 1993 the average had fallen below 150 Dobson units (WMO 1994: xxxi).

In 1988 an international panel of atmospheric experts scrutinised the ozone measurements made by satellites and ground-based instruments, and concluded that global ozone levels had diminished over the previous 17 years (WMO 1990). Natural events such as the solar cycle and severe volcanic eruptions explained only part of the observed reduction. The ozone layer is now thinner in most regions of the world. In 1993 global average depletion was estimated to be almost 3% per decade (DoE 1993: 3). Figure 3.2 shows the trend in global ozone levels from 1979 to 1994.

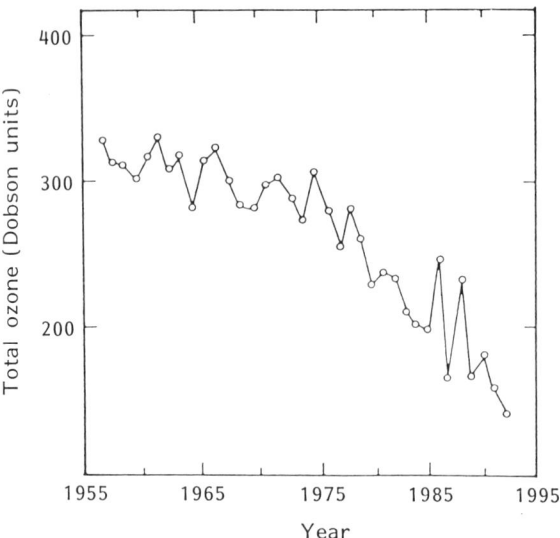

Figure 3.1. Average ozone levels recorded in the month of October (spring-time) in the Antarctic, 1957 to 1992 (Halley Bay 76 S). Reproduced from 'Scientific Assessment of Ozone Depletion: 1994' WMO Global Ozone Research and Monitoring Project, Report No. 37, 1994, page xxxi

2.1.3 THE ANTARCTIC OZONE 'HOLE'

In 1985 the British Antarctic Survey observed that ozone was largely destroyed over the Antarctic, covering a region as big as the United States and as deep as Mount Everest (DoE 1991: 4). This 'ozone hole' is the dramatic reduction in ozone which occurs over Antarctic for two to three months each southern spring (September to November). Strong winds during winter isolate Antarctica from the rest of the world, preventing replenishment of ozone from other regions. At the same time the very cold temperatures produce tiny aerosol particles of ice with sulphur and nitrogen compounds in the stratosphere. Complex chemical reactions occur on these particles, converting chlorine from inactive to active forms.

When sunlight returns to Antarctica in spring, the sun's UV rays free the chlorine and bromine (primarily ClO and BrO), starting a quick and massive destruction of ozone molecules. This creates a widespread 'hole' or severe thinning of the ozone layer (MfE 1994). About 25% of springtime ozone loss over Antarctica is probably due to reactions involving bromine (UNEP 1992a: 11). The ozone hole eventually diminishes when the stratosphere warms up sufficiently to disperse the polar stratospheric clouds and break up the winds which isolate the region. Ozone-rich air then flows in and helps to replenish the ozone depleted areas (DoE 1993: 4).

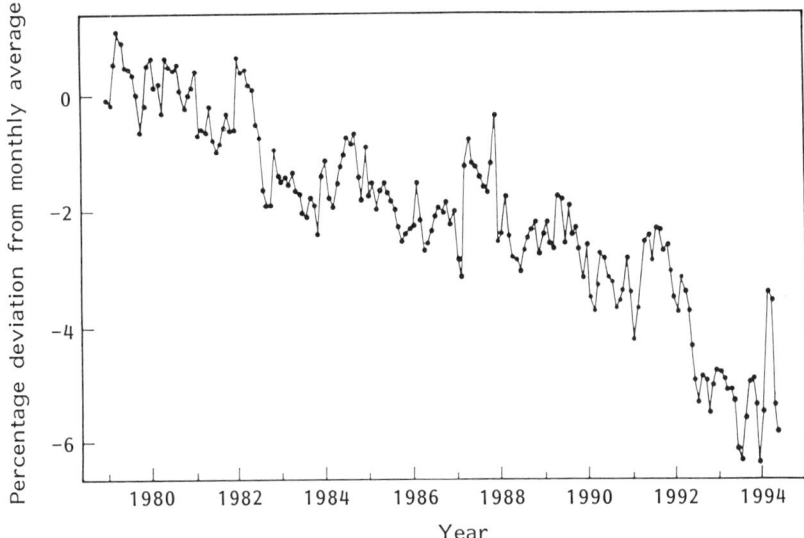

Figure 3.2. Trend in global ozone concentration, 1979–1994, between 60 N and 60 S. Adjusted to exclude 1–2% changes due to the solar cycle. Reproduced from 'Scientific Assessment of Ozone Depletion: 1994' WMO Global Ozone Research and Monitoring Project, Report No. 37, 1994, page xxx

Figure 3.3 shows the proportion of the southern hemisphere where ozone levels were less than 212 Dobson units in 1984 and 1992. This shows the progressive growth in the area of the ozone hole.

The Antarctic ozone holes in 1992, 1993 and 1995 were deeper and longer-lasting than any previously recorded. In just three weeks of 1993, for example, more than 70 million tonnes of ozone were destroyed over an area of 25 million square kilometres (MfE 1994). The severe depletion in 1992 and 1993 was due in part to the 1991 eruption of Mt Pinatubo, whose vast sulphate aerosol enhanced the effects of the chlorine and bromine already in the atmosphere (SORG 1993). But the effect of Mt Pinatubo has now diminished and ozone losses are still substantial. In 1994 ozone depletion of 20% to well over 50% occurred over very large areas of Antarctica (NOAA 1994). Depletion now occurs during summer (January and February) as well as spring.

The World Meterological Organisation (WMO) Scientific Assessment of Ozone Depletion in 1994 stated that a substantial ozone hole was expected to occur each austral spring for many more years (WMO 1994: xiv). The report also noted new results reaffirming the key role of anthropogenic bromine and chlorine in ozone loss in polar regions (WMO 1994: 3.1).

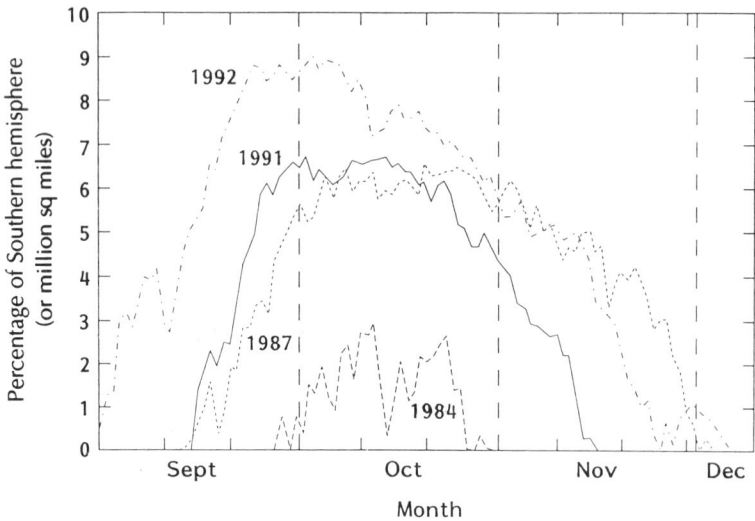

Figure 3.3. Proportion of the southern hemisphere where total ozone was less than 212 Dobson units, for the years 1984, 1987, 1991 and 1992. This approximately defines the area of the Antarctic ozone hole. Reproduced from 'Stratospheric Ozone 1993' HMSO, London, November 1993, page 8. Crown copyright is reproduced with the permission of the Controller of HMSO

2.1.4 DEPLETION IN THE SOUTHERN HEMISPHERE

Ozone depletion now occurs over many parts of the southern hemisphere. Losses diminish with increasing distance from the south pole, but extend as far north as Brazil. Data in the 1994 WMO report indicate that ozone losses in southern mid-latitudes currently average about 8% year-round (WMO 1994: xxiii).

NASA scientists in the USA have mapped out the average trend in ozone decline from 1978 to 1991, in percentage decline per decade. Figure 3.4 shows the decade trends for May to August (winter) in the southern hemisphere. Winter ozone levels diminished by 5–9% per decade over southern Chile and southern Argentina, 4–6% over New Zealand, 3–4% over Uruguay, 2–3% over Paraguay, 1–4% over southern Australia, southern Brazil and South Africa, and 1–2% over Namibia, Botswana, southern Zimbabwe, southern Mozambique, southern Madagascar and southern Bolivia. In countries nearer to the sub-tropics, UV levels are already high, so even small increases in UV radiation would be undesirable (section 3.1.2).

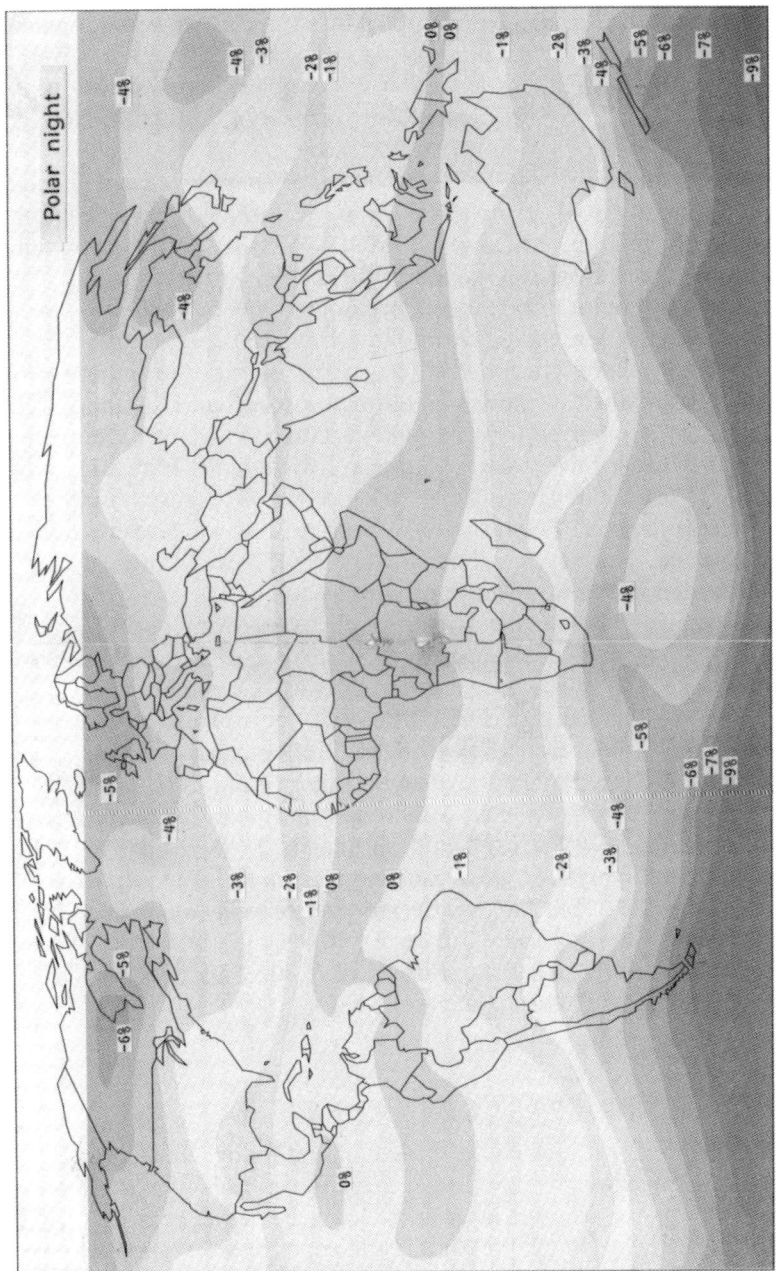

Figure 3.4. Average trends in ozone in different regions of the world, May to August from 1978 to 1991, measured in percentage of decline per decade. Reproduced from Stolarski, NASA, in 'Action on Ozone' updated September 1993, edited by N. Paulton, UNEP, Nairobi, pages 12–13

2.1.5   DEPLETION IN THE NORTHERN HEMISPHERE

Meterological conditions in the Arctic are different from the Antarctic and appear to have prevented an Arctic 'hole' from forming. However, several authoritative reports have not ruled out the possibility, commenting that there is no Arctic ozone hole 'at present' or 'so far' (DoE 1993: 4; NOAA 1992: 17).

At northern hemisphere mid-latitudes during winter/spring the ozone levels were about 10% lower in 1993 than in the late 1960s; during summer/fall they were about 5% lower (WMO 1994: xxiii). Ozone depletion in the northern hemisphere has generally been less intense than in the southern hemisphere, but it affects a greater number of countries.

Figure 3.5 shows the percentage decline in ozone levels per decade across the northern hemisphere, from 1978 to 1991, during the months of December to March when ozone losses are heavier over mid latitudes. By 1991 in Europe the per decade winter ozone decline was 6% over Germany, Denmark, Netherlands, Luxembourg, Austria, Switzerland, Italy and most of France, 5% over Greece, former Yugoslavia, Hungary, Bulgaria, Romania, Czechoslovakia, Poland, 4–7% over the former Soviet Union, 4–6% over Sweden, Norway and Finland, and 3–4% over Portugal, Eire and Turkey. Further ozone losses have occurred since 1991.

Figure 3.5 shows that in the Americas in 1991 winter ozone had diminished by 4–7% per decade over the USA, 2–5% over southern Canada and Alaska, and 3–4% over northern Mexico. Ozone levels in north Africa and the middle East declined by 3–5% over Tunisia, northern Algeria and Iran, 3–4% over Israel, Syria, Jordan, Iraq, 2–4% over Morocco, Canary Islands, Egypt. Libya and northern Saudi Arabia. In Asia ozone had diminished by 6–7% per decade over northern Japan, northern China and Mongolia, 4–6% over southern Japan, 3–5% over Pakistan, Afghanistan and Nepal, and 2–5% over northern India and southern China.

In early 1995, record ozone losses occurred over Siberia (more than 35% loss), large parts of Europe, North America and Asia (5–15% loss) (BNA 1995). The WMO reported that the acrosol from Mt Pinatubo no longer contributed to these ozone losses (Naughton 1995).

2.1.6   FUTURE OZONE LOSSES

The Montreal Protocol has slowed the growth of anthropogenic chlorine and bromine in the atmosphere (WMO 1994: 2.1). When emissions eventually cease, recovery of the ozone layer will be very slow because some compounds have extremely long lifetimes in the atmosphere (section 5.1).

The level of anthropogenic chlorine and, to a lesser extent, bromine in the stratosphere is expected to continue growing for the next few years, so

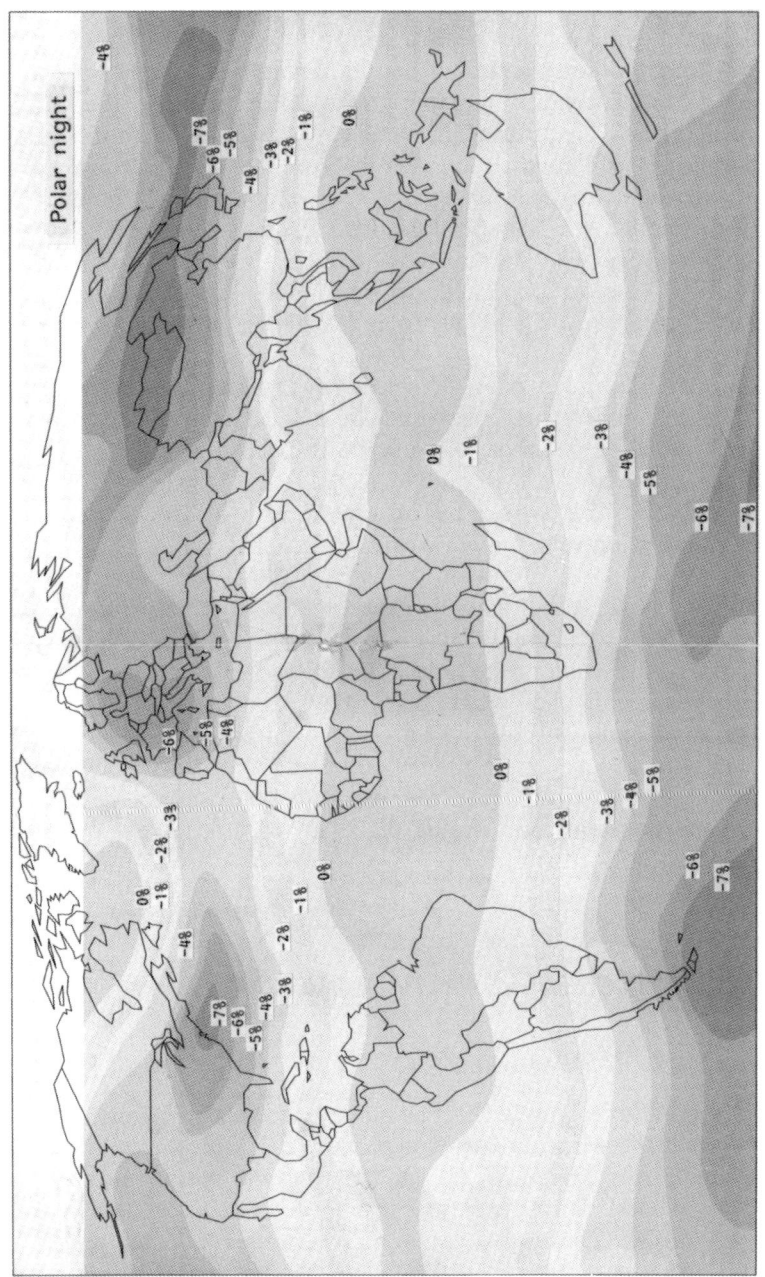

Figure 3.5. Average trends in ozone in different regions of the world, December to March from 1978 to 1991, measured in percentage of decline per decade. Reproduced from Stolarski, NASA, in 'Action on Ozone' updated September 1993, edited by N. Paulton, UNEP, Nairobi, pages 12–13

ozone losses are expected to increase for the remainder of the decade (WMO 1994: xiii). The 1994 WMO Scientific Assessment predicts that ozone depletion might reach its peak around the year 1998 (WMO 1994: xxiii). This is based in the assumption that all nations will comply fully with the Montreal Protocol controls already agreed. Unfortunately, the CIS is failing to comply, and illegal use of controlled substances is an emerging problem in industrialised countries. Trade in black market CFCs and other prohibited ozone depleting substances has been uncovered in a number of countries (for example, see New Scientist 1995: 7). So ozone depletion will more probably reach its peak after 1998.

The 1994 WMO report predicted the maximum size of ozone losses (compared to levels in the 1960s) likely to be reached in the next few years:

- In northern mid-latitudes, ozone losses of 12–13% during winter/spring, and 6–7% in summer/fall. This would be associated with increases in surface UV radiation of about 15% and 8% respectively.
- In southern mid-latitudes, losses are expected to reach roughly 11% on a year-round basis. This would be associated with a 13% increase in surface UV radiation year-round (WMO 1994: xxiii).

These estimates assume full compliance with the Montreal Protocol, so actual losses are likely to be greater. If there are major volcanic eruptions or extremely cold Arctic winters the ozone losses and UV increases would probably be larger in individual years (WMO 1994: xxiii).

During the 1980s and 1990s official estimates of future ozone losses were found frequently to underestimate the actual losses. According to the UK's Department of the Environment, for example, the springtime losses observed over northern latitudes during the 1980s were at least double the amount predicted (DoE 1991: 4).

3.1   PREDICTED INCREASES IN UV RADIATION

3.1.1   UV RADIATION MEASUREMENTS

The amount of UV-B reaching the earth's surface depends on factors such as the ozone column (the vertical density of ozone which varies naturally with latitude and time of year), angle of the sun (varying with season and time of day), cloud cover, and pollution from sulphate aerosols and tropospheric (ground level) ozone. The global biologically effective UV falling on a horizontal surface occurs primarily during the midday hours, about 50% during the four hours centred around the noon-time zenith (WHO 1994: 45).

The total ozone column tends to be greater in spring than in autumn (except in the tropics) so more UV-B normally reaches the earth's surface in early autumn than early spring (WHO 1994: 18). In the tropics the ozone column and angle of the sun remain relatively constant over the year, so UV-B levels vary little with season (WHO 1994: 19).

Over the last century UV radiation reaching the earth in some industrial regions may have been reduced by 6–18% due to sulphate aerosol pollution and by 3–10% due to increases in ground level ozone pollution (Environmental Effects Panel 1994: 1.18). Such changes make the interpretation of recent UV trends complex. High in the Swiss alps, studies detected UV-B increases of 0.7 ± 0.3% per year between 1981 and 1989 (Blumthaler and Ambach 1990). Measured UV in Lauder, New Zealand, increased by about 0.6% per year from 1981 to 1990, anti-correlated with ozone column data (Zheng and Basher 1993). Enhanced UV levels have also been detected in Canada, Germany, Austria, Greece, and parts of the USA (Environmental Effects Panel 1994: 1.14). These changes were probably due to a combination of ozone depletion and changes in pollution or cloud cover.

The WMO 1994 Scientific Assessment concluded that there is overwhelming experimental evidence that, all other things being equal, decreases in atmospheric ozone result in UV-B increases at the earth's surface, in quantitative agreement with predictions from radiative transfer models (WMO 1994: 9.1). Figure 3.6 illustrates the calculated changes in UV-B at latitude 45 °N for the period 1979 to 1994. The rate of increase in UV is not constant but is anti-correlated with ozone changes, including perturbations due to the 11-year solar cycle (WMO 1994: 9.15). At latitude 45° the trend is approximately +0.5% per year in both hemispheres. At latitude 55° the trends are significantly larger, particularly in the southern hemisphere (WMO 1994: 9.16). Gradients are larger at shorter wavelengths, and continue to increase at higher latitudes.

In 1992/3 large increases in UV-B were measured at northern middle and high latitudes compared with previous years, despite variability in cloudiness. The WMO commented that, for the first time, greatly enhanced UV was seen for extended periods of time in heavily populated latitudes (WMO 1994: 9.14).

Figure 3.7 shows the relationship between changes in column ozone and erythemally-weighted UV radiation; the relationship is not linear (WMO 1994: 9.14). The World Health Organisation reports that a 10% reduction in ozone could lead to a 15–20% increase in UV exposure, depending on the biological process being considered (WHO 1994: 6). UNEP's Environmental Effects Panel has calculated the probable trends in biologically effective UV doses, assuming no cloud cover or aerosol pollution (Table 3.1). The Panel found that significant increases were expected in the mid latitudes of both hemispheres (Environmental Effects Panel 1994: 1.18).

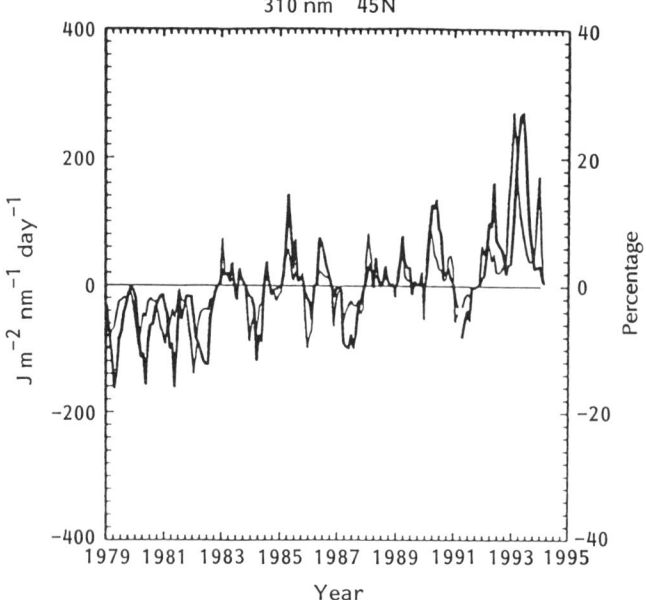

Figure 3.6. Calculated deviations from the 1978–1993 average monthly values of the daily spectral irradiance (310 nm) at latitude 45 N. Thick line and left scale give absolute irradiance changes; thin line and right scale give percentage change. Reproduced from 'Scientific Assessment of Ozone Depletion: 1994' WMO Global Ozone Research and Monitoring Project, Report No. 37, 9.17

3.1.2 IMPACTS OF INCREASED UV RADIATION

The WHO and UNEP have published scientific assessments of the likely impacts of predicted increases in UV-B radiation resulting from ozone depletion (WHO 1994; Environmental Effects Panel 1994). Figure 3.8 summarises some of the effects.

Additional UV-B radiation is likely to have a significant impact on human health in future (Environmental Effects Panel 1994: iv):

(a) The Panel estimates that a sustained 1% decrease in stratospheric ozone will result in an increase of approximately 2% in the incidence of non-melanoma skin cancer (Environmental Effects Panel 1994: iv).

(b) A 1% increase in ozone depletion may be associated with a 0.6–0.8% increase in eye cataracts.

(c) Additional UV exposure suppresses certain immune responses. This may enhance the risk of certain infections and might reduce the

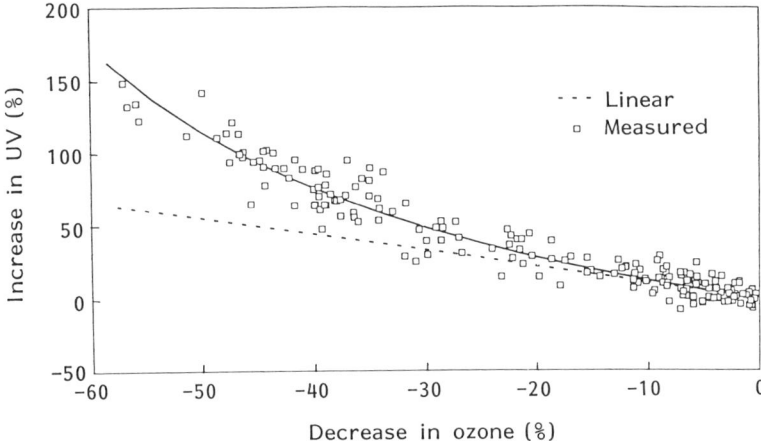

Figure 3.7. Relationship between erythemally weighted UV radiation and ozone column changes. Measurements from South Pole, 1 Feb 1991 to 12 Dec 1992. Reproduced from Booth and Madronich in 'Scientific Assessment of Ozone Depletion: 1994' WMO Global Ozone Research and Monitoring Project, Report No. 37, 9.15

efficacy of vaccination programmes (Environmental Effects Panel 1994: iv; WHO 1994: 178).

The Canadian government has calculated that increased UV from ozone depletion has already increased the risk of skin cancer in the Canadian population by more than 7% (Canada Gazette 1994). Actual UV exposure will vary from one geographical region to another, depending on factors such as cloud cover and air pollution. Outdoor workers would generally receive higher exposures than other population groups (WHO 1994: 50). The World Health Organisation recommends avoiding outdoor activities when UV levels are highest, or using UV-protective sunglasses, clothing and sunscreen (WHO 1994: 7). However, some sunscreens may not be effective in preventing immune suppression (WHO 1994: 244).

Some plants have mechanisms for ameliorating the effects of increased UV radiation and may acclimate to increased levels. But in some species and cultivars, physiological and developmental processes, such as photosynthesis, reproduction and disease resistance, are adversely affected by UV-B. Plant growth can be directly affected (WHO 1994). Sensitive crops include types of maize, soybean, oats, barley, sugar beet and rice, and horticultural crops such as tomato, cucumber, melon, cauliflower and broccoli (Krupa and Kickert 1989, Environmental Effects Panel 1994). About half of the

Table 3.1. Calculated trends in biologically effective UV exposures, 1979–1993, at various latitudes, giving annual erythemal dose, and percent increase in doses for DNA damage, erythema induction in humans, and skin cancer induction in laboratory mice. Annual UV exposures were calculated from stratospheric ozone (SBUV and SBUV/2) satellite instrument measurements, assuming no aerosol pollution or cloud cover. Percentage changes are expressed relative to the 1979–93 averages. Uncertainties are one standard deviation. Reproduced from 'Environmental Effects of Ozone Depletion: 1994 Assessment' Report of the Environmental Effects Panel, November 1994, UNEP, I-20.

Latitude	Erythema annual dose (MJ m^{-2})	Percentage increase during 1979–1993		
		Erythema induction in humans	DNA damage	Skin cancer in mice
65 N	0.47	5.6 ± 2.1	10.3 ± 3.8	7.9 ± 2.9
55 N	0.68	6.6 ± 1.8	11.6 ± 3.3	8.9 ± 2.5
45 N	1.01	8.6 ± 1.9	14.5 ± 3.3	11.1 ± 2.4
35 N	1.46	7.9 ± 1.9	13.0 ± 3.1	9.8 ± 2.3
25 N	1.94	6.0 ± 1.9	9.6 ± 3.0	7.1 ± 2.2
15 N	2.35	4.0 ± 1.5	6.3 ± 2.4	4.7 ± 1.7
5 N	2.58	2.6 ± 1.7	4.1 ± 2.7	3.0 ± 2.0
5 S	2.58	3.2 ± 1.5	5.1 ± 2.3	3.0 ± 1.7
15 S	2.34	1.8 ± 1.4	2.8 ± 2.1	2.0 ± 1.6
25 S	1.95	5.4 ± 1.5	8.7 ± 2.4	6.4 ± 1.8
35 S	1.49	8.0 ± 1.5	13.2 ± 2.4	9.8 ± 1.8
45 S	1.06	8.0 ± 3.9	13.6 ± 3.2	10.2 ± 2.4
55 S	0.73	10.7 ± 2.8	18.8 ± 5.1	14.2 ± 3.8
65 S	0.51	20.4 ± 4.4	37.2 ± 8.2	27.1 ± 6.1

species of conifer seedlings so far studied can be adversely affected by UV-B (DoE 1993: 2).

In agriculture, raised UV levels will necessitate the selection or breeding of more UV-tolerant cultivars. Secondary effects in plants, such as changes in plant form or timing of developmental phases, may be as important as the direct damaging effects of UV-B. Changes at the ecosystem level cannot be easily quantified. Nevertheless, changes will be of importance in agricultural production and in natural ecosystems (Environmental Effects Panel 1994: v).

The production of phytoplankton in the Antarctic shows a direct reduction due to ozone-related increases in UV-B (Environmental Effects Panel 1994: vi). The most severe effects of UV radiation on fish, shrimp, amphibians and other water organisms are on reproductive capacity and the early developmental stages. Small increases in UV-B exposure could result in significant reductions in the populations of certain organisms in the food

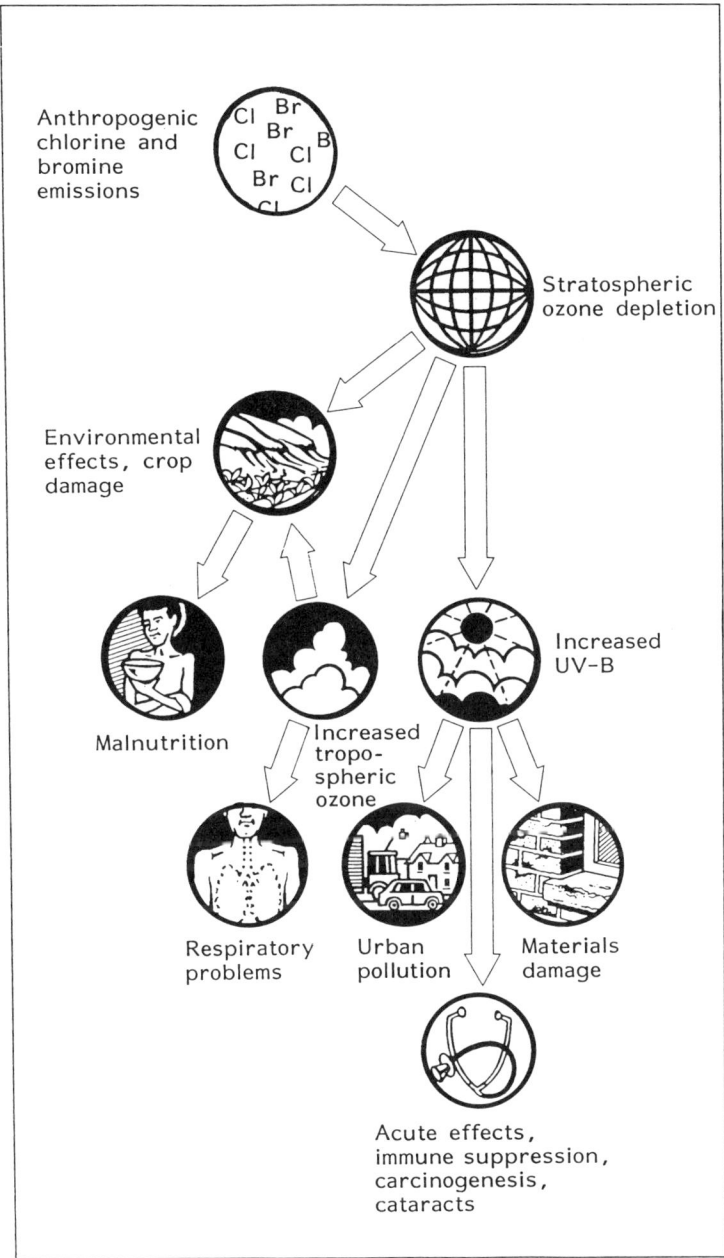

Figure 3.8. Environmental impacts of ozone layer depletion. Based on figure in 'Action on Ozone' updated September 1993, edited by N. Paulton, UNEP, Nairobi

chain (Environmental Effects Panel 1994: vi). Increased UV is predicted to reduce ocean fish stocks by several millions of tonnes (Environmental Effects Panel 1994: 4.1). This has implications for human welfare and fishing industries because more than 30% of the animal protein used as human food comes from the sea (Environmental Effects Panel 1994: v).

If additional UV radiation decreases the productivity of marine and terrestrial ecosystems as predicted, it would probably reduce the uptake of atmospheric carbon dioxide and alter both sources and sinks of greenhouse and important trace gases such as carbon monoxide and carbonyl suphide. This could potentially reinforce the atmospheric build-up of such gases (Environmental Effects Panel 1994: vi).

Additional UV-B would result in increased production and destruction of polluting gases such as ground level ozone and hydrogen peroxide, which are known to have adverse effects on human health, plants and building materials. In polluted regions with high NO_x tropospheric ozone is expected to increase, reaching potentially harmful concentrations earlier in the day (Environmental Effects Panel 1994: viii). Many polymers (plastics and rubber) and other materials of commerical significance are adversely affected by UV, reducing their useful lifetimes outdoors. The damage ranges from discolouration to loss of mechanical integrity, and will add to the cost of using or maintaining these outdoor materials in future (Environmental Effects Panel 1994: viii).

UNEP's Environmental Effects Panel concluded in 1994 that the increases in UV-B radiation already observed, and expected in the future, would have consequences of significant magnitude. This applied even to the most favourable scenario of ozone depletion. The Panel noted that many of the changes induced by increased UV radiation would be so complex that it was not possible to quantify them. The report strongly endorsed continued efforts to prevent ozone losses (Environmental Effects Panel 1994: ix).

4.1 COMPOUNDS INVOLVED IN OZONE DEPLETION

'For about a billion years, the natural ozone system worked smoothly, but now human beings have upset the delicate balance'. National Oceanic and Atmospheric Administration 1992 'Our Ozone Shield' Reports to the Nation, Washington DC.

Both natural and human-made gases can break up ozone molecules. For millennia, ozone molecules have been broken up by naturally occurring emissions such as methyl chloride from the oceans (NOAA 1992: 6, 8). The natural emissions of chlorine and bromine are believed to be part of a relatively stable rate of natural ozone break-up and replenishment estab-lished over billions of years (MfE 1994; NOAA 1992). Surprisingly small additional quantities of gases—primarily from human activities—have been able to disrupt the balance.

Ozone depletion is a net reduction in the level of ozone, arising when ozone molecules are broken up faster than they can be replenished. Severe volcanic eruptions (Mt Pinatubo in 1991 and El Chichon in the early 1980s) and shifting weather patterns have contributed to the perturbation of ozone levels in certain years. But the sustained downward trend and size of ozone losses clearly indicate that human activities are primarily responsible (WMO 1992, SORG 1993, WMO 1994).

The Stratospheric Ozone Review Group (SORG) which advises the UK Department of the Environment, pointed out that the most obvious change since the early 1980s has been the continued steady rise in the chlorine and bromine loadings of the atmosphere (SORG 1993: 10). The WMO Scientific Assessments in 1991 and 1994 concluded that ozone losses are largely due to human-made chlorine and bromine (WMO 1992: xi, WMO 1994: xiv). When compounds such as CFCs, carbon tetrachloride, halons and methyl bromide reach the stratosphere intense UV-C radiation severs their chemical bonds, releasing chlorine and bromine. The chlorine and bromine act as catalysts, repeatedly destroying ozone molecules by stripping off oxygen atoms (MfE 1994).

4.1.1 NATURAL AND ANTHROPOGENIC EMISSIONS

Severe volcanic eruptions contribute to temporary, net ozone reductions because they throw additional quantities of sulphur high into the atmosphere. Sulphate aerosol particles enhance the action of chlorine and bromine, probably because they provide a suitable surface for reactions (SORG 1993: 2).

Several emissions from natural sources give a natural background level of chlorine and bromine in the atmosphere. The oceans generate significant quantities of methyl chloride (CH_3Cl), contributing the majority of the natural chlorine in the atmosphere (which totals about 600 pptv) (WMO 1994: xxix; MfE 1994). (See Figure 3.9).

Natural bromides from the oceans—dibromomethane (CH_2Br_2), dibromochloromethane ($CHBr_2Cl$) and bromochloromethane (CH_2BrCl)—contribute approximately 6 pptv of bromine in the atmosphere (Figure 3.10) (SORG 1993: 25). SORG estimated that naturally generated methyl bromide contributes about 3–8 pptv bromine (SORG 1993: 26). More recent research indicates that the oceans may contribute a smaller net amount or that the oceans may be a net sink (Lobert *et al* 1995). The 1994 WMO Scientific Assessment examined whether the oceans might start to emit a large additional quantity of methyl bromide if anthropogenic emissions were eliminated. The Assessment concluded that ocean 'buffering' is realistically

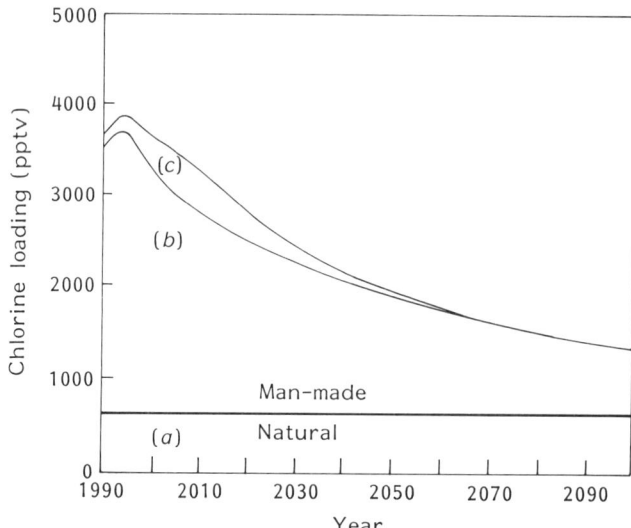

Figure 3.9. Current and projected atmospheric chlorine loading from natural and anthropogenic sources. Reproduced from 'Stratospheric Ozone 1993' HMSO, London, November 1993, page 25. Key: (*a*) natural methyl chloride, (*b*) anthropogenic CFCs, carbon tetrachloride and methyl chloroform, (*c*) anthropogenic HCFCs consumed at the maximum amount allowed under the 1992 amendments to the Montreal Protocol

limited to only about 2 or 3% of the atmospheric concentration (WMO 1994: 10.14)—an insignificant fraction.

The oceans emit not only methyl chloride and bromides but a number of other gases that play a role in ozone chemistry, such as methane (CH_4), carbon monoxide (CO), carbonyl sulphide (COS), and significant quantities of nitrous oxide (N_2O) (WMO 1992). Emissions of gases often vary with location, temperature, ocean circulation, and other processes.

Some of the oceanic gases, for example methyl chloride, are also generated by land-based natural processes and/or anthropogenic activities. Other sources of nitrous oxides include denitrification processes in soils, deforestation, nitrogenous fertilizers, and the burning of biomass (vegetation), coal, oil and aircraft fuel. Nitrous oxides in the stratosphere are important in ozone chemistry (WMO 1994: 2.20).

Other natural sources of methane include termites and wetlands, while anthropogenic sources account for about two-thirds of the gas and include fossil fuels, biomass burning, landfills, agricultural animals and rice paddies (WMO 1994: 2.19). Methane is an important greenhouse gas that contributes to the oxidising capacity of the troposphere and affects the reactive

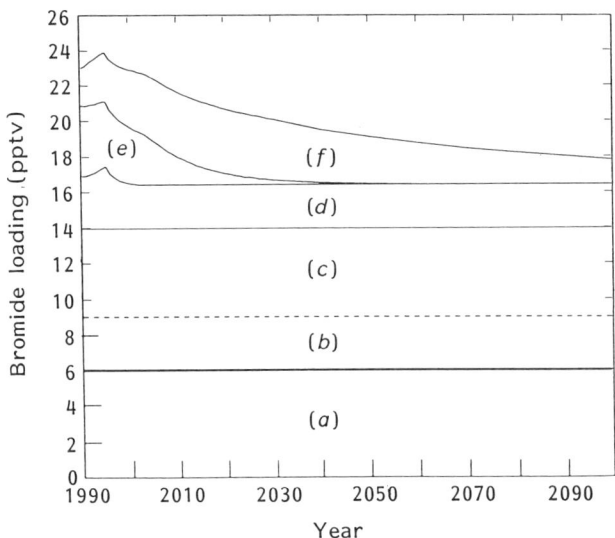

Figure 3.10. Current (1990) and projected atmospheric bromine loading from natural and anthropogenic sources. Reproduced from 'Stratospheric Ozone 1993' HMSO, London, November 1993, page 26. Key: (*a*) natural sources (CH_2Br_2, CH_2BrCl and $CHBr_2Cl$), (*b*) natural methyl bromide from the oceans, (*c*) anthropogenic and/or natural methyl bromide, (*d*) minimum proportion of anthropogenic methyl bromide (3 pptv in 1990) showing the projected freeze in industrialised countries, and assuming no growth in use in developing countries, (*e*) anthropogenic halon-1211, (*f*) anthropogenic halon-1301

state of many other gases. In the stratosphere it is a source of hydrogen and water vapour, and a sink for atomic chlorine.

Carbonyl sulphide generates sulphur aerosol particles in the atmosphere (SORG 1993: 2), enhancing the damaging effect of chlorine and bromine. The major sources of carbonyl sulphide are believed to be the oceans (20–40%), soils (20%), biomass burning (10%), and other anthropogenic activities (20%) (WMO 1992: 1.16).

Termite mounds are estimated to emit 100 000 tonnes per year of chloroform ($CHCl_3$) (WMO 1992: 1.15). Most chloroform emissions are natural, but approximately 40% are believed to come from anthropogenic sources.

Like the oceans, biomass burning generates a range of gases that can play a role in ozone chemistry: methyl chloride, methyl bromide, nitrous oxide, methane and carbonyl sulphide. About 80% of biomass burning is estimated to occur in the tropics, due to activities such as deforestation and savanna burning (WMO 1994: 10.8). WMO considered that less than 20% of biomass burning could be described as natural (WMO 1994: 10.11).

Table 3.2 shows the estimated atmospheric concentrations of many of the natural and anthropogenic compounds contributing to ozone chemistry. Natural emissions of the gases cannot normally be controlled or prevented. However, action could be taken to reduce anthropogenic emissions of compounds such as methane, nitrous oxides and carbonyl sulphide. Reductions in greenhouse gases that are not controlled by the Montreal Protocol, methane and nitrous oxide, are due to be introduced as a result of the international Framework Convention on Climate Change. The Montreal Protocol has focused initially on controlling solely anthropogenic chlorinated and brominated compounds such as CFCs and halons; the examination of methyl bromide marks a shift towards examining compounds that have complex anthropogenic and natural sources.

Table 3.2. Natural and anthropogenic compounds contributing to ozone chemistry, showing estimated atmospheric concentrations (1992) and increase in concentration per annum (early 1990s). *Sources*: Compiled from data in WMO 1994, sections 2 and 10; WMO 1992: 1.4, 1.16; SORG 1993: 26.

Gas	Sources Natural	Anth.	Concentration (pptv)	Increase (pptv/yr)
Substances involved in ozone chemistry:				
Methane (CH_4)	✓	✓	1 714 000	5000
Nitrous oxide (N_2O)	✓	✓	310 000	550
Carbonyl sulphide (COS)	✓	✓	c. 510	–
Sulphur from severe volcanic eruptions	✓		Variable	–
Sources of chlorine:				
Methyl chloride (CH_3Cl)	✓	✓	600	–
CFC-12 (CCl_2F_2)		✓	503	13.0
CFC-11 (CCl_3F)		✓	268	2.5
CFC-113 (CCl_2FCClF_2)		✓	82	2.5
CFC-114 ($CClF_2CClF_2$)		✓	20	0.5
CFC-115 ($CClF_2CF_3$)		✓	–	–
Methyl chloroform (CH_3CCl_3)		✓	160	3.5
Carbon tetrachloride (CCl_4)		✓	132	(−1)
Chloroform ($CHCl_3$)	✓	✓	c. 10	–
HCFC-22 (CHF_2Cl)		✓	102	7.0
HCFC-142b (CF_2ClCH_3)		✓	3.5	c. 1.0
HCFC-141b ($CFCl_2CH_3$)		✓	0.3	c. 0.7
Sources of bromine:				
Methyl bromide (CH_3Br)	✓	✓	11	–
Halon-1211 (CF_2ClBr)		✓	2.5	c. 0.1
Halon-1301 (CF_3Br)		✓	2	0.2
Halon-2402 ($C_2F_4Br_2$)		✓	c. 2	–

4.1.2 ANTHROPOGENIC CHLORINE

Before man-made chlorine compounds were used, the abundance of chlorine in the atmosphere (mainly from the oceans) was probably about 600 pptv (MfE 1994). By 1992 the concentration of chlorine in the atmosphere had been raised to about 3500 pptv (WMO 1994: 10.3). See Figure 3.9. CFCs (CFC-11, 12 and 113) contribute about 57% of the chlorine currently in the atmosphere, while carbon tetrachloride, methyl chloroform and HCFC-22 contribute a further 12%, 10% and 3% respectively (WMO 1994: xxix). Phase-out schedules for these four major anthropogenic sources of chlorine have been agreed under the Montreal Protocol.

CFCs have carried millions of tonnes of extra chlorine into the stratosphere (NOAA 1992: 8). Their Ozone Depletion Potentials, indictors of relative potency, are estimated to range from approximately 0.4 to 1.0 (see Table 3.3 and section 5.1) (WMO 1994: 13.17). CFCs have been used for flexible and rigid foams (used in furniture, bedding, carpet underlay, packaging), as insulation material (in cold stores, refrigerators, freezers, ice makers and buildings), as refrigerants (in domestic refrigerators, freezers,

Table 3.3. Comparison of anthropogenic chlorine and bromine compounds, showing best-estimate Ozone Depletion Potentials (ODP), internationally agreed reduction and phase-out (PO) dates, estimated emissions to the atmosphere (in 1991), and estimated lifetimes in the atmosphere. *Source*: Compiled from WMO 1994: 10.11, 10.23, 13.6, 13.17, DoE 1993: 6, MBTOC 1994: 2, SORG 1993: 23, 26.

Compound	Estimated ODP	Control dates*	Estimated emissions (tonnes/year)	Lifetime (years)
Anthropogenic chlorine:				
CFCs	0.4–1.0	75% cut 1994 PO 1996	643 200	50–1700
Carbon tetrachloride	1.2	85% cut 1995 PO 1996	80 000	42
Methyl chloroform	0.1	50% cut 1994 PO 1996	608 000	5.4
HCFCs	0.01–0.1	35% cut 2004 65% cut 2010 90% cut 2015 PO 2020	237 000	1.4–19.5
Anthropogenic bromine:				
Halons	5.0–13.0[†]	PO 1994	12 500[†]	20–65[†]
Methyl bromide	0.3–0.8	25% cut 2001 50% cut 2005 PO 2010	89 750[‡]	c. 1

*Reductions and phase-out apply to industrialised countries; a limited number of essential uses may continue after the PO date. Developing countries normally have an additional ten years in which to achieve phase-out.
[†]Data for halon 1211 and 1301 only.
[‡]Mid-point of range, refer to Table 3.4.

air conditioning, commercial and transport refrigeration and freezing systems), as solvents (in the electronics industry, precision and general engineering and in dry cleaning) and in aerosols such as asthma inhalers (DoE 1993: 10).

International controls have reduced the rate of growth, but the atmospheric concentration of CFCs is still rising (WMO 1994: 2.1). Large quantities of CFCs remain in refrigeration units and air conditioning systems. The vast majority of CFC production will be phased out in industrialised countries by 1996. Developing countries have an additional 10 years in which to achieve this. Some CFCs are so long-lived (see Table 3.3) that they will continue to have a detrimental effect on the ozone layer throughout and beyond the next century.

Carbon tetrachloride (CCl_4, sometimes known as CFC-10) is slightly more potent than CFC-11. It has been used as a solvent, and for the manufacture of CFCs, rubberised paints, pesticides and pharmaceuticals. Due to its high toxicity its use has already been reduced or prohibited in many countries, including the UK (DoE 1991: 6). Production is due to be phased out in industrialised countries by 1996.

HCFCs (hydrochlorofluorocarbons) were mainly introduced as interim substitutes for CFCs. They have lower Ozone Depletion Potentials (ODPs) estimated to range from about 0.01 to about 0.1 (Table 3.3). This means they have approximately 1%–10% of the potency of CFC-11, the benchmark substance. They are due to be phased out in industrialised countries by 2020, although the European Union has agreed to phase out by 2014, and Sweden and Denmark will phase out by 2002. Future Montreal Protocol meetings are likely to discuss an earlier international phase-out date.

Methyl chloroform (1,1,1-trichloroethane) is estimated to have an ODP of about 0.1, giving it about 10% of the potency of CFC-11. It will be phased out by 1996 in industrialised countries. It has been used as a solvent, a degreasing agent for cleaning metals and plastics, and for adhesives and correction fluids.

4.1.3   ANTHROPOGENIC BROMINE

Anthropogenic bromine in the stratosphere is an important catalyst for ozone depletion, alone and in coupled reactions with chlorine (SORG 1993: 21). Bromine has a more significant impact in the northern hemisphere than in the southern hemisphere. In the Antarctic, bromine is probably responsible for about 20–30% of the springtime ozone losses caused by anthropogenic chlorine and bromine (UNEP 1992a: 11). In the Arctic, bromine probably accounts for about 50% of the winter ozone losses caused by anthropogenic substances. Bromine is probably also responsible for about

50% of the winter ozone losses at mid-latitudes in the northern hemisphere (UNEP 1992a: 11).

Although bromine is responsible for a substantial proportion of damage, the concentration of bromine in the atmosphere is a tiny fraction, less than 1%, of the chlorine (WMO 1994: 3.20).

The two main anthropogenic sources of bromine are halons and methyl bromide. Halons have been widely used because they are effective fire extinguishers. Their ODPs range from approximately 5.0 to 12.0 (Table 3.3). In the UK, about 30% of each year's production was emitted to the atmosphere while the rest was temporarily 'stored' in fire-fighting equipment (DoE 1991: 6). Production of halons was largely phased out in industrialised countries by 1994 (WMO 1994: 10.3).

Methyl bromide is known to reach the upper troposphere and stratosphere (WMO 1994: 10.6). The 1994 WMO report estimated methyl bromide's ODP to lie between 0.3 and 0.8, suggesting it may have at least 30% of the potency of CFC-11 (see Section 5.1) (WMO 1994: 10.23). Manufactured methyl bromide is used as a fumigant and as a chemical feedstock. It is also generated as an unintended by-product of leaded fuel combustion, biomass burning and certain industrial processes (see Section 4.1.4). It is possible that further sources and/or sinks might be identified.

Anthropogenic halons contributed about 6 pptv bromine in the atmosphere in 1990, according to SORG's estimates (see Figure 3.10) (SORG 1993: 26). The atmospheric concentration of bromine from all sources was estimated to be 21 pptv (± 0.8 pptv) in 1991/2 (WMO 1994: 2.9). The 1994 WMO Scientific Assessment data indicated that anthropogenic methyl bromide probably contributed approximately 4.5 pptv (2.0–7.5 pptv) bromine in the early 1990s.

SORG estimated that if the anthropogenic emissions of bromine (halons and methyl bromide) were phased out, the bromine loading in the troposphere could be reduced from more than 20 pptv to 15 pptv or even to 10 pptv, depending on the precise size of the natural background level of methyl bromide (SORG 1993: 26).

4.1.4   EMISSIONS OF ANTHROPOGENIC METHYL BROMIDE

Table 3.4 shows the estimated emissions of methyl bromide from anthropogenic sources. Taking the mid-points of the ranges, the fumigant would contribute roughly 52% of anthropogenic emissions. Biomass burning, combustion of leaded fuel, and industrial processes would account for perhaps 33%, 13% and 2% of anthropogenic emissions respectively, in the early 1990s. The proportion from leaded petrol is probably now smaller.

Table 3.4. Estimated emissions of methyl bromide from anthropogenic sources (early 1990s). *Source*: Compiled from WMO 1994: section 10, MBTOC 1994: 45, 48.

Anthropogenic sources of methyl bromide	Estimated emissions to atmosphere (tonnes/year)	Mid-point of estimated emissions (tonnes/yr)	Approximate percentage (%)
Fumigant	34 000–59 000	46 500	52%
Biomass burning	10 000–50 000	30 000	33%
Leaded fuel	500–22 000	11 250	13%
Industrial processes	2000	2000	2%
Total anthropogenic emissions	46 500–133 000	89 750	100%

4.1.4.1 Fumigant Use and Emissions

The chemical industry sold about 75 600 tonnes of manufactured methyl bromide in 1992; about 96% of this was used as a fumigant (TEAP 1995: 67). Of the estimated 70 946 tonnes sold worldwide as a fumigant in 1991, about 23 100 tonnes (33%) were used in the USA, 15 040 (21%) in the European Union (12 countries), and about 8700 tonnes (12%) in Japan (TEAP 1995: 67, CMR 1994, MBGC 1994). About 14 500 tonnes (19%) of methyl bromide was estimated to be used in developing countries in 1992 (TEAP 1995: 67).

Soil fumigation is estimated to emit 30–80% of the fumigant to the atmosphere, depending on factors such as application method and soil type (MBTOC 1994: 48). Depending on the circumstances, emissions of methyl bromide range from about 51% to 88% for durable commodities, about 85–95% for perishable commodities, and about 90–95% for structures (MBTOC 1994: 48). Low level emissions can continue for some time after fumigation and airing in the case of certain commodities, such as grains, nuts and timber. For example, a Japanese study calculated that 30% of the methyl bromide used during fumigation of timber was emitted slowly from the timber after airing had finished (MBTOC 1994: 63). Dried fruit is calculated to emit 68% of the methyl bromide after airing has finished; while nuts were estimated to release 36% after airing (MBTOC 1994: 61–62).

The 1994 WMO Scientific Assessment estimated that worldwide use of the fumigant led to emissions of 24 000–64 000 tonnes of methyl bromide to the atmosphere per year (in 1991/2) (WMO 1994: 10.11), based on the assumption that 50% of the fumigant on average may be emitted to the atmosphere (WMO 1994: 10.8). This included emissions from all uses: soil, quarantine, stored commodities and structures. Subsequently, the Montreal Protocol's Methyl Bromide Technical Options Committee (MBTOC) refined the range of emissions to 34 000–59 000 tonnes of methyl bromide (1992) (MBTOC 1994: 45). This was based on a more detailed estimate of emissions from each area of use, which concluded that on average about 64% of the fumigant is emitted to the atmosphere (MBTOC 1994: 48).

4.1.4.2 Emissions from Biomass Burning

The 1994 WMO report estimated that biomass burning emits 10 000–50 000 tonnes of methyl bromide per year (WMO 1994: 10.11). Biomass burning includes burning of tropical forests, temperate forests, savannah, grasslands, fuel wood and crop remains (stubble) following harvest. The report noted that almost all biomass burning is initiated or controlled by human activities, estimating anthropogenic emissions to be more than 80% (WMO 1994: 10.8, 10.11). Emissions in the southern hemisphere, where the largest savannah areas are burned and a significant amount of deforestation occurs, are about twice as large as those in the northern hemisphere (WMO 1994: 10.8). Some anthropogenic biomass burning would be extremely difficult to control; and the use of wood for fuel is necessary for human survival in some regions. However, it would be feasible to curtail certain practices such as deliberate forest and savannah burning by commercial companies.

4.1.4.3 Emissions from Leaded Fuel

Combustion of ethylene dibromide, an additive in leaded petrol, has been estimated to emit either 500–1500 tonnes or 9000–22 000 tonnes methyl bromide per year (WMO 1994: 10.11). The discrepancy in estimates was largely due to the quantity of ethylene dibromide assumed to be used in the USA in 1991. Use of leaded petrol has been reduced markedly. In 1995 use of leaded fuel in the USA has fallen to about 2% of the market and is due to be phased out in 1996. Leaded petrol is no longer used in Japan and is being phased out in other industrialised countries. However, use may not be diminishing in developing countries.

4.1.4.4 Emissions from Industrial Processes

A small proportion of manufactured methyl bromide is used as a chemical feedstock or process agent: 2648 tonnes (3.5%) in 1992 (TEAP 1995: 67). It is used in the manufacture of other pesticides and integrated circuits, and as an extractant (WMO 1994: 10.9; TEAP 1995: 84). Methyl bromide is also created as a by-product of certain chemical processes, which in some cases generate substantial emissions. One factory in the USA, for example, reported emissions of 335 tonnes in 1991 (US EPA 1992). The 1994 WMO assessment estimated that industrial sources emit 2000 tonnes of methyl bromide per year to the atmosphere (WMO 1994: 10.11).

5.1 COMPARISON OF CHLORINE AND BROMINE

A synergistic reaction occurs between chlorine and bromine, so bromine inflicts more damage when both are present in the atmosphere (WMO 1992: 4.15). Each chlorine atom destroys thousands of molecules of ozone (UNEP 1993: 4). Each atom of bromine from halons and methyl bromide is substantially more destructive than each atom of chlorine. This is mainly because a greater fraction of bromine remains in active forms, and bromine catalysis is most efficient in the lower stratosphere where the ozone concentration is greatest (WMO 1994: 10.20, WMO 1992: 4.15).

The 1992 UNEP Scientific Assessment estimated that, atom for atom, bromine was approximately 40 times more efficient than chlorine in destroying ozone (UNEP 1992: 4). The 1994 WMO assessment raised this figure by 25%, estimating bromine to be about 50 times more efficient (WMO 1994: xxi). The report indicated that this may be a conservative figure because bromine is about 100 times more destructive than chlorine at an altitude of about 20 km, the region of peak observed ozone loss (WMO 1994: 13.15). Here, bromine's contribution to the overall ozone loss rate is nearly as important as chlorine's (WMO 1994: 10.20), even though the concentration of bromine is very small.

The level of total bromine in the atmosphere is about 21 pptv, compared with about 3500 pptv for total chlorine (WMO 1994: 2.9). Using conservative estimates, 2 pptv of additional bromine in the atmosphere could cause as much damage to ozone as about 100 pptv of chlorine (WMO 1994: 10.3).

The potency of ozone depleting substances is usually compared by calculating the steady-state Ozone Depletion Potential (ODP). The ODP provides a relative indicator of the expected impact on ozone per unit mass of gas, compared to that from CFC-11, the benchmark substance, over an infinite period of time.

The current estimated ODPs for ozone depleting gases vary from 0.01–0.02 for HCFC-123, to 12.0–13.0 for halon-1301 (WMO 1994: 13.17).

The 1994 WMO scientific assessment calculated that it was improbable that methyl bromide's ODP would be less than 0.3 or greater than 0.8 (WMO 1994: 10.23). This suggests it has at least 30% of the potency of CFC-11, measured over an infinite period of time. The WMO assessment took account of the uncertainties in the data. They stated that there was no single process whose current estimated uncertainty could reduce the ODP below 0.3. Smaller values would only be possible if two improbable situations occurred simultaneously and several parameters were at the extremes of their error limits (WMO 1994: 10.22).

A steady-state ODP of at least 0.3 would make methyl bromide more potent than some other compounds for which phase-out dates have already been set, such as methyl chloroform which has an estimated ODP of 0.1 (see Table 3.3). In the highly improbable event of methyl bromide's ODP being reduced by a factor of 10, it would remain higher than HCFC-123 and HCFC-225ca (ODP 0.01–0.02).

Methyl bromide's lifetime (response time) is estimated to be 0.8–1.7 years (WMO 1994: 10.1), and new data suggest it is at the lower end of this range. A few other ozone depleting compounds have relatively short lifetimes: chloroform, for example has a life of about 0.5 years, and methyl chloroform has an estimated life of 5.0–5.8 years (WMO 1994: 13.6). In contrast, the lifetime of CFC-11 is estimated to be 45–55 years, and CFC-115 is approximately 1700 years (WMO 1994: 13.6).

The convention of using steady-state (long term) ODPs tends to mask methyl bromide's high potency. Methyl bromide's short life means that ODPs calculated over short periods of time provide a more realistic indicator of its impact on ozone. Methyl bromide's ODP calculated over a 10-year period is at least 2.7, and over a 5 year-period is at least 12.0 (WMO 1994: 10.22). Over methyl bromide's realistic lifetime of approximately 1 year the adjusted ODP is estimated to be much greater than 12.0.

A short lifetime means that a compound inflicts its damage rapidly on ozone. As a consequence, early reductions in the use of methyl bromide and other compounds with a short life will produce improvements rapidly in the ozone layer, while elimination of compounds with long lifetimes (eg. CFCs) will bring improvements in the long term. Early reductions in the emissions or use of compounds with short lifetimes would shorten the period of high ozone losses. Action on compounds with short lifetimes will also reduce the impact of substances already in the atmosphere, because of synergistic reactions between bromine and chlorine.

6.1 SCIENTIFIC ASSESSMENTS OF METHYL BROMIDE AND OZONE DEPLETION

Scientific investigations into methyl bromide's impacts on stratospheric ozone started more than 20 years ago. A number of studies from the 1970s onwards attempted to model or measure methyl bromide's role in ozone chemistry or ozone depletion (for example Wofsy, McElroy and Yung 1975, Robbins 1976, Penkett et al 1981, Molina, Molina and Rowland 1982, Prather, McElroy and Wofsy 1984). Some studies examined emissions of the fumigant to the air or atmosphere (for example Daelemans 1978, NAS 1978, Brown and Rolston 1980); others raised the possible role of automobile emissions and leaded petrol (Harsch and Rasmussen 1977, Hao 1986) and biomass burning (Crutzen and Andreae 1990). One of the early studies in 1975 estimated that natural sources—the oceans—might contribute about 80% of the bromine in the atmosphere, while anthropogenic sources—the agricultural fumigant and leaded petrol—might contribute about 20% (approximately 10% each) (Wofsy *et al* 1975).

For several decades the predominant view was that anthropogenic methyl bromide was relatively unimportant in ozone depletion, especially compared to CFCs and the potent halons. In 1984, for example, a review of methyl bromide's environmental impacts carried out for the US Department of Agriculture noted that the fumigant and leaded petrol emitted methyl bromide to the atmosphere. It said the fumigant was considered to be a very minor source of bromine, and that bromine was not believed to be a major cause of ozone depletion (Curley 1984: III-3). A study published in 1985 threw doubt on this perspective. and recommended a more careful examination of bromine chemistry and anthropogenic emissions (Penkett *et al* 1985).

A number of factors eventually put anthropogenic methyl bromide on to the agendas of scientific panels. Emissions from the fumigant increased substantially when sales rose between the 1970s and the 1990s. Global sales of the fumigant increased from less than 16000 tonnes in 1975 to about 41500 tonnes in 1984 and about 73000 tonnes in 1992 (WMO 1994: 10.10, TEAP 1995: 66). Sales increased at the rate of 5.5% per annum between 1988 and 1992 (TEAP 1995: 66). Leaded petrol was of lesser interest to scientists because use of bromine for petrol additives in the USA fell from 100000 tonnes in 1978 to 24000 tonnes in 1991 (WMO 1994: 10.9). Leaded petrol is being phased out in many countries because of lead emissions.

Scientific scrutiny was also directed to anthropogenic methyl bromide as a result of improved information about the potency of bromine in breaking up ozone molecules. It was clear that bromine played a substantial role in polar ozone losses and a significant role in mid-latitude losses. By 1990 controls had been placed on CFCs and halons, and scientists turned their attention to the shorter-lived chlorine and bromine compounds (UNEP 1992a: 5). In 1990 and 1991, two official scientific assessments from the UK and the WMO drew attention to the importance of methyl bromide in ozone depletion (SORG 1990, WMO 1992). Observations found that bromine was directly involved in the catalytic destruction of ozone in the lower stratosphere, and much of the bromine was known to come from methyl bromide (UNEP 1992a: 5).

The 1991 WMO Scientific Assessment estimated that if the abundance of methyl bromide in the atmosphere could be reduced by as little as 1.5 pptv this would achieve a rapid reduction in chlorine/bromine loading, comparable to accelerating the CFC phase-out schedule by up to 3 years (WMO 1992: xvii). The total abundance of methyl bromide in the atmosphere—

from natural and anthropogenic sources—was approximately 11 pptv in the early 1990s (WMO 1994: 10.4). If anthropogenic sources contributed as little as 14% of the total, their elimination would benefit the ozone layer significantly.

Evidence of methyl bromide's impact on the ozone layer was reviewed again in 1992 for the Montreal Protocol (UNEP 1992a). UNEP's expert panel estimated that 15–35% of methyl bromide emissions were anthropogenic, based on 1990 data (UNEP 1992a: 9). The UNEP Panel noted that all studies showed there was significantly more methyl bromide detected in the atmosphere of the northern hemisphere than in the southern hemisphere—about 30% more in the north. They concluded this most probably indicated an excess source in the northern hemisphere (UNEP 1992a: 6).

The UNEP assessment calculated that anthropogenic methyl bromide could have accounted for about 5–10% of the observed ozone losses globally. It predicted that if use of methyl bromide continued growing at the recent rate of 5–6% per year, it could be responsible for about 15% of global ozone loss by the year 2000 (UNEP 1992a: 12).

In 1993 the Stratospheric Ozone Review Group (SORG), which advises the UK Department of the Environment, pointed out that increasing emissions of bromine would account for an increasingly larger fraction of ozone depletion as time progressed, if sources were not controlled (SORG 1993: 25). They reported that at least 3 pptv (30%) or more of methyl bromide emissions were estimated to be anthropogenic (SORG 1993: 25). SORG calculated that anthropogenic bromine emissions (from halons and methyl bromide) had raised the abundance of bromine in the atmosphere to more than 20 pptv from the natural background level of about 9–14 pptv (SORG 1993: 26). They concluded that a reduction in the emissions of man-made methyl bromide would reduce atmospheric bromine loading significantly and rapidly (SORG 1993: 21).

UNEP in 1993 commented that 'Stricter control measures have to be devised for methyl bromide and its consumption phased out as early as possible'. (UNEP 1993a: 24).

The 1994 WMO Scientific Assessment for the Montreal Protocol examined in detail the current data on methyl bromide and concluded that 'methyl bromide continues to be viewed as a significant ozone-depleting compound' (WMO 1994: xv). The report noted that a cessation of the emissions of anthropogenic methyl bromide would have a rapid impact on the extent of stratospheric ozone loss (WMO 1994: 10.23). This is primarily due to its short lifetime and high potency.

Data in the 1994 WMO report indicated that anthropogenic methyl bromide may have contributed about 4.5 pptv (2.0–7.5 pptv) of atmospheric bromine in the early 1990s.

A number of uncertainties, particularly about methyl bromide's chemical reactions in the stratosphere, had mostly been resolved by 1994 (WMO 1994: 10.23). Some uncertainties remained, for example in the size of certain sources and sinks. The 1994 Scientific Assessment took account of the uncertainties when calculating estimates, and expressed 'considerable confidence' in its findings (WMO 1994: 10.23).

The 1994 WMO Assessment identified four areas where governments could introduce further controls to reduce ozone losses (WMO 1994: xxiii). There are listed in Table 3.5. The Assessment calculated that if emissions of methyl bromide from agricultural and industrial sources were to be eliminated in the year 2001, the integrated future chlorine/bromine loading above the 1980 level (which is related to the cumulative future loss of ozone) was predicted to be 13% less over the next 50 years, compared to compliance with the existing Montreal Protocol controls (WMO 1994: xxiii). For comparison, rapidly eliminating the emissions from other remaining ozone depleting compounds would give reductions in integrated future chlorine/bromine loading of 10% for halons, 5% for HCFCs, and 3% for CFCs (WMO 1994: xxiii). Of the four areas, acting on methyl bromide would have the single greatest impact, representing 42% of the protective action identified by the Scientific Assessment (Table 3.5).

Table 3.5. Remaining areas where controls could be established under the Montreal Protocol to reduce future ozone losses. For each area, the 1994 WMO Scientific Assessment estimated the reduction in integrated future chlorine/bromine loading (above the 1980 level) over the next 50 years relative to controls under the Protocol in 1992. *Source*: WMO 1994: xxiii, TEAP 1995: 1–2.

Action identified by the UNEP/WMO Scientific Assessment	Estimated reduction in integrated future ozone losses	Percentage of identified reductions	TEAP conclusions about the feasibility of further controls
1. Eliminating agricultural and industrial methyl bromide emissions in the year 2001	13%	42%	Technically and economically feasible
2. Preventing release of the halons in existing equipment	10%	32%	Technically feasible, but costly
3. Eliminating HCFC emissions by 2004	5%	16%	Technically and economically feasible
4. Preventing the release of the CFCs in existing equipment	3%	10%	Technically feasible, but not economically feasible
Total	31%	100%	

7.1 LEGISLATIVE CONTROLS DUE TO OZONE LAYER IMPACTS

Under the Montreal Protocol, an international agreement to protect the ozone layer, methyl bromide was officially listed as an ozone depleting compound in 1992. The 1995 Montreal Protocol meeting agreed that industrialised countries will phase out methyl bromide by 2010. Since 1993 some countries have approved national reduction or phase-out schedules for methyl bromide because of its effects on ozone. The European Union and Canada have introduced regulations to cut consumption by 25% in 1998. The USA, Austria, Denmark, Norway, Sweden, Indonesia and several other countries have agreed early phase-out dates.

A summary of adopted legislation and policies is presented in Table 3.6. Details of the legislation and policy issues are given below.

7.1.1 USA: PHASE-OUT LEGISLATION

The USA manufactures and uses more methyl bromide than any other country. It manufactured about 28 000 tonnes (40% of global manufacture), and used about 25 500 tonnes in 1991 (about 36% of worldwide consumption of the fumigant and industrial process chemical) (CMR 1994, TEAP 1995: 67). In 1993 the USA finalised regulations to phase out production and importation of methyl bromide by the year 2001 (Federal Register 1993b). The regulation was made under the Clean Air Act, which requires ozone depleting compounds to be listed as Class I or Class II substances. Methyl bromide has been added to the Class I list of compounds which 'cause or contribute significantly to harmful effects in the stratospheric ozone layer', defined as compounds with an Ozone Depletion Potential (ODP) of 0.2 or above (Clean Air Act Amendments 1990). Class I substances have to be phased out within seven years.

In the improbable event that methyl bromide's ODP were calculated to be less than 0.2, it would be re-classified as a Class II ozone depleting compound. The Clean Air Act would still require phase-out, but over a longer period of time.

Section 6.11 of the Clean Air Act requires warning labels for products manufactured with ozone depleting compounds and traded between states in the USA. Products manufactured with CFCs, for example, must carry labels stating: 'Warning: manufactured with [name of ozone depleting compound] a substance which harms public health and environment by destroying ozone in the upper atmosphere'. As a result of strong opposition by methyl bromide manufacturers, fumigation companies, some growers and the US Department of Agriculture the federal rule does not require labels for products treated or grown with methyl bromide. There are also considered

Table 3.6. Summary of legislation and national policies for phasing out or reducing consumption of methyl bromide at national level. Situation in late 1995. Note: In some cases regulations allow exemptions for essential uses or for quarantine and pre-shipment uses. *Source*: Miller 1995a.

Country	Measure and year adopted	Agreed phase-out or reductions
USA	Regulation 1993	Phase-out by 2001
Austria	Agreed policy	Phase-out by 2000
Denmark	Regulation 1994	Phase-out by 1998 – eliminate certain uses by 1995 – phase out main use (glasshouse tomatoes) by 1996
Sweden, Norway, Finland	Nordic environmental strategy agreement	Phase-out by 1998
Indonesia	Decree 1994	Phase-out by 1998
Italy	Parlimentary law 1993	Phase-out by 2000 agreed by Chamber of Deputies and Senate, but date suspended and under review
European Union	Regulation 1994	25% cut in 1998 – requirement to take all precautions to prevent leakage – regular review of permitted uses
Canada	Regulation 1994	25% cut in 1998 – permits required to import methyl bromide for quarantine or pre-shipment uses
Industrialised countries	Montreal Protocol amendment 1995	25% cut by 2001 50% cut by 2005 Phase-out by 2010 with some exemptions
Italy*	Ordinance 1994	Fields may be fumigated only in alternate years
Germany*		Soil treatments for food crops have been phased out
Netherlands*		All soil treatments have been phased out
Switzerland*		Soil uses not permitted

*Introduced because government agencies were primarily concerned about water contamination, food residues and/or occupational and community health.

to be some technical difficulties in defining 'manufactured with' in the context of agricultural production (Federal Register 1993). However, lawyers representing the Natural Resources Defense Council and Friends of the Earth, with other parties, are suing the US Environmental Protection Agency to implement labelling provisions under the Clean Air Act, to allow consumers a choice in purchasing agricultural products which utilise an ozone depleting compound (NRDC *et al* 1994).

In the USA, ozone depleting compounds such as CFCs are subject to special taxes. Due to industry opposition, the current US administration has no plans to introduce taxes on methyl bromide. A coalition of US environmental and rural organisations is pressing for a tax, on the grounds that rising prices would make alternatives more attractive, and that the revenue (up to $US 1 billion over 5 years) could be used to develop further alternatives (MBAN 1994).

7.1.2 DENMARK AND SCANDINAVIA: REDUCTIONS AND PHASE-OUT

In June 1994 Denmark agreed a regulation to phase out use of methyl bromide by 1998, under the Chemical Substances and Products Act (Danish EPA 1994). The regulation requires certain commerical uses to cease by January 1995. Soil fumigation for tomatoes grown in glasshouses has to be phased out by 1996. Fumigations for glasshouse tomatoes constitute the major use in Denmark, accounting for 68% of the methyl bromide used in 1991/92 (Nordic Council 1993: 44).

The remaining uses have to be phased out by January 1998. These comprise:

– soil disinfestation for other glasshouse crops,
– disinfestation of herbs, imported seeds, nuts, dried fruit, raw tobacco, and furs,
– disinfestation of flour mills and warehouses for flour and grain,
– disinfestation of aircraft, rat control on ships,
– quarantine for wooden packing materials for export,
– control of wood-boring beetles,
– disinfestation of museum items and historic buildings (Danish EPA 1994: 5).

Quarantine and pre-shipment uses will be phased out (Host Rasmussen 1994). The phase-out dates were selected on the basis of technological developments to date, and anticipated developments of alternative substances and methods (Danish Market Committee 1993). The Nordic Council of Ministers has published two reports on existing and potential alternatives to methyl bromide (Nordic Council 1993, Nordic Council 1995). As part of a Nordic environmental strategy agreement, Norway, Sweden and the other Scandinavian countries have made a commitment to phase out methyl bromide by 1998.

Use of methyl bromide in Denmark was reduced voluntarily from 33 tonnes in 1991/2 to 17 tonnes in 1993, primarily in anticipation of the regulation. Use was reduced to 12 tonnes in 1994 (Danish EPA 1995), representing a 64% reduction within three years.

An interesting legal challenge was made to the Danish regulation. The European Commission, led by the trade and industry Directorate DG III, attempted to prevent the draft regulation being adopted on the grounds that it would be an unjustified barrier to trade within the European Union, violating Article 30 of the Treaty. The Commission said Denmark has no legal basis for introducing stronger environmental measures than the European Union (Bangemann 1993). The Danish government responded that the regulation was compatible with the Treaty, and was based on significant environmental concerns. They stated that 'it is of the greatest importance that no substances which deplete the ozone layer are used anywhere longer than absolutely necessary' (Danish Market Committee 1993). The regulation was adopted, setting a legal precedent for Member States to introduce environmental measures that go further than the common EU position.

7.1.3 OTHER NATIONAL RESTRICTIONS

A number of other countries have made commitments to reduce or phase out the use of methyl bromide rapidly. The Austrian government has agreed a policy to phase out use by the year 2000 (Aichinger 1995). Indonesia has issued regulations to phase out use by 1998, specifying expiry dates for registered products (Menteri Pertanian 1994).

Italy is the largest European user, consuming an estimated 7000 tonnes per annum in 1993 (MBTOC 1994: 114). In 1993 the Italian Senate and Chamber of Deputies passed a Parliamentary Proposal (equivalent to a Private Members' Bill) to phase out methyl bromide by the end of 1999 (Law 549/92 of 28 December 1993). The law includes a provision for certain uses of methyl bromide to continue beyond 2000 under exceptional circumstances. A tax was due to be placed on methyl bromide. The Ministry of Environment was allocated $3.5 million (for 1994/5) to finance the law, and it was to be enforced by a newly established national agency for environmental protection. However as a result of pressure from industry and the European Commision the phase-out date has been temporarily suspended and is under review (Agra Europe 1994, Ministry of Environment 1995). The position of the Italian administration in early 1995 was that it would support a 50% cut in methyl bromide use by 2005.

The European Union, including Italy, has adopted regulations to reduce consumption and production by 25% in 1998, except for quarantine and pre-shipment uses (Council of EU 1994: 5, 7). The regulation requires Member States to take 'all precautionary measures practicable' to prevent methyl bromide leakage from fumigations (Council of EU 1994: 11). It also notes the need for a regular review of the permitted uses of methyl bromide and other ozone depleting substances.

Canada has agreed a regulation to cut consumption by 25% in 1998, except for quarantine and pre-shipment uses, under the Environmental Protection Act. The regulation requires permits for importing the methyl bromide used for quarantine or pre-shipment (Canada Gazette 1994: 4015).

7.1.4 THE MONTREAL PROTOCOL

The Montreal Protocol has established phase-out schedules for a number of chlorinated and brominated compounds, such as CFCs, HCFCs, carbon tetrachloride and halons (UNEP 1992b). In 1992 the Montreal Protocol meeting considered an interim UNEP Scientific Assessment on the effects of methyl bromide on the ozone layer (UNEP 1992a), and debated what action to take. The USA and some other countries argued that action to restrict methyl bromide was necessary, because it would make a significant contribution to ozone layer protection. The USA proposed a phase-out in the year 2000, while some other nations favoured either a freeze, or a freeze and reduction (Federal Register 1993b: 65032). The Methyl Bromide Global Coalition, an industry lobby group, claimed that action was premature because scientific knowledge was too uncertain and controls would have an unjustified economic impact. Reductions were blocked by a number of governments, in particular Israel and France which manufacture methyl bromide.

The 1992 Montreal Protocol meeting agreed a compromise, as follows:

- To add methyl bromide to the official list of ozone depleting substances (UNEP 1992b: 45).
- To freeze methyl bromide production and consumption in industrialised countries in 1995 (at 1991 levels), except for the amounts used for quarantine and pre-shipment treatments (UNEP 1992b: 38).

Governments unanimously adopted a non-binding Resolution, noting the serious environmental concerns raised in the Scientific Assessment and urging nations to make every effort to reduce emissions of methyl bromide. Nations agreed to discuss reductions and a possible phase out of methyl bromide at the Montreal Protocol meeting in 1995 when further assessments of ozone effects and alternatives would be available (UNEP 1992b: 74).

At the Montreal Protocol meeting in 1993 seventeen countries signed a voluntary Declaration, stating their 'firm determination' to:

- reduce their consumption of methyl bromide by at least 25% by the year 2000 at the latest, and
- phase out totally the consumption of methyl bromide as soon as technically possible (UNEP 1993b: 62).

The Declaration was signed by two of the world's major methyl bromide producing countries, the USA and Israel. Other signatories were: the UK, Italy, Germany, Austria, Belgium, Finland, Iceland, Denmark, Liechtenstein, Netherlands, Sweden, Switzerland, Canada, Botswana and Zimbabwe (UNEP 1993b: 62, UNEP 1994b).

The Montreal Protocol's Technology and Economic Assessment Panel (TEAP) believed that placing controls on methyl bromide would be more cost-effective than additional controls on other ozone depleting compounds, such as destroying CFCs in old equipment (TEAP 1995: ES.4). During the Montreal Protocol meeting in 1995, governments agreed international schedules to reduce and phase out almost all uses of methyl bromide in industrialised countries.

TEAP examined the lessons for methyl bromide from the successful CFC and halon phase-outs. TEAP noted that users of methyl bromide have been reluctant to commercialise and implement alternatives because they believe methyl bromide uses to be more important and more difficult to eliminate than CFCs or halons (TEAP 1995: 21). TEAP believes that the challenges faced by users of methyl bromide, while substantial, are no more difficult than the challenges already overcome by users of the other ozone depleting compounds (TEAP 1995: ES.4).

In 1987 the CFC and halon users claimed that no single chemical offered the same advantages, and that it would not be possible to develop acceptable alternatives. Numerous reports from industry, governments and independent institutes predicted few prospects for alternatives (TEAP 1995: 21). TEAP found that these initial perspectives failed to appreciate the potential for technical innovation, the power of market forces, the efficiency of public/private partnerships and the leadership of specific companies that pledged early phase-outs. TEAP reported that in many cases the alternatives were cost-saving, no-cost or low-cost. The initial pessimism gave way to attitude shifts, innovative development and profitable commercialisation (TEAP 1995: 22).

TEAP noted that innovation and the development of alternatives was particularly rapid when:

- Chemical suppliers started to support ozone layer protection, such as DuPont and ICI after 1987. In contrast, methyl bromide producers still question the conclusions of the scientific assessments (TEAP 1995: 23).
- Leadership companies set voluntary goals to halt use. So far, few methyl bromide customers are publicly demanding rapid change (TEAP 1995: 23).
- National regulations encourage change. The Netherlands phase-out of soil treatments is a successful example. A few other countries are starting to set early phase-out dates.

When controls were agreed for CFCs and halons in 1987 very few alternatives had been identified (TEAP 1995: 21). Users of methyl bromide are in a superior position at this stage, because potential or existing alternatives have been identified for a substantial proportion of use. A detailed UNEP report about alternatives to methyl bromide has identified feasible alternatives, either currently available or at an advanced state of development, for a substantial proportion of methyl bromide use, including some quarantine procedures. The report failed to identify alternatives for less than 10% of use (MBTOC 1994: 3).

7.1.5 POLICY APPROACHES

7.1.5.1 Results of Decisions on CFCs

In 1974 Molina and Rowland first suggested that CFCs might cause ozone depletion (Molina and Rowland 1974). This was hotly debated over the following decade. A scientific assessment in 1985 predicted that if CFC production continued to increase, ozone levels would drop by about 5% by the year 2050 (NOAA 1992: 12). CFCs were regarded as ideal chemicals, indispensible to modern life, and forming the centre of a multi-billion dollar industry. Officials feared that a ban would disrupt many segments of society. In 1985 scientists were still uncertain whether ozone levels had actually started to drop. Some researchers recommended further research before taking action, while others recommended early action to avoid the predicted risks to human health and the environment.

By 1987 scientific expeditions had demonstrated that additional chlorine and bromine had shifted the fragile chemical balance in the Antarctic (NOAA 1992: 16). Nevertheless, there were still many uncertainties about the mechanisms. Scientists at that time were not sure that depletion would eventually become significant. Many governments decided that the risk was too great to wait for further research.

The Protocol controls on ozone depleting substances were tightened further in 1990 and 1992. Measurements in the atmosphere show that the Protocol has so far reduced the annual rate of growth of chlorine in the atmosphere from 110 ppt/year in 1989 to about 60 ppt/year in 1992. Growth of bromine from halons was reduced to about 0.2–0.3 ppt/year in 1992 compared to 0.6–1.1 ppt/year in 1989 (WMO 1994: 2.1). The Protocol has been effective in preventing a very high degree of ozone depletion. If controls had not been introduced and strengthened, the abundance of chlorine/bromine in the atmosphere would probably have reached almost 6000 pptv by the year 2025 (Figure 3.11) (WMO 1994: 13.12). The 1992 controls will reduce levels to about 2400 pptv by 2025. However, this will still be about four times greater than the natural background level of about 600 pptv.

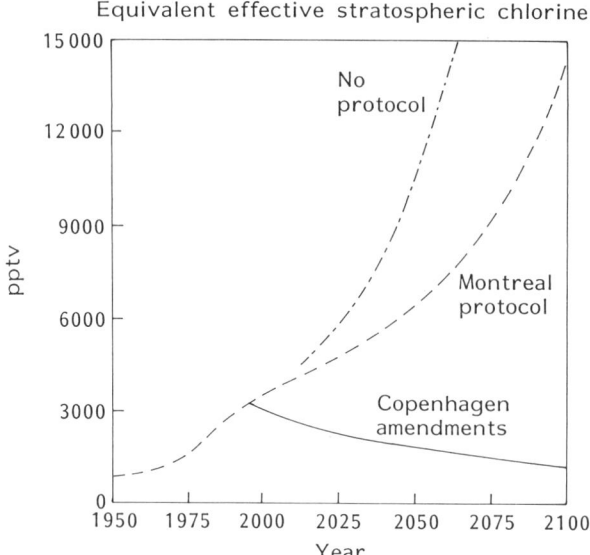

Figure 3.11. Estimated impact of the Montreal Protocol and 1992 Copenhagen amendments on stratospheric chlorine/bromine loading (calculated as the equivalent effective stratospheric chlorine). The 'No protocol' scenario assumes 3% per year increase in global emissions of CFCs and methyl chloroform, which is less than the known trends prior to the Montreal Protocol agreement. Reproduced from 'Scientific Assessment of Ozone Depletion: 1994' WMO Global Ozone Research and Monitoring Project, Report No. 37, 1994, 13.12

Despite the achievements, the reduction and phase-out schedules agreed under the Protocol were too slow to prevent significant ozone thinning occurring now and over the next decade. UNEP stated in 1993 that humans are bound to pay a price for the delay in signing a treaty (UNEP 1993: 24). More recently UNEP has warned that further progress under the Protocol is necessary: '. . . we cannot afford to be complacent. The line that divides complacency from catastrophe is very thin . . .' (UNEP 1994c). If global use of methyl bromide were to increase at the rate of 5% per annum, for example, the abundance of chlorine/bromine in the atmosphere would reach 6000 pptv within several decades, negating the achievements of the Montreal Protocol.

7.1.5.2 Policy Significance of Remaining Uncertainties

Although there are uncertainties in data relating to ozone depletion, ample data demonstrate that it is a real phenomenon and that anthropogenic compounds play a substantial role. The conclusions of the recent scientific assessments of the WMO and the Environmental Effects Panel were supported by hundreds of published, peer reviewed papers by leading experts in the field (WMO 1994, Environmental Effects Panel 1994).

Since 1991 four authoritative scientific assessments have concluded that anthropogenic methyl bromide contributes significantly to ozone losses (WMO 1992, UNEP 1992a, SORG 1993, WMO 1994). Scientists are not yet able to explain or quantify certain aspects of the process. Concerning control measures for methyl bromide, UNEP has concluded 'the sooner action could be initiated by every country, the better'. (UNEP 1995: 2). On the other hand, a number of methyl bromide industry representatives and scientific consultants have said that action to phase out or reduce the use of methyl bromide would be premature while some uncertainties remain (British Pest Control Association 1993: 1, European Methyl Bromide Association 1995: 4, Methyl Bromide Global Coalition 1994b: 1–3, Dead Sea Bromine 1993: 38).

More research can always be carried out to increase understanding on a topic. In almost every area of science, and especially those where research started barely twenty years ago, scientists can state that there are large areas of uncertainty, so this should be expected in the case of methyl bromide. The key issue is whether policy makers have sufficient, reliable scientific information about the consequences of action and non-action.

For policy-makers the most important calculations of the WMO Scientific Assessment are the integrated future chlorine/bromine loading scenarios. These estimate the consequences of different actions, such as continuing or eliminating emissions by certain dates. The scenarios are based on emissions of methyl bromide, taking full account of the fact that emissions are significantly lower than usage/sales, and thus providing a better indicator of ozone losses than the ODP. The estimate that eliminating agricultural and industrial emissions of methyl bromide would produce a 13% reduction in future chlorine/bromine loading is very significant (see Table 3.5). If methyl bromide had only half that impact on ozone, its phase-out would still be the second greatest remaining step that governments could take to reduce ozone losses.

New information may reduce methyl bromide's estimated lifetime of 1.3 years (0.8–1.7 years) to about 1 year or slightly less. This would reduce the ODP, but scientists estimate that it would most probably remain within the calculated range. It is not uncommon for the WMO assessments to refine estimated lifetimes or ODPs in the light of improved information. The lifetime for CFC-11 for example, was estimated to be 42–105 years, and was recently revised to 50 years ±5 (WMO 1992: 8.7, WMO 1994: 13.6).

Policy makers in the Montreal Protocol need a general indication of potency but do not need to know a precise ODP. In 1991 models suggested the ODP for halon-1211 lay between 16.3 and 49.6; semi-empirical data suggested an ODP of 3.9–4.4; WMO recommended a best estimate ODP of 4 (WMO 1992: 6.15). Halon-1211 was clearly a potent substance, and a phase-out date had already been set. The 1994 Assessment calculated that it was improbable that methyl bromide's ODP would be less than 0.3, so it is likely to be more potent than HCFCs and methyl chloroform, which have been scheduled for phase-out. But if methyl bromide's ODP were reduced ten-fold it would remain more potent than some HCFCs. Methyl bromide's short-term ODP will remain very high.

In policy terms, it is not necessary to know precisely the proportion of methyl bromide that is natural. It is the additional chlorine/bromine that causes problems for the ozone layer, by breaking up ozone faster than it can be replenished. Eliminating the additional emissions would eventually restore equilibrium to the rate of ozone break-up and replenishment. The 1991 WMO Scientific Assessment calculated that if the abundance of methyl bromide in the atmosphere could be reduced by as little as 1.5 pptv, this would achieve a rapid reduction in chlorine/bromine loading, equivalent to accelerating the CFC phase-out by up to three years (WMO 1992: xvii). From a policy perspective, if anthropogenic methyl bromide contributes as little as 1.5 pptv then its elimination would clearly benefit the ozone layer. The 1994 WMO data indicate that anthropogenic methyl bromide may contribute 2.0–7.5 pptv.

It is normal to have a spectrum of opinion among scientists about any issue, and methyl bromine is no exception. The WMO report and conclusions received input from 226 scientists from 29 countries, and was reviewed by a further 147 scientists (WMO 1994: ii). Scientific consultants and researchers funded by the methyl bromide industry were also actively involved in the process, so their technical points and arguments were considered during the assessment. The main conclusions of the scientific assessment were clear (see Table 3.5).

Humans cannot take any action to eliminate the long-lived anthropogenic chlorine and bromine compounds already released to the atmosphere, nor to prevent volcanic eruptions and similar natural events. The stratosphere is already overburdened with chlorine and bromide, making further ozone thinning inevitable. The existence of natural emissions does not justify continued release of additional emissions. On the contrary, it means it is important to prevent further emissions wherever feasible.

Delays in eliminating preventable sources of ozone depleting compounds are expected to incur costs in agriculture, forestry, fisheries, tourist industries, health and social services in many countries as a result of increases in UV radiation (Environmental Effects Panel 1994, WHO 1994). In purely economic terms, it would not be cost effective to sit back and wait for more refined data. The US Environmental Protection Agency has carried out a cost-benefit analysis, estimating that the total costs of phasing out methyl bromide in the USA would be $1.7–$2.3 billion, whereas the costs of continued use would be between $14 and $56 billion for the period 1994 to 2010 (US EPA 1993).

8.1   LOCAL ENVIRONMENTAL IMPACTS

Use of methyl bromide sometimes produces residues in soil, certain crops, air or water. A proportion of methyl bromide is broken down and/or metabolised into a variety of residues, depending on factors such as the levels of moisture and organic matter (WHO 1995). In aqueous solution it hydrolyses slowly to form methanol and hydrobromic acid. It is an effective methylating agent, reacting with amines to form methylammonium bromide derivatives. Methyl bromide also reacts with sulphur compounds under alkaline conditions to give mercaptans, thioethers, and disulphides (IPCS 1994: 8).

8.1.1   SOIL AND CROP RESIDUES

When applied to soil, methyl bromide diffuses and some is degraded primarily by complex reactions with organic matter, or primarily by hydrolysis if little organic matter is present (Arvieu 1983, Arvieu and Cuany 1985). The principal residue is generally inorganic bromide. There is limited data identifying or quantifying the other diverse residues and metabolites formed under different conditions.

The efficacy of methyl bromide depends on the concentration of the fumigant in the gas phase (ACP 1992: 8). Methyl bromide dosages accrued in soil and between soil and sheeting during commercial fumigations can vary from 250 to 10 000 mg hour/litre (concentration time product), even within the same glasshouse (ACP 1992: 7). The doses required to kill the more tolerant stages of organisms are 200–600 mg hour/litre for insects and higher plants, 600 mg h/l for nematodes, and 2500–4000 mg h/l for fungi, bacteria and viruses (ACP 1992: 7).

A well-conducted fumigation destroys injurious pests, but also destroys a wide variety of non-target organisms, such as insects, earthworms and soil microflora, altering the trophic structure of the soil environment (Van Rhee 1977, WHO 1995, Sassaman *et al* 1986). The elimination of most soil organisms can be a disadvantage if unwanted pests enter a fumigated field on new plants or from deeper soil levels or neighbouring land, because they may multiply rapidly in the absence of competing species or predators. More sustainable agricultural methods would seek to protect specific soil organisms and micro-organisms that recycle and release plant nutrients, aerate the soil or help plants to withstand pest attacks (Conway and Pretty 1991, Pimentel and Lehman 1993).

Methyl bromide *per se* has been detected in various soil types up to three weeks after fumigation, the highest levels generally being found in the upper layers (0–40 cm) of the soil (Lepschy, Stark and Sub 1979). In sandy soils which have little organic matter methyl bromide has a longer half-life. It is also relatively persistent when in deeper soil levels (Herzel and Schmidt 1984).

Inorganic bromide occurs naturally in the soil, usually at less than 10 mg/kg (WHO 1995). Soil residues resulting from fumigation vary greatly; measurements from a few mg to 218 mg/kg have been reported (ACP 1992). Raised levels of bromide in some cases have been found to persist in soil for four years (Fallico and Ferrante 1991).

Naturally-occurring bromide is generally bound to soil particles. In contrast, the bromide residues from fumigation are at first only slightly bound and are free to move in the soil (WHO 1995). Residues can therefore be taken up by plants or washed out by water (WHO 1995).

Certain crops such as cotton, celery, pepper and onion do not reach adequate growth when grown in fumigated soil (Bromine and Chemicals Ltd 1990). A few crops, such as carnations, can be damaged by inorganic bromide in the soil (Kempton and Maw 1974).

Crops grown in fumigated soil often contain higher organic bromide residues; uptake depends on factors such as plant species, soil type, climate and agricultural practices (Basile and Lamberti 1981). A study comparing residues in tomatoes grown in fumigated and untreated soil found bromide levels were about two times higher (55 mg/kg) in the crop grown on fumigated soil. Residues decreased with each successive harvest but were still higher than in control plants after the fourth harvest (Fallico and Ferrante 1991). A study on lettuce, which accumulates bromide easily, found that the crop grown on unfumigated soil contained less than 10 mg inorganic bromide per kg, while most lettuces harvested from fumigated soil contained considerably more, 30% containing more than 500 mg/kg and 2% in excess of 2000 mg/kg (Roughan and Roughan 1984).

Residues as high as 500 or 2000 mg/kg are no longer permitted in many countries. The UK, for cxample, has set legal limits (MRLs) ranging from 20 to 100 mg/kg, depending on the foodstuff (ACP 1992: 107). Growers have reduced residue levels substantially by practices such as leaching after fumigation, altered timing of fumigation, and integrated pest control (WHO 1995).

8.1.2 WATER RESIDUES

Leaching soil after fumigation to keep crop residues within permitted limits or to prevent phytotoxicty has been practised in countries such as the UK, Netherlands, Belgium and the USA. Deliberate leaching, heavy rain and irrigation systems can in some situations transport residues into surface or groundwater.

Residues may reach groundwater under specific conditions. For example, leaching a glasshouse area where clay occurred in the deep soil layers was found to raise inorganic bromide residues to 280 mg/litre in groundwater (Guns 1989). Methyl bromide *per se* can contaminate groundwater in special circumstances, such as a high water table, low temperatures and a low density of the underlying strata (Herzel and Schmidt 1984).

A recent review by the IPCS notes that relatively high levels of inorganic bromide (up to 72 mg/litre) can be found in drainage water from greenhouses and could adversely affect aquatic organisms (IPCS 1994: 12). For different fish species a no-observed effect concentration (NOEC) of 25 mg inorganic bromide/litre was determined on long-term exposure (IPCS 1994: 12).

Concentrations of methyl bromide have occasionally been found in drainage water, at levels of up to 9.3 mg per litre (IPCS 1994: 12). For methyl bromide the lowest median effect concentration (EC_{50}) or median lethal concentration (LC_{50}) values reported are 0.3 mg/litre for fish, 2.8 mg/litre for algae, 1.7 mg/litre for daphnids. The no-observed-effect concentrations (NOEC) in long-term studies were 0.06 mg/litre for fish and daphnids (IPCS 1994: 11). Methyl bromide's octanol/water partition coefficient is reported to be 1.19, so it probably does not have any significant tendency to bioaccumulate (WHO 1995).

8.1.3   AIR CONTAMINATION

MBTOC has estimated that 30–85% of the methyl bromide applied during soil fumigation is emitted to the air (MBTOC 1994: 48), depending on factors such as fumigation technique, sheeting permeability and organic matter in the soil. Air concentrations up to 900 ppb have been measured in adjacent fields up to 100 metres away from fumigated fields (Brodberg *et al* 1992: 15). Another study found values of 1–4 mg methyl bromide per m^3 at distances of up to 20 m from a greenhouse a few hours after injection; one tenth of this value was found four days later (WHO 1995: 4). A California study found that methyl bromide in the breathing zone of operators carrying out soil fumigation ranged from 134 to 1096 ppb (24 hour time-weighted average) depending on the type of work (Brodberg *et al* 1992: 15).

In the past, cases of fatal human exposures from relatively high concentrations led many governments to introduce controls on use. Limits for occupational exposure now vary from 1 mg/m^3 (8 hour time-weighted average) in the Netherlands, to 20 mg/m^3 in the UK, and 60 mg/m^3 in Italy (IPCS 1994: 34, Van Haasteren 1995). Methyl bromide may be applied only by licensed operators in a number of countries. However, in some countries, particularly in developing countries, it is still applied by untrained personnel.

Excessive exposures can occasionally occur, despite precautions such as operator training. In California, for example, state authorities recorded 148 cases of systemic illness, 52 eye injuries and 60 cases of skin injury due to methyl bromide exposure, for the period 1982 to 1990 (Brodberg *et al* 1992: 4). From 1984 to 1990 methyl bromide ranked eighth among pesticides causing acute illness in California (Pease 1993).

Non-fatal human poisoning has resulted from exposure to concentrations of 389 mg/m^3 (100 ppm) (IPCS 1994: 13). It can damage the nervous system, lung, nasal mucosa and kidney. Central nervous system effects include blurred vision, mental confusion, numbness, tremor, speech defects. Topical exposure can cause skin irritation, burns and eye injury. Exposure to relatively high concentrations causes convulsions, coma and death (IPCS 1994: 13).

The IPCS reports that there are no data on the direct effects of methyl bromide on birds and wild mammals. However, methyl bromide is highly toxic to all animal species by all routes of exposure (IPCS 1994: 12).

9.1 PHASE-OUT DUE TO LOCAL ENVIRONMENTAL IMPACTS

The regulatory authorities in a few countries have required major uses of methyl bromide to be phased out as a result of concerns about occupational safety and residues in water or food (Lower House 1981, Ketzis 1992, Schuepp 1994). These cases demonstrate that the development of other methods of pest control, though disruptive and in certain areas very difficult, has been technically feasible for diverse crops, pests, climates and conditions. Examples of regions where methyl bromide has been phased out are described below. Table 3.7 gives examples of some alternative methods of pest control used in place of methyl bromide.

9.1.1 THE NETHERLANDS: PHASE-OUT FOR SOIL USES

Methyl bromide was officially approved as a soil fumigant in the Netherlands in 1965 (Lower House 1981). By 1980 horticulture had become heavily dependent on the fumigant; annual consumption of methyl bromide was 2500 tonnes on about 3530 hectares (glasshouse and open fields) producing high-value crops such as strawberries, tomatoes and cut flowers. Studies on the levels of bromide ion residues in food crops led to a regulation in 1978 requiring soil to be leached after fumigation with copious amounts of water (350–500 litres of water per m^2) (Lower House 1981: 1).

In 1979 an inter-Ministerial working group started to examine the long-term acceptability of methyl bromide. Phase-out was agreed in 1980/81 because of concerns about occupational safety, water contamination, air

Table 3.7. Examples of alternative methods of pest control used in place of methyl bromide for horticultural crops. *Source*: Miller 1995a

Alternative methods of pest control for soil	Examples of crops and countries where alternative is used
Solarisation – plastic sheeting placed on soil to trap heat from sun	Melon – Spain Tomato – Italy north. Lettuce, vegetables – Italy centre. Eggplant, vegetables – Italy south. Tomato – Greece. Tomato – Portugal. Tomato, eggplant, melon – Israel. Olives – Greece. Greenhouse tomato – France. Strawberry, tomato – Japan. Strawberry, nut trees – USA. Tomato – Morocco. Used in more than 40 countries eg. Egypt, India, Pakistan, Jordan, Australia.
Solarisation & chemical treatment	Strawberry – USA. Strawberry – Israel.
Solarisation & IPM	Strawberry, vegetables – Italy. Tomato – Florida.
Soil substitutes – eg. mix of gravel with grain husks, peat, clay granules, rockwool, composted tree bark	Strawberry, tomato – Netherlands. Nurseries – Zimbabwe. Tomato – Denmark. Nurseries for flowers, trees & vegetables – USA.
Waste organic materials – waste materials applied to soil eg. paper industry waste, winemaking waste, twig chippings	Cut flowers – Italy Strawberry, tomato – Senegal. Horticulture – USA. Horticulture – Israel.
Grafting – use of pest resistant rootstock	Premium tomatoes – Israel.
Steam – diverse methods of heating soil with steam	Fruit & vegetables – Italy. Tomato, lettuce – UK. Cucumber – Netherlands. Also used in France, Belgium, Germany, Switzerland.
Integrated Pest Management (IPM) – mixture of cultivation practices, biological controls and low pesticide use	Tomato – Denmark. Tomato, grapes, apples – USA. Fruit & vegetables – Canada. Tomato – Mexico. Tree nurseries – China.
Chemical treatments	Tomato, tree nurseries – USA.
IPM with other fumigant	Tomato – Florida.

contamination close to residential neighbourhoods, and residues in vege-tables, fruit and milk (Lower House 1981). Water contamination was a particular problem due to low-lying land, leaching and intensive use of methyl bromide. During the fumigation season, for example, levels of 3–40 mg bromide/litre were measured in surface water used by cattle in pasture areas downstream from the main glasshouse area. Outside the fumigation season the levels varied from 0.3 to 1.5 mg/litre (RIVM 1980). Residues of methyl bromide *per se* (0.004–0.37 mg/litre) were detected in drinking water in situations where plastic water supply pipes ran under fumigation sites. 1500 tonnes of methyl bromide were estimated to be emitted to the air in the Westland area during the fumigation season (Lower House 1981: 12).

Reduction and phase-out measures in the Netherlands included: reducing doses by using less permeable sheeting for fumigations, issuing permits for methyl bromide on a case-by-case basis, and no longer permitting the fumigant where soil could be steamed (Lower House 1981, International Workshop 1992). All soil fumigations ceased by 1992 (van Haasteren 1994). Phase-out was helped by publicly financed research, skilled agricultural advisers, relatively skilled growers, initial limited grants for investing in steam facilities, and industry/government cooperation (International Work-shop 1992). Horticultural production is of considerable importance to the Netherlands economy. The value of the industry in 1980 was DFL 11 185 million, with employment of about 87 400 (Lower House 1981). The government estimated that the phase-out would have considerable impact on business, increasing glasshouse costs, for example, by DFL 13 500 to 30 000 per hectare (Lower House 1981: 7).

The alternatives adopted by growers included mobile steam facilities, rockwool or clay granule substrates, crop rotation, resistant cultivars and selected pesticides. In a number of cases methyl bromide was replaced with sophisticated high-yield systems which required increased investment. Nevertheless, official statistics show that the production of horticultural crops was maintained and increased over the period of phase-out. The production value of horticultural crops doubled from DFL 5955 million in 1980 to DFL 12 757 million in 1991 (International Workshop 1992: 17). Table 3.8 illustrates yields for three major horticultural crops before and after soil fumigation was phased out. The yield increases were largely due to the adoption of soil substitutes (artificial substrates) that avoided the need for using methyl bromide (Miller 1995a).

9.1.2 GERMANY: PHASE-OUT FOR FOOD CROPS

The German government prohibited use of methyl bromide for food crops in 1985 because it led to bromide residues in excess of strict national limits. In the late 1980s additional concerns about groundwater contamination led to

Table 3.8. Trends in yields of major horticultural crops in the Netherlands, comparing periods before and after methyl bromide was phased out. *Source*: Compiled from Kwantitatieve Informatie 1994/5.

Year	Tomato yield (kg/m^2)	Pepper yield (kg/m^2)	Cucumber yield (kg/m^2)
Period when methyl bromide was permitted:			
1976	15	14	36
1985	25	16	48
1991	38	22	53
Period when methyl bromide was not permitted:			
1992	40	23	55
1993	44	23	55

restricted use for non-food crops (Miller 1993). The predominant alternatives used by German horticulturalists are integrated pest management (IPM) methods that combine the use of resistant varieties, intercropping, crop rotations and selected pesticides, for field crops such as strawberries. Growers sometimes exchange fields with cereal farmers in order to rotate crops. Alternatives for greenhouses crops include steam, soil substitues and IPM (Ketzis 1992). Alternative pest control methods adopted in Germany are generally less capital intensive than those adopted in the Netherlands, and are more suitable for a wide range of countries. Table 3.9 shows tomato yields in several European countries, illustrating the fact that both Germany and the Netherlands achieved good yields using different systems to replace methyl bromide (Miller 1995b).

9.1.3 *ITALY: REGIONAL PHASE-OUT AND NATIONAL RESTRICTIONS*

Concerns, about contamination of Lake Bracciano in Italy in 1983, led the regional government to prohibit use of methyl bromide in the vicinity of the lake (Regional Ordinance 1983). As a result, a significant area of very intensive horticulture no longer uses methyl bromide. Erzal, a regional body for agricultural development, purchased steam machinery and organised subsidised soil treatment (Cori and Triolo 1994). More recently, researchers have developed a cheap alternative, combining solarization (laying plastic film on soil to trap the sun's heat) with IPM techniques such as composting, cultivation techniques and selected use of pesticides (Correnti and Di Luzio 1994). Yield has not fallen and in some cases has been increased. The cost of the combined solarization/IPM method is very low compared to methyl bromide or steam (MBTOC 1994: 100).

Italy is estimated to use more methyl bromide than any other European

Table 3.9. Comparison of tomato yields and use of methyl bromide in various European countries. *Source*: Miller 1995b.

Country	Tomato yield (100 kg/hectare 1992)	Using methyl bromide	Estimated use of methyl bromide for tomato crops (tonnes 1993)
Countries using low-yield systems predominantly:			
France	670	yes	450
Germany	640	no	0
Greece	514	yes	600
Spain	476	yes	< 1430
Italy	465	yes	2800
Portugal	447	yes	–
Denmark	200	yes	21
Countries using high-yield systems predominantly:			
Netherlands	4332	no	0
Belgium	3425	yes	200
UK	2600	yes	40
Average (12 EC countries)	559	yes	

country. The Emilia-Romagna region in particular uses considerable quantities (about 670 tonnes in 1990) (Cori and Triolo 1994). As part of a regional programme to reduce environmental contamination from pesticide residues, a local code of production limits soil treatments with methyl bromide to once in every two years. As a result, growers on 20 000 hectares have reduced use of methyl bromide by about 50% (MBTOC 1994: 100). In June 1994 Italy's Health Ministry issued a National Ordinance restricting fumigation of fields to one year in two (Ministry of Health 1994). National sales of methyl bromide were reduced in 1994 as a result of the Ministerial Ordinance, according to chemical industry sources (Agrofarma 1994: 9, Agrow 1995: 8).

A review of the cases where methyl bromide has been reduced or phased out in Italy concluded that it would be feasible for methyl bromide to be reduced by more than 60% in Italy within a few years, and that it is technically feasible to entirely replace methyl bromide (Cori and Triolo 1994). Alternatives have been adopted in diverse parts of Italy, with different climates, pests and social cultures.

9.1.4   SWITZERLAND: SOIL USES NOT APPROVED

Methyl bromide was never registered in Switzerland for production of food crops because of concerns about food residues and groundwater contamination (Schuepp 1994). However, it was used for non-food crops such as flowers and tree nurseries. In 1986 the Swiss Federal Office of Public Health commissioned an assessment of the public health implications of methyl bromide (Guillemin, Hillier and Bernhard 1990). Concerns about the safety of horticulturalists and the public, groundwater and food residues led to voluntary withdrawals and restrictions (Bernhard, Guillemin and Hillier 1989, Scheupp 1994). Methyl bromide is not now permitted for any kind of soli treatment (Pflanzenbehandlungsmittel Verzeichnis 1992/3). Steam, IPM resistant varieties and crop rotations are used instead (Ketzis 1992). The Federal Agricultural Institutes promote knowledge of soil pests, resistant varieties and IPM, providing growers with information on available resistant varieties, effective crop rotation and integrated methods of pest control. Handbooks such as 'Pflanzenschutz im Integrierten Gemusebau' (Plant Protection in Integrated Vegetable Production) describe production systems growers can use to control pests such as Verticillium wilt, Fusarium spp. and nematodes, which are treated with methyl bromide in other countries.

REFERENCES

ACP, 1992, Evaluation on methyl bromide. Advisory Committee on Pesticides. Ministry of Agriculture, Fisheries and Food, London.
Agra Europe, 1994, Phased end to use of methyl bromide. EC Food Law, July, p. 15. London.
Agrofarma, 1994, Annual report 1993–94, October. Agrofarma, Rome.
Agrow, 1995, Speciality products gain ground in Italy. Agrow, **226**, p. 8, 17 February. (Richmond: PJB Publications).
Aichinger H, 1995, Austrian Ministry of Environment, Vienna. Personal communication.
Arvieu JC, 1983, Some physico-chemical aspects of methyl bromide behaviour in soil. *Acta Horticulturae*, **152**, 267–74.
Arvieu JC and Cuany A 1985, EPPO Bull. **15**, 87–96.
Bakker DJ, 1993, Emissions of methyl bromide in fumigation of soil, structures and commodities: A review of Dutch studies over the period 1981–1992. TNO report IMW-R93/309, TNO Institute of Environmental Sciences, Delft.
Bangemann M, 1993, Letter to the Environment Minister of Denmark, re: Notification 93/0113/dk, 15 July, DG III, European Commission, Brussels.
Basile M and Lamberti F, 1981, Bromide residues in edible organs of plants grown in soil treated with methyl bromide, *Med. Fac. Landbouw*, **46**, 337–41.
Bernhard CA, Guillemin MP and Hillier RS, 1989, Occupational and para-occupational exposure to methyl bromide during soil fumigation in Switzerland, *Acta Horticulturae*, **255**, 327–55.
Blumthaler M and Ambach W, 1990, Indication of increasing solar ultraviolet-B radiation flux in Alpine regins, *Science*, **248**, 206–8.
BNA, 1995, WMO says unusually low ozone levels. Bureau of National Affairs, International Environment Daily, 15 February.
British Pest Control Association, 1993, Methyl bromide may not threaten ozone. Press release, 12 October, Derby, UK.

Brodberg R, Dong MH, Fong H, Formoli T, Haskell D, Meinders D, Ross JH, Sanborn JB and Thongsinthusak T, 1992, Estimation of exposure of persons in California to pesticide products containing methyl bromide. California Department of Pesticide Regulation, Worker Health and Safety Branch, Sacramento.

Bromine and Chemicals Ltd, 1990, Methylbromide soil fumigation — use and application instructions, London.

Brown BD and Rolston DE, 1980, Transport and transformation of methyl bromide in soils, *Soil Science* **130** (2), 68–75.

Canada Gazette, 1994, Ozone-depleting Substances Regulations amendment, 14 December, Department of the Environment, Department of National Health and Welfare, Ottawa.

CDPR, 1992, Estimation of exposure of persons in California to pesticide products containing methyl bromide. California Department of Pesticide Regulation, Sacramento.

Clean Air Act Amendments, 1990, Public Law 101-549. Title VI Stratospheric ozone protection, Washington DC.

CMR, 1994, Methyl bromide production and consumption estimates. Chemical Marketing Reporter, January, p. 37.

Conway GR and Pretty JN, 1991, Unwelcome harvest: Agriculture and pollution. Earthscan, London.

Cori L and Triolo L, 1994, Review of cases of methyl bromide reduction and elimination in Italy. SAFE Alliance, London.

Correnti A and Di Luzio PB, 1994, Attivit dell' ENEA nell'ambito degli interventi per la salvaguardia igienico sanitaria del lage di Bracciano. Sviluppo di attivit agricole compatibili nei territori prospicienti il lago. ENEA Internal Technical Report, Italian Committee of Innovation Technology, Energy and Environment, Rome.

Crutzen PJ and Andreae MO, 1990, Biomass burning in the tropics: Impact on atmospheric chemistry and biogeochemical cycles, *Science*, **250**, 1669.

Curley WH, 1984, Methyl bromide: a profile of its fate and effect in the environment. Contract No. 53-3294-3-39, US Department of Agriculture Animal and Plant Health Inspection Services, USA.

Council of EU, 1994, Council Regulation (EC) No. 3093/94 of 15 December on substances that deplete the ozone layer. Official Journal L, 22 December, **333**, 1–20, Brussels.

Daelemans A and Siebring H, 1977, Distribution of methyl bromide over the phases in Soil, *Mac. Fac. Landbouww*. **422** (2), 1729–38.

Daelemans A, 1978, in Wegman *et al* (ed.), 1981, Methyl bromide and bromide-ion in drainage water after leaching of glasshouse soils, *Water, Air, Soil Pollution*, **16**, 3–11.

Danish EPA, 1994, Statutory order from the Ministry of the Environment prohibiting the use of certain ozone depleting substances, No, 478, 3 June, Copenhagen.

Danish EPA, 1995, Danish Ministry of Environment, Copenhagen. Personal communication.

Danish Market Committee, 1993, Report of meeting, 8 November, Copenhagen.

Dead Sea Bromine, 1993, *Bromon*, No. 20, September, Dead Sea Bromine Co. Ltd, Beer Sheva, Israel.

DoE, 1991, The ozone layer, UK Department of the Environment, London.

DoE, 1993, The ozone layer, UK Department of the Environment, London.

European Methyl Bromide Association, 1995, Position paper on methyl bromide, 5 April, EMBA, CEFIC, Brussels.

Environmental Effects Panel, 1994, Environmental effects of ozone depletion: 1994 assessment. United Nations Environment Programme, Nairobi.

Fallico R and Ferranti M, 1991, *Zentralblatt für Hygiene und Umweltmedizin* **191**, 555–62.

Federal Register, 1993a, Production of stratospheric ozone, Proposed rule, 58 (51), 15014, 18 March, Washington DC.

Federal Register, 1993b, Protection of stratospheric ozone, Final rule, 58 (236), 65018–82, 10 December, Washington DC.

Guillemin, MP, Hillier RS and Bernhard CA, 1990, Occupational and environmental hygiene assessment of fumigations with methyl bromide, *Annals of Occupational Hygiene*, **34** (6), 591–607.

Guns MF, 1989, Monitoring of the groundwater bromine content and plant residues in methyl bromide fumigated glasshouses. Third International Symposium in Soil Disinfestation, Leuven, Belgium, 26–30 September 1988, *Acta Horticulturae*, **255**, 337–46.

Harsch DE and Rasmussen RA, 1977, Identification of methyl bromide in urban air, *Analyt. Lett.*, **10** (13), 1041–7.

Hao WM, 1986, Industrial sources of atmospheric N_2O, CH_3Cl and CH_3Br, thesis, Harvard university.

Herzel F and Schmidt G, 1994, The persistence of the fumigant methyl bromide in soil, *Wasser u. Boden*, **12**, 589–91 (in German).

Host Rasmussen M, 1994, Danish Ministry of Environment, Copenhagen. Personal communication.

International Workshop, 1992, Alternatives to methyl bromide for soil fumigation. Proceedings of International Workshops 19–23 October, Rotterdam and Rome/Latina.

IPCS, 1994, Methyl bromide (bromomethane) health and safety guide, *Health and Safety Guide No. 86*, International Programme on Chemical Safety, World Health Organisation, Geneva.

Kempton RJ and Maw GA, 1974, Soil fumigation with methyl bromide: the phytotoxicity of inorganic bromide to carnation plants, *Ann. Appl. Biol.* **76**, 217–29.

Ketzis J, 1992, Case studies of the virtual elimination of methyl bromide soil fumigation in Germany and Switzerland and the alternatives employed, in International Workshop 1992 op. cit.

Krupa SV and Kickert RN, 1989, The greenhouse effect: impacts of ultraviolet-B (UV-B) radiation, carbon dioxide and ozone (O_3) in vegetation, *Environmental Pollution*, **61**, 263–393.

Lepschyet J, Stark H and Sub A, 1979, Behaviour of methyl bromide (Terabol) in soil, *Gartenbauwissenschaften*, **44**, 84–8 (in German).

Lobert JM, Butler JH, Montzka SA, Geller LS, Myers RC and Elkins JW, 1995, A net sink for atmospheric CH_3Br in the East Pacific ocean, *Science*, **267**, 1002–5.

Lower House, 1981, Report of the Netherlands Parliamentary session 1980–1981, 16 400, Chapter XIV, **50**, 1–23, Amsterdam.

MBAN, 1994, Methyl bromide briefing kit. Methyl Bromide Alternatives Network, Pesticide Action Network, San Francisco.

MBTOC, 1994, 1994 Report of the Methyl Bromide Technical Options Committee. Report for the 1995 Assessment of the Montreal Protocol on Substances that Deplete the Ozone Layer. United National Environment Programme, Nairobi.

Menteri Pertanian, 1994, Pembatasan penggunaan dan izin pestisida berbahan aktif metil bromida. Surat Keputusan Menteri Pertanian, No. 322/Kpts/TP.270/4/94 of 28 April, Republic of Indonesia.

MBGC (Methyl Bromide Global Coalition), 1994a, Methyl bromide: annual production and sales for the years 1984–1992, MBGC, Washington DC.

Methyl Bromide Global Coalition, 1994b, Methyl Bromide Global Monitor, 1 (1), spring/summer, MBGC, Washington DC.

MfE, 1994, Ozone depletion: repairing the damage. New Zealand Ministry for the Environment, Wellington.

Miller MK, 1993, Alternatives to methyl bromide for soil and post-harvest fumigation. SAFE Alliance, London.

Miller MK, 1994, National and international legislation on methyl bromide. Report No. 13/94, Environmental Policy Research, Napier.

Miller MK, 1995a, Methyl bromide use and alternatives. The economic and trade implications of replacing methyl bromide in developing countries. Preliminary case study report for Environment Canada, Agriculture Canada, International Development Research Center of Canada, Friends of the Earth, US Environmental Protection Agency, Development Corporation of Switzerland.

Miller MK, 1995b, Preliminary cost benefit assessment on replacing methyl bromide in Europe. Technical report No.11/95, Environmental Policy Research, Napier.

Ministry of Environment, 1995, Servizio I.A.R., Ministero dell'Ambiente, Rome. Personal communication.

Ministry of Health, 1994, Ordinanza del Ministro della Sanita 16 giugno 1994: Misure cautelative concernenti i presidi sanitari a base di bromuro di metile. Official Gazette, 152, 1 July, Rome.

Molina MJ and Rowland FS, 1974, *Nature*, **249**, 810–2.

Molina LT, Molina MJ and Rowland FS, 1982, Ultraviolet absorption cross-sections of several brominated methanes and ethanes of atmospheric interest, *J. Phys. Chem.*, **86**, 2672–6.

NAS, 1978, Chloroform, carbon tetrachloride and other halomethanes: an environmental assessment. National Academy of Sciences, Washington DC.

Naughton P, 1995, UN reports record ozone depletion ozone depletion over north. Reuter, Geneva, 14 February.

New Scientist, 1995, 28 January, p.7.

NOAA, 1992, Our ozone shield. Reports to the nation. National Oceanic and Atmospheric Administration, Boulder, Colorado.

NOAA, 1994, Southern Hemisphere Winter Summary 94/4. 15 December. National Oceanic Atmospheric Administration, Boulder, Colorado.

Nordic Council, 1993, Methyl bromide in the Nordic countries—current use and alternatives. Nord 1993: 34, Nordic Council of Ministers, Copenhagen.

Nordic Council, 1995, Alternatives to methyl bromide. TemaNord 1995: 574, Nordic Council of Ministers, Copenhagen.

NRDC et al, 1994, NRDC et al v. US EPA: Petitioners' statement of issues to be raised. No. 94-1079, United States Court of Appeals for the District of Colombia Circuit, Washington DC.

Pease B, 1993, Preventing pesticide-related illness. California Policy Seminar.

Penkett SA, Prosser NJ, Rasmussen RA and Khalil MA, 1981, *Journal of Geophysical Research*, **86**, 5172–8.

Penkett SA, Jones BM, Rycroft MJ and Simmons DA, 1985, An interhemispheric comparison of the concentrations of bromide compounds in the atmosphere, *Nature*, **318**, 550–3.

Pimentel D and Lehman H (ed.), 1993, *The Pesticide Question: Environment, Economics and Ethics* (New York: Chapman & Hall).

Prather MJ, McElroy MB and Wofsy SC, 1984, Reductions in ozone at high concentrations of stratospheric halogens, *Nature*, **312**, 227–31.

Regional Ordinance, 1983, No. 288, 3 August. Municipalities of Trevignano and

Anguillara, Rome province, Italy.

RIVM, 1980, Methyl bromide. Report No. 149/80, National Institute of Public health and Environmental Protection, the Netherlands.

Robbins DE, 1976, Photodissociation of methyl chloride and methyl bromide in the atmosphere, *Geophysical Research Letters*, **3**, 213–6; also UV absorption cross-sections for methyl bromide and methyl chloride, *Geophysical Research Letters*, **3**, 757–8.

Roughan JA and Roughan PA, 1984, Pesticide residues in foodstuffs in England and Wales: Part I. Inorganic bromide ion in lettuce grown in soil fumigated with bromomethane. *Pesticide Science*, **15**, 431–8.

Sassaman, JF *et al*, 1986, Pesticide background statements: Methyl bromide and chloropicrin. In Vol. II, Fungicides and fumigants. Agriculture Handbook, No. 661. Forest Service, US Department of Agriculture.

Schuepp H, 1994, Eidgenossische Forschungsanstalt [Federal Research Station for Fruit-growing, Viticulture and Horticulture], Wadenswil, Switzerland. Personal communication.

Shanklin J, 1994, True story of CFCs. Letters to the editor. *Chemistry & Industry*, 18 April, p. 282, London.

SORG, 1990, Stratospheric ozone 1990. UK Stratospheric Ozone Review Group. Third report. HMSO, London.

SORG, 1991, Stratospheric ozone 1991. UK Stratospheric Ozone Review Group. Fourth report. HMSO, London.

SORG, 1993, Stratospheric ozone 1993. UK Stratospheric Ozone Review Group. Fifth report. HMSO, London.

Swedish EPA, 1994a, Report on ozone depleting substances. No. 4278, Swedish Environmental Protection Agency, Solna.

Swedish EPA, 1994b, Phase out all substances that deplete the ozone layer. Press release, 1 February, Swedish Environmental Protection Agency, Solna.

TEAP, 1995, Supplement to the 1994 assessments. Part I: Synthesis of the reports of the Scientific Assessment Panel and Technology and Economic Assessment Panel on the impact of HCFC and methyl bromide emissions, March, United Nations Environment Programme, Nairobi.

UNEP, 1992a, Methyl bromide: its atmospheric science, technology, and economics. Synthesis report. Methyl Bromide Interim Scientific and Technology and Economic Assessment, United Nations Environment Programme, Nairobi.

UNEP, 1992b, Report of the fourth meeting of the Parties to the Montreal Protocol on Substances that Deplete the Ozone Layer. 23–25 November, Copenhagen. UNEP/OzL.Pro.4/15, UNEP, Nairobi.

UNEP, 1993a, Action on ozone. United Nations Environment Programme, Nairobi.

UNEP, 1993b, Report of the fifth meeting of the Parties to the Montreal Protocol on Substances that Deplete the Ozone Layer. 17-19 November, Bangkok. UNEP/OzL.Pro.5/12, UNEP, Nairobi.

UNEP, 1994a, 1994 report of the Technology and Economic Assessment Panel, United Nations Environment Programme, Nairobi.

UNEP, 1994b, Corrigendum to the report of the fifth meeting of the Parties to the Montreal Protocol on Substances that Deplete the Ozone Layer. UNEP/OzL.Pro.5/12/Corri.1, UNEP, Nairobi.

UNEP, 1994c, International scientists report record ozone depletions, but future improvements expected. Press release, 6 September. United Nations Environment Programme, Nairobi.

UNEP, 1995, Report of the eleventh meeting of the Open-Ended Working Group of the Parties to the Montreal Protocol, 8–12 May. UNEP/OzL.Pro/WG.1/11/10, 13

May, UNEP, Nairobi.

US EPA, 1992, Toxic Release Inventory 1991. US Environmental Protection Agency, Washington DC.

US EPA, 1993, The cost and cost-effectiveness of the proposed phase-out of methyl bromide. US Environmental Protection Agency, Washington DC.

Van Haasteren, 1994, The Netherlands Ministry of Housing, Physical Planning and Environment, The Hague. Personal communication. Methyl bromide can only be used as a soil treatment in emergency situations if special permission is granted. No permission has been given since 1992, although some requests have been made.

Van Haasteren, 1995, The Netherlands Ministry of Housing, Physical Planning and Environment, The Hague. Personal communications. A new limit came into force in December 1994 and is based on the conclusions of Met Nederlandstalige Samenvatting [Dutch Expert Committee for Occupational Standards] 1990. Health-based recommended occupational exposure limits for methylbromide. Report 13/90, The Hague.

Van Rhee JA, 1977, Effects of soil pollution on earthworms, *Pedobiologia*, **17**, 201–8.

WHO, 1994, Ultraviolet radiation: an authoritative scientific review of environmental and health effects of UV, with reference to global ozone layer depletion, *Environmental Health Criteria*, **160**, World Health Organisation, Geneva.

WHO, 1995, Methyl bromide, *Environmental Criteria*, **166**. International Programme on Chemical Safety, World Health Organisation, Geneva.

WMO, 1990, Report of the International Ozone Trends Panel: 1988. World Meteorological Organisation, Global Ozone and Monitoring Network, report No. 18, WMO, Washington DC.

WMO, 1992, Scientific assessment of ozone depletion: 1991. World Meteorological Organisation Global Ozone Research and Monitoring Project, report No. 25, WMO, Geneva.

WMO, 1994, Scientific assessment of ozone depletion: 1994. World Meteorological Organisation Global Ozone Research and Monitoring Project, report No. 37, UNEP, Nairobi.

Wofsy SC, McElroy MB and Yung YL, 1975, The chemistry of atmospheric bromine, *Geophysical Research Letters*, **2** (6), 215–8.

Zheng X and Basher RE, 1993, Homogenisation and trend detection analysis of broken series of solar UV-B data, *Theor. Appl. Climatol.*, **47**, 189–203.

4

Effects on Target Organisms

OLIVER C. MACDONALD* and CHRISTOPH REICHMUTH[†]
*Central Science Laboratory, MAFF, Harpenden, Herts, UK

[†]Federal Biological Research Institute for Agriculture and Forestry, Institute of Stored Product Protect, Berlin, Germany

C.H. Bell, N. Price and B. Chakrabarti: The Methyl Bromide Issue
© 1996 John Wiley & Sons Ltd

1.1 INTRODUCTION

After the discovery of the insecticidal properties of methyl bromide by Le Goupil (1932) methyl bromide quickly became adopted for plant quarantine treatments (Bond 1984). Later it was used on a large scale to disinfest stored products, flour mills, warehouses, ships and railway cars (Bond 1984, Monro 1969, Peters 1942). By 1938 it was probably the most widely used fumigant in the US, the main use of methyl bromide at the time, being for the treatment of dried fruit, but other products such as grain, live plants, fruits and vegetables were also fumigated.

Given sufficient concentration and duration of exposure methyl bromide appears to be toxic to virtually all forms of life. Against insects, which are probably the most frequently targeted organisms, methyl bromide is not as toxic as some other chemicals that have been used as fumigants. However, other properties of the gas serve to make it an effective fumigant (Bond 1984). Short lethal exposure periods of less than two days are a particularly important quality of this chemical compared to phosphine, the other main fumigant in use today, or modified atmospheres containing high concentrations of carbon dioxide or nitrogen and low levels of oxygen, where several days or even weeks may be required for the control of pests.

2.1 MODE OF ACTION

The mode of action of methyl bromide, while having been subject to much study, has never been clearly elucidated and it is likely that its toxicity is caused by damage to multiple sites. Insects exposed to lethal concentrations of the gas show no immediate symptoms of poisoning. Unlike phosphine and hydrogen cyanide which are respiratory poisons and cause rapid paralysis in insects, the onset of symptoms in insects, and for that matter mammals, exposed to methyl bromide, can be delayed for many hours (von Oettingen, 1955, Winteringham and Barnes 1958). Insects exposed to methyl bromide typically show irritation, followed by uncoordinated movement and spasms and then a decline in activity that is eventually followed by death. However, some species may behave differently. *Tribolium confusum* Jaquelin du Val has been reported as showing no signs of irritation, merely a decline in activity in reponse to poisoning by methyl bromide, while *Tenebroides mauritanicus* (L.) becomes paralysed following a period of hyperactivity (Bond 1956). The response of mammals is similarly varied. Rats became extremely excitable when exposed to a $1\,mg\,l^{-1}$ concentration for 20 hours. Any stimulus resulted in wild activity which finally subsided into muscular tremors. Eventually the animals became exhausted and relaxed. Rabbits responded differently exhibiting no obvious symptoms until they collapsed.

Those rabbits that survived test treatments invariably became paralysed. Higher concentrations than $1\,mg\,l^{-1}$ resulted in signs of discomfort in rats. Lower concentrations, around $0.24\,mg\,l^{-1}$ however had no effect on either rats or guinea pigs even after six months of routine exposures (Irish *et al* 1940).

Various workers have tried to determine the specific mode of action of methyl bromide. Respiration itself does not appear to be directly affected. In *T. mauritanicus* no effect of methyl bromide was found on the respiration of either adults or larvae until individuals became paralysed, after which the respiration rate slowly declined until the insects died (Bond, 1956). Methyl bromide has nevertheless been shown to affect repiratory enzymes. Dixon and Needham (1946) showed that hexokinase, which catalyses the phosphorylation of glucose by adenosine triphosphate (ATP), was inhibited *in vitro*. Further work by Winteringham *et al* (1958), however, showed that inhibition of respiratory SH enzymes was not responsible for the lethal effect of methyl bromide in *Musca domestica* L., though they did show that very high levels of methyl bromide caused a substantial depletion of phosphoglycerate and ATP. Miller and Haggard (1943) found high levels of intracellular bromide ions in rats killed by methyl bromide and suggested that poisoning was attributable to the release of free bromide ions.

A number of studies have indicated that methyl bromide acts on SH groups. Blackburn *et al* (1944) found that thiol groups in bisulphited wool protein were methylated by methyl bromide. Lewis (1948) showed that methyl bromide reacted with a range of groups in proteins but the reaction with SH groups *in vitro* was particularly rapid and proposed that the toxicity of methyl bromide was due to the disruption of enzyme function. Loveday and Winteringham (1951) showed that methyl bromide reacted with free SH groups in insects. This was followed by a suggestion by Winteringham and Barnes (1955) that the toxicity of the gas was due to the general methylation of SH dependent and possibly other proteins, and therefore that no selective action in the toxicity of methyl bromide was likely to be found.

More recent work has shown that glutathione is methylated and sulfhydryl groups alkylated by methyl bromide. There is also a reduction in cytotoxicity if reduced glutathione is added to cells exposed to methyl bromide, thus supporting earlier theories that methyl bromide reacts with protein SH groups (Nishimura *et al* 1980). Shivanandappa and Rajendran (1987) also showed that exposure to methyl bromide reduced glutathione levels in the Khapra beetle, *Trogoderma granarium* Everts.

It would appear that methyl bromide exerts at least part of its toxic effect by the methylation of SH groups in enzymes systems that are responsible for cell respiration. However, wider interference with SH groups in other systems causes widespread toxic effects (Stark 1994). The claims by Cheetham (1990) and Nishimura *et al* (1980) that methyl bromide is a general cellular toxin is widely supported, not just for animal studies but also in fungi (Parameswaran and Ruetze 1984), and thus has considerable basis.

3.1   TOXICITY OF METHYL BROMIDE

3.1.1   CTP

The toxic effect of a poisonous gas depends on its concentration c — mostly in a mixture with air — and the period t during which an organism is exposed to the toxic substance. According to Haber's rule, first published in 1924, the mathematical product of concentration and time (ct) is a constant, k, the ct-product, or CTP, giving an appropriate measure to express the relative toxicity of poisonous gases.

$$ct = \text{constant } k \tag{1}$$

If the concentration c in the formula is reduced, the exposure time has to be extended accordingly to give the same value for k. Thus half of the concentration ($c/2$) will lead to the same mortality within twice the exposure time ($2t$) and *vice versa*. Different mortalities require different CTPs. Peters and Ganter (1935) called k the efficiency number in experiments with adult *Sitophilus granarius* (L.) and hydrogen cyanide. Peters (1936, 1942), suggested the name for the unit as gram-hour-unit for k with the units g.m^{-3} for c and hours for t. The constant k is widely known as the CTP with units mg l^{-1} h. The obvious advantage of this rule lies in its applicability under practical conditions where the exposure time often has to be adapted to the circumstances and the lethal concentration can be calculated in advance.

For methyl bromide and some other fumigants the CTP remains constant over a wide range of lethal concentrations and exposure periods. However, the relationship does break down at extreme concentrations and exposure periods, Brown (1954) gave 2 mg l^{-1} as the lower limit at which the CTP relationship still holds. Howe and Hole (1966) determined higher CTPs for the control of *S. granarius* at 25 °C and 70% r.h. with 3 mg l^{-1} than with 6 mg l^{-1} or 9 mg l^{-1}. Bennett (1969) found that the CTP required to control pupae of *Tribolium castaneum* (Herbst) was constant with concentrations of methyl bromide down to 5 mg l^{-1}, whereas at 2.8 mg l^{-1} significantly higher CTPs were required to achieve the same mortality. For seven species of Coleoptera attacking stored products, Bell (1988) determined the minimum concentrations of methyl bromide at which Haber's rule applied, or efficacy threshold, to be between 0.6 and 2.0 mg l^{-1} at 15 °C, and between 1.3 and 4.0 mg l^{-1} at 25 °C depending on species. The order of the efficacy thresholds largely reflected the order of susceptibility of the different species, i.e. less susceptible species had higher efficacy thresholds. Strains of insects that had higher than usual tolerance to methyl bromide had slightly higher efficacy thresholds than their stock counter parts at 25 °C. However, this difference was not apparent at 15 °C. Previous exposure to a high concentration of methyl bromide appears to lower the efficacy threshold (Bell 1981, 1982). One of the few detailed studies of the fumigation of fungi showed decreasing CTPs required to achieve the LD_{99} for a whole range of species as the concentration of methyl bromide was increased from 5 to $10 \, mg \, l^{-1}$, (Munnecke *et al* 1978). It is unclear though whether this is because Haber's rule does not apply to the treatment of fungi or whether the efficacy threshold is substantially higher for fungi than for insects.

A certain minimal exposure time, below which Haber's rule does not hold, is necessary to allow a gas to penetrate into the organism and reach the lethal site. This time is quite short for methyl bromide. In one study on the methyl bromide tolerant diapausing larval stage of *Ephestia elutella* (Hübner), the minimum time was found to be substantially less than 1 hour (Bell and Glan-ville 1973).

Methyl bromide appears to be toxic to any organism in sufficient dosage. While most of its uses are against insects, mites and fungi, a wide range of other classes of organisms have also been controlled. More advanced organisms are generally more sensitive to methyl bromide than are more primitive ones. Table 4.1 indicates the typical CTPs required to kill the most tolerant stages of various organisms (MAFF/PSD Advisory Committee 1992).

4.1 EFFECT ON TARGET ORGANISMS

4.1.1 MICRO-ORGANISMS

Most work on micro-organisms has concentrated on plant pathogenic organisms and their control by soil fumigation. However, methyl bromide

Table 4.1. Susceptibility of different organisms to methyl bromide

Organism	Concentration Time Product $(mg \, l^{-1} \, h)$
Higher Plants (excluding seeds)	200–600
Mites	
Insects	50–200
Nematodes	600–1000
Fungi	
Bacteria	2500–4000
Viruses	

has also been tested for the sterilisation of a range of other materials infected with both bacteria and fungi. High levels of success were reported in many of these studies. Complete elimination of pathogens is difficult, though, as many micro-organisms produce resting stages that are highly resistant to treatment.

In 1950 Kolb and Schneiter reported that methyl bromide at area dosages of 3400 to 3900 $g\,m^{-2}$ killed spores of anthrax, *Bacillus anthracis*, in 24 hours in the presence of moisture. Dry spores however could survive at least 72 hours. A shorter treatment of only one hour was insufficient to eliminate the disease from any sample tested. Three species of bacteria, *Pasteurella multicoda, P. anatipestifer and Escherichia coli*, were eliminated from poultry litter by fumigation with 300 $g\,m^{-2}$ for 48 hours. High levels of control were achieved against *E. coli* with 100 $g\,m^{-2}$ for 48 hours, while satisfactory control of *Pasteurella* spp. required 150 $g\,m^{-2}$ for 96 hours (Bendheim and Shoshan, 1979). Spores of *Bacillus larvae*, a bacterium found in bee hives, were killed by fumigation with 2550 $g\,m^{-2}$ methyl bromide (Faucon *et al* 1982), though the safety of using such high concentrations of gas for fumigating hives was questioned, as was the possibility of toxic residues being absorbed by wood and wax. Tests with jute bags contaminated with potato ring rot (*Corynebacterium sepedonicum* = *Clavibacter michiganensis* sub sp *sepedonicus*) showed that high concentrations of methyl bromide (5% $\approx 200\,mg\,l^{-1}$) could reduce the presence of the bacterium to no more than a trace in 18 hours. Complete elimination of the bacterium, however, required a treatment of 15% methyl bromide for 48 hours (Richardson and Monro 1962).

With fungi, Majumder (1954) used methyl bromide to prevent spoilage in shipments of tapioca and wheat that had absorbed large quantities of water and had been found to be heavily infested with yeast, *Aspergillus* spp. and *Penicillium* spp. Treatments of 80 $mg\,l^{-1}$ for 24 or 48 hours were used. The number of microbes per g of dry matter was reduced from 125×10^6 to 200 after 24 and 150 after 48 hours in the tapioca and from 2.4×10^6 to 120 and 72 respectively in the wheat samples. Control infestations meanwhile increased 2–3 fold over 48 hours. Methyl bromide proved to be a more effective control than ethylene dibromide. Sufficiently high levels of control have been achieved for methyl bromide fumigation (240 $mg\,l^{-1}$ for 72 hours at 5–10 °C) to be accepted as a quarantine treatment of the oak-wilt fungus, *Ceratocystis fagacearum* (Bretz), in oak logs (Rütze and Liese 1983, Liese and Rütze 1985). No deterioration in the quality of the timber was observed. Unger *et al* (1992) showed in both laboratory and field experiments that several stages of the mould *Serpula lacrimans* infesting wood and walls of old buildings could be controlled within 72 hours with a dose of 150 $mg\,l^{-1}$ at about 20 °C. However, the fumigation of seeds to control fungi associated with them has not proved successful due to the phytotoxicity of methyl bromide (Kennedy 1961, Raju 1984).

Many reports attest to the efficacy of methyl bromide for the control of soil borne plant pathogens. The incidence of diseases of apple seedlings (Benson *et al* 1978), iris bulbs (Haas 1976), tomatoes (Clerjeau *et al* 1973), soybeans (An *et al* 1990) and cucumbers (Summer *et al* 1983) have all been reduced by fumigation with methyl bromide. Most field studies report only the dosage of gas applied to the soil and measure the efficacy of the treatment by determining the reduction in disease incidence in a subsequent crop compared to a control. Only a few studies have looked specifically at the reduction in the numbers of disease propagules either by soil sampling (An *et al* 1990) or by investigating the effect of pathogen samples buried in the soil prior to treatment (Haas 1976). Such studies still offer only a partly quantitative description of the effect of methyl bromide on micro-organisms as the CTPs to which soil organisms are exposed is dependent not just on the applied dose but on a number of other factors as well. Soil type and wetness as well as application method and prior treatment of the soil will affect the penetration of gas through the soil, while the amount of organic matter will affect the rate at which the gas is absorbed. The type of sheeting and how it is used will also influence the rate at which gas is lost, either by penetration of the sheet or round its edges. Soil moisture and type as well as host tissues have been shown to have a considerable effect on the response of fungi to fumigation (Munnecke *et al* 1959, Munnecke *et al* 1971).

The most thorough study of the effect of methyl bromide on soil borne plant pathogens was carried out by Munnecke *et al* (1978) who looked at ten different species of phytopathogenic fungi. In most cases these were mycelia cultured on agar but parallel tests of some species infecting plant material or as sclerotia or chlamydospores were also conducted. Sclerotia were more resistant to fumigation than the mycelia for the two species tested and for *Phytophthora cinnamomi* mycelia were more resistant than chlamydospores. Little difference was found in the response of mycelia growing on agar and that growing on roots, though mycelia of *Verticillium albo-atrum* on stems was more tolerant than that on agar. Table 4.2 shows the CTPs required to obtain the LD_{90} for mycelia of nine of the pathogens tested by Munnecke *et al* (1978) at a range of gas concentrations. The increasing doses required for the same level of kill at different gas concentration suggest that either the parameters chosen for these tests fell outside the range at which the CTP relationship applies or possibly that Haber's law does not hold for the treatment of fungi. Whatever the reason for the variations in the CTPs the results suggest that high concentrations of gas may be necessary for soil fumigation to effectively control fungi.

4.1.2 NEMATODES

Nematodes have been treated in a similar manner to phytopathogenic fungi in many studies of soil fumigation. As with most soil fumigation work, these investigations are generally qualitative and say little of the dose response of

Table 4.2. CTPs (mg l^{-1} h) required for LD$_{90}$ values at specific concentrations (ml l^{-1}) of methyl bromide (Munnecke *et al* 1978)

Species	CTP required at concentration				
	5	10	15	20	30
Armillaria mellea	250	220	206	196	180
Pythium ultimum	385	215	150	118	84
Phytophthora cinnamomi	210	155	132	116	99
Phytophthora citrophthora	230	165	138	122	105
Phytophthora parasitica	250	190	168	148	126
Rhizoctona solani	610	350	255	200	150
Sclerotium rolfsii	510	317	240	200	150
Verticillium albo-atrum	365	350	345	330	330
Whetzelinia sclerotiorum	395	330	300	280	252

the species under study. Table 4.3 lists some of the studies that have been carried out on the control of soil borne nematodes.

Many nematodes, however, are borne on seeds or other propagative plant material. A number of studies have therefore investigated the fumigation of plant material, especially seeds, for the control of plant parasitic species. Because of the phytotoxic effects of methyl bromide, achieving a dose that will kill the target organisms without damaging the host requires careful control of the doses used. These studies have therefore generally been more quantitative and have investigated the effect of a range of doses.

The stem eelworm (*Ditylenchus dipsaci*, (Kuhn) Filipjev) has been one of the most studied plant parasitic nematodes. Early work showed that CTPs between 600 and 800 mg l^{-1} h could be used to control *D. dipsaci (Anguillulina dipsaci)* on onion (Goodey 1945), teazel (Goodey 1949) and red clover (Goodey 1949, Bingefors and Hahlin 1955). Page *et al* (1959) and Hague and Clark (1959) reported studies on artificially desiccated eelworm larvae extracted from infested stems. No larvae survived fumigation with methyl bromide at CTPs above 850 mg l^{-1} h. By comparison, treatments up to 1250 mg l^{-1} h on teazel (Coolen *et al* 1972), or 2222 mg l^{-1} h on onion (Goodey 1945), had little effect on the yield of treated seed when sown under commercial conditions. Further studies by Gostick (1963) showed that complete control of *D. dipsaci* infesting lucerne seed could be obtained at CTPs of around 800 mg l^{-1} h between 10 and 20 °C but at 25 °C and above, this was reduced to 600 mg l^{-1} h. Powell (1974) studied *D. dipsaci* infesting field beans. A CTP of 800 mg l^{-1} h gave complete control in lightly infested seed, with 285–374 nematodes extracted from batches of 100 seeds. However in heavily infested seed, with up to 5806 nematodes per batch, this dose only gave 98.5% control. Complete control in this case was only achieved with a CTP of 3000 mg l^{-1} h.

Table 4.3. Results of studies on soil fumigation of nematodes

Nematode species	Plant infested	Treatment rate	Control
Meloidogyne javanica	Cucumber Eggplant	$40\,\mathrm{g\,m^{-2}}$	Good (Stephan *et al* 1991)
Meloidogyne incognita	Tobacco	$26.8\,\mathrm{g\,m^{-2}}$	Highly Effective (Hussaini 1989)
Meloidogyne arenaria	Peanut	$25.2\,\mathrm{g\,m^{-2}}$	No significant effect (Dickson and Hewlett 1988)
Meloidogyne incognita	Tobacco	$22.5\text{–}45\,\mathrm{g\,m^{-2}}$	Highly Effective (Hussaini 1985)
Meloidogyne incognita	Turnips Maize Southern Peas	$39.2\,\mathrm{g\,m^{-2}}$	Reduced to very low levels (Rhode *et al* 1980)
Meloidogyne spp.	Watermelon	$16.8\,\mathrm{g\,m^{-2}}$	Increased yield by $> 37\,000\,\mathrm{kg\,ha^{-1}}$ (Brust and Scott 1994)
Pratylenchus vulpus	Roses	$90\,\mathrm{g\,m^{-2}}$	Effective down to 90 cm (Coolen *et al* 1972)
Hoplolaimus columbus	Cotton	$1.68\text{–}13.4\,\mathrm{g\,m^{-2}}$	Significant effects at all doses (Noe 1980)
Trichodorus spp.	Strawberry	$33.6\,\mathrm{g\,m^{-2}}$[†]	All nematodes virtually eliminated (Reidel and Ellis 1983)
Xiphinema spp.		$20\,\mathrm{g\,m^{-2}}$	Eradicated to a depth of 90 cm (Basile *èt al* 1986)
Radolphus similis		$48.7\,\mathrm{g\,m^{-2}}$	Eradicated to a depth of 8 feet (O'Bannon and Tomerlin 1968)

[†] 67% Methyl bromide 33% Chloropicrin mixture.

4.1.3 PLANTS

In addition to the control of pests, soil fumigation is also used for weed control. As seeds with low moisture contents are most tolerant (Powell 1975, Cobb 1958) soil to be treated is normally watered at least two weeks prior to fumigation to encourage the germination of seeds (Dead Sea Bromine Group 1991). While a number of studies have been conducted on the effect of treatment on stored seeds and on the phytotoxic effects of plant disinfestation treatments, few studies appear to have been carried out on the efficacy of methyl bromide soil fumigation on weeds. Hussaini investigating the effect of fumigations to control both weeds and nematodes in tobacco nurseries in India showed that both methyl bromide at 15.5 ml m^{-2} (Hussaini 1989) and a 1:1 mixture of methyl bromide and ethylene dibromide applied at 22.5 and 45.0 g m^{-2} (Hussaini 1985) gave more than 95% control of weeds.

Extensive studies have been carried out on the effects of fumigation on stored seed. Powell (1975) investigated the effect of fumigation at CTPs of 200 and 400 mg h^{-1} h on forty varieties of vegetable, cereal, fodder and grass seeds at a range of moisture contents. At these CTPs, although germination was rarely reduced by more than 10%, many varieties showed a delay in germination. While this was typically only 1–2 days, for some seeds it was up to a week. If this were repeated in a field situation, possibly with a greater delay caused by the higher doses used in soil fumigation, it would allow crops a greater chance to become established before weeds whose seeds had survived treatment started to appear. Cobb (1958) investigated the effect on the germination of seeds of 35 different crops, at different moisture contents, of three different fumigation conditions, 80 mg l^{-1} for 12 hours and 48 hours, both at 10 °C, and 64 mg l^{-1} for 12 hours at 22 °C. All the seeds withstood at least a single treatment with little if any loss of viability at the lowest moisture content tested for each species. For almost all the seeds though, an increase in moisture content resulted in greater injury when treated. Alfalfa, barley and ryegrass were exceptions to this. In general leguminous seeds appear to be more tolerant of fumigation than are cereals (Thompson 1966).

4.1.4 MOLLUSCS

A number of snail species are pests of quarantine concern, there being a risk of their introduction to new areas through movement in traded plant material. The use of methyl bromide to control the apple snail, *Pomacea canaliculata* (L.) has been studied in Japan (Sadoshima *et al* 1993). The most tolerant stage was found to be the young snails in diapause. Complete kill required a dose of 300 g m^{-3} for 27 hours at 10–19 °C in large scale mortality tests. In the US, methyl bromide has been investigated for the disinfestation of rosemary seed infested with juvenile *Helicella candidula* (Studer) and *H. conspurcata* (Draparnaud) snails (Roth and Kennedy 1973). At NAP *H. conspurcata* survived $96 \, \text{mg} \, l^{-1}$ for 8 hours, however under a reduced pressure of 100 mm Hg *H. candidula* survivors were found up to a dose of $32 \, \text{mg} \, l^{-1}$ for 2 hours (CTP $\approx 60 \, \text{mg} \, l^{-1} \, \text{h}$) but not at $48 \, \text{mg} \, l^{-1}$ for 2 hours (CTP $\approx 87 \, \text{mg} \, l^{-1} \, \text{h}$). All fumigations were carried out at 22.2–22.8 °C.

4.1.5 ARTHROPODS

Arthropods are by far the most extensively studied group of organisms in relation to methyl bromide fumigation. Within the group, insects have been studied more extensively than any other organisms, although mites have received some attention and fumigation has also been used in New Zealand to prevent the import of black widow spiders, *Latrodectus* spp. with shipments of grapes (New Zealand, Ministry of Agriculture and Fisheries 1990).

Tables 4.4–4.7 show the response to methyl bromide of insects from eight different orders. Tables 4.4 and 4.5 contain a collection of data describing the response of a wide range of insects and mites of different stages and infesting a wide range of produce to methyl bromide. Table 4.4 summarises reports where dose response parameters have been determined, while Table 4.5 summarises reports in which the doses required to cause complete, or near complete, mortality have been described. Many papers deal only with adult insects. At 25 °C, the LD_{50} for this stage ranges between $10 \, \text{mg} \, l^{-1} \, \text{h}$ to $50 \, \text{mg} \, l^{-1} \, \text{h}$ and the $LD_{99.9}$ between $20 \, \text{mg} \, l^{-1} \, \text{h}$ to $70 \, \text{mg} \, l^{-1} \, \text{h}$, respectively (Table 4.6).

The groups most tolerant to methyl bromide tend to be stored product pests, especially mites and some of the Coleoptera, with LD_{99}s for some species of over $200 \, \text{mg} \, l^{-1} \, \text{h}$. The only other insects showing similar tolerance are the diapausing larvae of some moths (Tables 4.6 and 4.7). Most insects from other groups all have similar ranges of response, with LD_{99}s between about 25 and $70 \, \text{mg} \, l^{-1} \, \text{h}$. Of the four species of mite listed (Table 4.5), *Acarus siro* (L.), *Tyrophagus longior* (Gervais) and *Lepidoglyphus destructor* Schrank, all pests of stored products, were highly tolerant in comparison to most insect species (MAFF/PSD Advisory Committee on Pesticides 1992, Bowley and Bell 1981).

Some of the data in Table 4.5 come from experiments with many thousands of insects that have been carried out for the development of quarantine treatments for which some plant health authorities demand probit 9 security. Again, some of the Coleoptera appear to require higher doses for complete control than do species from most other classes.

Table 4.4. Summary of literature reports of dose/mortality regression responses of various insects to fumigation with methyl bromide

Species	Family or superfamily	Stage	Commodity	LD_{50}	D_{95} (mg l^{-1} h)	LD_{99}	Temp (°C)	Reference	Note
Mites									
Tetranychus urticae	Tetranychidae	A		34.5	112.7	184.1	11	Powell & Gostick 1971	also at 18 & 5 °C
		E		28.3	95	157.1	11	Powell & Gostick 1971	also at 18 & 5 °C
		L		25.6	147.6	365.1	11	Powell & Gostick 1971	also at 18 & 5 °C
Coleoptera									
Acanthoscelides obtectus	Bruchidae	A	Cowpea	30.5	34		25	Fisk & Shepard 1938	
Callosobruchus chinensis	Bruchidae	A	Cowpea	28	30.6			Adu & Mithu 1985	
		E	Cowpea	20.4	95.5			Adu & Mithu 1985	
		P	Cowpea	21.4	63.4			Adu & Mithu 1985	
		L	Cowpea	53				Adu & Mithu 1985	
Leptinotarsa decemlineata	Chrysomelidae	A	Potatoes	45	116		10	Powell & Mills 1981	
Sitophilus granarius	Curculionidae	A		7.4		8.4	25	Shepard et al 1937	
		A		27.5			25	Fisk & Shepard 1938	
Sitophilus oryzae	Curculionidae	A		4		6.2	25	Shepard et al 1937	
		A		20			25	Fisk & Shepard 1938	
Tribolium castaneum	Tenebrionidae	P		2.9			26	Rajendran 1990	
Tribolium confusum	Tenebrionidae	A		11.2		14.4	25	Vincent & Lindgren 1975	
		E		29.5			25	Fisk & Shepard 1938	
		A		50			25	Fisk & Shepard 1938	
Trogoderma granarium	Dermestidae	L		68		160	25	El-Lakwah 1977a	
		L		55–66		200–240	15	El-Lakwah 1977a	
		L		288–290		460–480	10	El-Lakwah 1977b	
		A		153–173		240–280	10	El-Lakwah 1977b	
Trogoderma variabile	Dermestidae	A		44–79		65–102	10	Vincent & Lindgren 1975	
		P		56–103		78–96		Cobb 1958	
		L		28–72		43–102	21	Cobb 1958	Depending on Time/Dose
Diptera									
Liriomyza trifolii	Agromyzidae	P	Chrysanthemum cuttings			26–69	15	Mortimer & Powell 1984	Depending on age
		L	Chrysanthemum cuttings			27.8–34.5	15	Mortimer & Powell 1984	
		E	Chrysanthemum cuttings			34	15	Mortimer & Powell 1984	
Liriomyza huidobrensis	Agromyzidae	L		15.5		45	15	Macdonald 1996	
		EP		23		45–49	15	Macdonald 1996	
Ragoletis completa	Tephritidae	E	Peaches	18.2			21	Yokoyama et al 1992	
		L1	Peaches	9.6			21	Yokoyama et al 1992	
		L2	Peaches	7.2			21	Yokoyama et al 1992	
		L3	Peaches	17.4–23.7			21	Yokoyama et al 1992	

	Family	Host	Stage				Reference	Notes
Homoptera								
Aonidiella aurantii	Coccoidea	Lemons	Early grey adult	78		25	Yust & Busbey 1942	HCH Resistant strain
	Coccoidea	Lemons	Early grey adult	63		25	Yust & Busbey 1942	HCH Susceptible strain
Bemisia tabaci	Aleyrodidae	Poinsettias	E	6	26.5	15	Macdonald & Mills 1994	
		Poinsettias	Nymphs	18.2	52.1	15	Macdonald & Mills 1994	
		Poinsettias	A	18.9	56	15	Macdonald & Mills 1994	
Hymenoptera								
Systole albipennis	Chalcidoidea		L		28.6–37	27	Verma 1986	Depending on host
			P		54.2–57.9	27	Verma 1986	Depending on host
Bruchophagus mellipes	Chalcidoidea		P	55.6		27	Verma 1986	
			L	43.3		27	Verma 1986	
Isoptera								
Zootermopsis angusticollis	Termopsidae		Pseudergates	26	34.8	27	Scheffrahn & Su 1992	
Neotermes castaneus	Kalotermitidae		Alates	22.7	34.8	27	Scheffrahn & Su 1992	
			Pseudergates	38	48.8	27	Scheffrahn & Su 1992	
Neotermes jouteli	Kalotermitidae		Pseudergates	32.4	44.9	27	Scheffrahn & Su 1992	
Incisitermes snyderi	Kalotermitidae		Alates	27.2	47.2	27	Scheffrahn & Su 1992	
			Pseudergates	38.8	57.2	27	Scheffrahn & Su 1992	
Incisitermes minor	Kalotermitidae		Alates	34.2	54.5	27	Scheffrahn & Su 1992	
			Pseudergates	34.7	49.6	27	Scheffrahn & Su 1992	
Cryptotermes cavifrons	Kalotermitidae		Pseudergates	45.9	75	27	Scheffrahn & Su 1992	
Reticulitermes flaviceps	Rhinotermitidae		Pseudergates	13.1	21.8	27	Scheffrahn & Su 1992	
Reticulitermes hesperus	Rhinotermitidae		Pseudergates	11.4	16.5	27	Scheffrahn & Su 1992	
Prorhinotermes simplex	Rhinotermitidae		Pseudergates	19.5	22	27	Scheffrahn & Su 1992	
Coptotermes formosanus	Rhinotermitidae		Pseudergates	20.4	28.9	27	Scheffrahn & Su 1992	
			Alates	35.8	49.4	27	Scheffrahn & Su 1992	
Lepidoptera								
Ephestia kuehniella	Pyralidae		L (Diapausing)	35.3	65.2	10	Cox *et al* 1984	
			L	25.1	65.2	10	Cox *et al* 1984	
Grapholita molesta	Tortricidae	Nectarines	E	34–40		21	Yokoyama *et al* 1987	
			L1–L5	9.76–16.56		21	Yokoyama *et al* 1987	
Cydia pomonella	Tortricidae	Cherries	E	29.4–36	48–54.5	12	Maindonald *et al* 1992	Depending on cultivar
		Cherries	L3	25.7–26.9	48–54.5	12	Moffitt *et al* 1992	Depending on cultivar
Thysanoptera								
Frankliniella occidentalis	Thripidae		E	7.5	20.9	15	Macdonald 1993	
			L	18.4–21.3	44.8–53.1	15	Macdonald 1993	Depending on age
			P	13.6	20.9	15	Macdonald 1993	
Psocoptera								
Liposcelis entomophilus	Liposcelidae	Rice	E	12.9	43.5	27	Pike 1994	

[a]E = Eggs, L = Larvae (numbers refer to specific instars), P = pupae or pseudo-pupae, A = Adults

Table 4.5. Literature reports of doses of methyl bromide causing complete or near complete mortality to a range of insects and mites

Species	Family or superfamily	Stage[+]	Commodity	Dose (g m³)	Time (h)	CTP (mg l⁻¹ h)	n	Survival	Temp (°C)	Reference	
Mites											
Tetranychus urticae	Tetranychidae	E	Cutflowers				800	0	18	Wit & van der Vrie 1985	
		E				88		0	18	Anon 1975, Bond & Buckland 1978, Powell & Gostick 1971	also at 11 & 5 °C
		A				88		0	18	Wit & van der Vrie 1985	
		L				44		0	18	Anon 1975; Bond & Buckland, 1978, Powell & Gostick 1971	also at 11 & 5 °C
Acarus siro	Acaridae	E	Wheatfeed			430		0	15	Bowley & Bell 1981	also at 10 °C
Tyrophagus longior	Acaridae	E	Wheatfeed			650		0	15	Bowley & Bell 1981	also at 10 & 20 °C
Lepidoglyphus destructor	Tyroglyphidae	E	Wheatfeed			430		0	15	Bowley & Bell 1981	also at 10 °C
Coleoptera											
Curculio dentipes	Curculionidae	L	Chestnuts	50	4			0	21	Hah *et al* 1982	
Sitophilus granarius	Curculionidae	A		16	12	192		0	14.5	Hah *et al* 1982	
Acanthoscelides obtectus	Bruchidae	A		16	12	192		0	14.5	Hah *et al* 1982	
Silvanus gemmelatus	Curculionidae	L		16	12	192		0	14.5	Hah *et al* 1982	
		A		16	12	192		0	14.5	Hah *et al* 1982	
Tenebrio obscurus	Tenebrionidae	A		16	12	192		0	14.5	Hah *et al* 1982	
Trogoderma versicolor	Dermestidae	L		16	12	192		0	14.5	Hah *et al* 1982	
Anthonomus grandis	Curculionidae	A	Cotton			29–32	658	0	20–22	Roth & Kennedy 1972	Wide range of doses at 0.6–35 °C
		A	Cotton			32–36	101	1	20–22	Roth & Kennedy 1972	Wide range of doses at 0.6–35 °C
Asynonychus godmani	Curculionidae	E	Citrus			75.6		0.06%	21	Soderstrom *et al* 1991	
Leptinotarsa decemlineata	Chrysomelidae	A (diapausing)	Soil			166		0		Powell 1981, Powell & Mills 1981	

Order / Species	Family	Stage	Commodity						Min. for 100%	Reference
Diptera										
Agromyza pusilla	Agromyzidae	E	Gerbera	12	4		3–40	0	21	Blanton 1942
		L	Gerbera	12	4		1130	0	21	Blanton 1942
		P	Gerbera	20	4		853	0	21	Blanton 1942
Liriomyza trifolii	Agromyzidae		Gypsophila	40	3			0	15–18	Carmi 1985
Ragoletis completa	Tephritidae	L3	Stone fruits	40	2		220	0	21	Yokoyama et al 1992
		E	Stone fruits	18.2	2	51.3		0	21	Yokoyama et al 1992
Anastrepha ludens	Tephritidae	L	Grapefruit	40	2		116219	3	26.7	Williamson et al 1986
Anastrepha suspensa	Tephritidae	E	Carambolas	40	2		104303	1	23.1	Hallman & King 1992
Bactrocera xanthodes	Tephritidae		Watermelons	32	4		28000	0	21–26	Cowley et al 1991
Ceratitis capitata	Tephritidae	L	Stonefruits	48	1		550000	0	30	Armstrong & Couey 1984
		L	Strawberries	48	3		150000	0	15	Armstrong et al 1984
		L	Strawberries	48	3		300000	2	21	Armstrong et al 1984
Dacus curcurbitae	Tephritidae	L	Cucumbers	32	4		150000	0	21	Armstrong & Garcia 1985
		L	Cucumbers	48	2		150000	0	21	Armstrong & Garcia 1985
Dacus dorsalis	Tephritidae	L	Strawberries	48	2		750000	0	21	Armstrong et al 1988
		L	String Beans	35	2.5		500000	2	15	Tanaka et al 1986
		L	Stonefruits	48	1		500000		30	Armstrong et al 1988
		L	Stonefruits	32	1.5			0	30	Armstrong et al 1988
		L	Cucumbers	48	2		110000	0	21	Armstrong & Garcia 1985
		L	Cucumbers	32	4		110000	0	21	Armstrong & Garcia 1985
Thysanoptera										
Thrips tabaci	Thripidae	L	Cut flowers	30	1.5		160	0	17–19	Wit & van der Vrie 1985
Frankliniella occidentalis	Thripidae	L, A, E	Chrysanthemums			40		0	15	Macdonald 1993
Psocoptera										
Liposcelis entomophilus	Liposcelidae	Nymphs, A	Rice		4	50	100	0	27	Pike 1994

continued overleaf

Table 4.5. (continued)

Species	Family or superfamily	Stage[†]	Commodity	Dose ($g\,m^3$)	Time (h)	CTP ($mg\,l^{-1}\,h$)	n	Survival	Temp (°C)	Reference
Homoptera										
Eriococcus ironsidei	Coccoidea		Macadamia	55	2				20	Ironside et al 1978
Trialeurodes vaporariorum	Aleyrodidae	E, P, A	Cut flowers	0	0		2–400	0	18	Yokoyama et al 1992
Aleurocanthus woglumi	Aphididae		Citrus	16	2			0	26–29	Sanchez-Riviello & Rhode 1976
Myzus persicae	Aphididae					88		0	18	Powell & Gostick 1971
		A				28		0	18	Powell & Gostick 1971 also at 11 & 5°C
Hemiberlesia lataniae	Coccoidea			16	2.5	50–55			22.8	Witherell 1984
Bemisia tabaci	Aleyrodidae	Nymphs					1781	4	15	Macdonald & Mills 1994
Heteroptera										
Cimex lectularis	Cimicidae					256		100	0	Mackie 1938
Lepidoptera										
Sitotroga cerealella	Gelechiidae	P	Popcorn	32	8			0	4.4	Vincent et al 1980
		E, L, A	Popcorn	32	8			0	4.4	Vincent et al 1980
Plodia interpunctella	Pyralidae	E, L, A	Popcorn	32		8		0	4.4	Vincent et al 1980
Opogona sacchari	Lyonetiidae	L	Banana	25.9	3			0	0	Cintra et al 1978
Spodoptera exigua	Noctuidae	L5	Cut flowers	0	0		160	0	18	Yokoyama et al 1992
Spodoptera littoralis	Noctuidae	E	Cut flowers	0	0		280	0	18	Yokoyama et al 1992
		E		0	0	44		0	18	Powell & Gostick 1971
Grapholita molesta	Tortricidae	L1–L5	Nectarines		16–23	2160		0	21	Yokoyama et al 1987
		E	Nectarines			46.7	1178	0	21	Yokoyama et al 1987
Cydia splendana	Tortricidae	L	Chestnuts	50	4				21	Hah et al 1982
Cydia pomonella	Tortricidae	L3	Cherries	64	2		10839	0	12	Moffitt et al 1992
		L5	Walnuts	56	4		34306	1	15.6	Hartsell et al 1991
		L	Walnut sacking	5.5	4			0	21	Lindgren 1936
		E	Nectarines	48	2	22	93744	0	24	Yokoyama et al 1994 With added CO_2
Amyelois transitella	Pyralidae	E, L, P	Pistachio nuts	16	21		>400	0	26.6	Hartsell et al 1986
		L, P	Pistachio nuts	32	24		>400	0	15.5	Hartsell et al 1986
		E	Pistachio nuts	32	24		>400	0	15.5	Hartsell et al 1986 At 100 mm Hg
Clepsis spectrana	Tortricidae	L	Cut flowers	0	0		160	1	18	Wit & van der Vrie 1985

[†] E = Eggs, L = Larvae (numbers refer to specific instars), P = pupae or pseudopupae, A = Adults

However, comparisons between these data are less reliable as the experiments involve an unknown degree of overdosing and in many cases CTP values cannot be determined accurately as only the initial dose of methyl bromide was recorded and no measurements of the decline in concentration during the course of a treatment were reported. Tables 4.6 and 4.7 facilitate comparisons to some extent by presenting data at 25 °C and 30 °C respectively. It can be seen that the difference in tolerance is small for many insect stages between 25 and 30 °C, though for diapausing larvae tolerance reduces by about a third at the higher temperature.

5.1 FACTORS AFFECTING THE EFFICACY OF METHYL BROMIDE

Many different factors can affect the toxicity of methyl bromide to a given species. Physical parameters such as temperature and pressure will cause variations in the efficacy and the CTP relationship is known to hold true only between certain limits. Biological parameters are also important. The age and prior treatment of individuals can have an effect and many organisms have life stages that are more tolerant of methyl bromide.

5.1.1 TEMPERATURE

The effect of temperature on fumigation with methyl bromide has been the subject of much study. Although the boiling point of methyl bromide is 3.56 °C fumigation has been shown to be effective at temperatures as low as −10 °C (Bogs 1976, El-Lakwah 1977a). Vincent and Lindgren (1975, 1977) showed that all stages of both the cigarette beetle *Lasioderma serricone* (F.) and *Trogoderma variable* Ballion were more tolerant to methyl bromide at lower temperatures. The greatest differences in effect were noticed as the temperature was increased from 4.4° to 10 °C and from 10° to 15.6 °C, whereas an increase from 15.6° to 21.1 °C had relatively little effect. Likewise with the grain weevil, *Sitophilus granarius*, there was a decrease in the LD_{50} of methyl bromide of up to 44% between 5° and 15 °C but only 20% between 15° and 26 °C (Cherif *et al* 1985). Similar results have been described for 4th instar cadelles, *Tenebroides mauritanicus*, adults of the flour beetles, *Tribolium confusum* and *T. castaneum* (Bond and Buckland 1976, 1978) and larvae of Khapra beetle, *Trogoderma granarium* Everts (El-Lakwah, 1977a; b). Reductions in the efficacy of methyl bromide at lower temperatures have been recorded for the peach potato aphid *Myzus persicae* (Sulz.) and the red spider mite, *Tetranychus urticae* (Koch) (Powell and Gostick 1971), and for all stages of the leafmining agromyzid fly, *Liriomyza trifolii* (Burgess in Comstock) (Mortimer and Powell 1984).

Table 4.6. Comparison of literature data on the susceptibility of adults or the most tolerant stages of different stored product Coleoptera and Lepidoptera to methyl bromide at 25 °C (from Detmers 1993)

Insects	Stage	Time/dose	CTP (mg l⁻¹ h)				References
			LD_{50}	LD_{95}	LD_{99}	$LD_{99.9}$	
Coleoptera							
Anobiidae:							
Lasioderma serricorne	pupa					100	Brown 1959
Bostrichidae:							
Rhyzopertha dominica	old pupa					45	Brown 1959
	adult	5 h	20			40	Brown 1959
	adult					38	Brown 1959
	adult	3.5–4 mg/l	16–17		36–49		Brown 1959
Prostephanus truncatus	egg	9.7 mg/l	15.2	20.0		25.6	Detmers 1993
	egg	21.9 mg/l	16.1	21.8		28.4	Detmers 1993
	pupa	10 mg/l	39.8	48.9		58.6	Detmers 1993
	pupa	20.3 mg/l	42.2	50.5		50.3	Detmers 1993
Bruchidae:							
Callosobruchus chinensis	preimago					40	Brown 1959
Cucujidae:							
Cryptolestes minutus	pupa					125	Brown 1959
Cryptolestes ferrugineus	young adult	5 h	35	41			Barker 1967
Cryptolestes turcicus	young adult	5 h	37	43			Barker 1967
Oryzaephilus mercator	adult	5 h	29			43	Anon 1975
Oryzaephilus surinamensis	adult		29			50	Brown 1959
	adult	5 h				43	Brown 1959
	adult	4 mg/l	25–29		36–44		Bell 1988
Curculionidae:							
Sitophilus granarius	adult	9 mg/l	17	30		38	Howe & Hole 1966
	adult	5 h	26				Anon, 1975
	pupa	9 mg/l	29	43		60	Howe & Hole 1966
	old pupa				50	65	Brown 1959
Sitophilus oryzae	pupa	2 h	29	44			Khrone & Lindgren 1958
	preimago					85	Brown 1959
	adult	5 h	18			24	Anon 1975

Species	Stage	Concentration				complete control	Reference
Sitophilus zeamais	adult	3.5–4 mg/l	12–14		19–21		Bell 1988
	adult	5 h	16			27	Anon 1975
	adult	4 mg/l	14		22		Shivanandappa & Rajendran 1987
Dermestidae:							
Trogoderma granarium	larva	14–30 mg/l	68			110	Brown 1959
	diapausing larva					210	El-Lakwah 1977a
Trogoderma variabile	pupa	2–8 h	44–52	58–62	160		Vincent & Lindgren 1975
Ostomidae:							
Tenebroides mauritanicus	pupa	5 h	88		121		Bond & Munro 1961
Ptinidae:							
Ptinus tectus	pupa					100	Brown 1959
Tenebrionidae:							
Tribolium castaneum	young pupa	10 mg/l	81			125	Godden & Howe 1965
	pupa					100	Brown 1959
	young pupa	5 mg/l	69				Bennett 1969
	adult	5 h	42			59	Anon 1975
Tribolium confusum	adult	5 h	45		64		Bond & Munro 1961
Tribolium confusum	adult	4 h	42–50	54–61			Lindgren & Vincent 1965
Tribolium confusum	pupa	5 h	43			90	Brown 1959
	adult	5 h				56	Anon 1975
	adult	5 h	43 l, 61 r		66 l, 86 r		Bell 1988
Lepidoptera							
Pyralidae:							
Ephestia cautella	pupa	4–11 mg/l				50 l	Bell 1976
Ephestia elutella	pupa	4–11 mg/l				50 l, f	Bell 1976
	diapausing larva	4–11 mg/l	76 f		150 f	183 f	Bell 1977
Ephestia kuehniella	pupa	4–11 mg/l				53 f	Bell 1976
Plodia interpunctella	pupa	4–11 mg/l				43 l, f	Bell 1976
	diapausing larva	4–11 mg/l	26 f		67 f		Bell 1977

l = laboratory strain, f = strain from the field, r = methyl bromide resistant strain, t = tested at 26.7 °C

Table 4.7. Comparison of literature data on the susceptibility of adults or the most tolerant stages of different stored product coleoptera and lepidoptera to methyl bromide at 30 °C (from Detmers 1993)

Insects	Stage	Time/dose	CTP (mg l⁻¹ h)				References
			LD_{50}	LD_{95}	LD_{99}	$LD_{99.9}$	
Coleoptera							
Bostrichidae:							
Rhyzopertha dominica	adult	5 h	7–26	24–62			El-Nahal *et al* 1986
Prostephanus truncatus	egg	5–20 mg/l	8.3	12.1		16.9	Detmers 1993
	larva	5–21 mg/l	15	22		33.5	Detmers 1993
	pupa	6–20 mg/l	28.5	37.8		48.8	Detmers 1993
	adult	5–21 mg/l	17.3	22.4		28.0	Detmers 1993
Silvanidae:							
Oryzaephilus surinamensis	adult					40	Brown 1959
Curculionidae:							
Sitophilus granarius	old pupa	5 h	14–23	22–46		65	Brown 1959
	adult	5 h	6–18	24–54			El-Nahal *et al* 1986
Sitophilus oryzae	adult						Rajendran 1990
Dermestidae:							
Trogoderma granarium	diapausing larva	7–40 mg/l	44		93	118	El-Lakwah 1977
	larva					70	Brown 1959
Tenebrionidae:							
Tribolium castaneum	young pupa	10 mg/l	81				Godden & Howe 1965
	adult	5 h	15–49	28–82			El-Nahal *et al* 1986
Lepidoptera						complete control	
Pyralidae:							
Ephestia cautella	pupa	4–11 mg/l				40 l, f	Bell 1976
Ephestia elutella	diapausing larva	4–11 mg/l	41 f		115 f		Bell 1977
Plodia interpunctella	pupa	411 mg/l				40 l, f	Bell 1976

l = laboratory strain, f = strain from the field, r = methyl bromide resistant strain

Tables 4.6 and 4.7 show further examples of differences in the response of adult *Rhizopertha dominica* (F.), *Oryzaephilus surinamensis* (L.) and *Sitophilus granarius* pupal *Prostephanus truncatus* (Horn) and *Plodia interpunctella* (Hübner) and diapausing larvae of *Ephestia elutella*, when treated at 25 or 30 °C.

Soil fumigations have also been shown to be affected by temperature. Fumigations against the plant parasitic nematodes *Xiphinema index* and *Meloidogyne incognita* (Kofoid & White) Chitwood increased in effectiveness with rising soil temperature (Abdalla and Lear 1975).

Bell (1978) showed that the efficacy threshold, the minimum concentration at which Haber's rule holds for the warehouse moth, *E. elutella*, increased at higher temperatures suggesting that some insects may have a greater ability to detoxify methyl bromide at higher temperatures. At 15 °C, the concentration time relationship applied at concentrations as low as 1.3–1.9 mg l^{-1}, while at 25 °C the minimum concentrations were between 2.7 and 4.0 mg l^{-1}. Clearly temperature must be taken into account when planning any fumigation or developing treatment schedules.

5.1.2 PRESSURE

Fumigation at reduced pressure can improve the penetration of gas into commodities. It may also have a direct effect on toxicity, possibly simply by increasing the stress to which an organism is exposed. Roth and Kennedy (1973) showed that *Helicella* spp snails could survive fumigation with 96 mg l^{-1} at normal atmospheric pressure but were controlled by a treatment of 32 mg l^{-1} for 3 hours when fumigated at a reduced pressure of 100 mm Hg. Calderon and Leesch (1983) showed that the susceptibility of adult *Tribolium confusum* was increased 1.6 fold by reducing the pressure at which they were fumigated from atmospheric to 100 mm Hg.

5.1.3 HUMIDITY

The influence of relative humidity on the susceptibility of insects to methyl bromide has been investigated without leading to any clear conclusion. Mortality has been found to be much more dependent on temperature changes—which often lead to humidity changes—than on humidity changes *per se* (El-Lakwah 1977c, Schacher and Knülle 1980, Bond, 1984). In dry conditions, micro-organisms may form spores that are resistant to methyl bromide (Kolb and Schneiter 1950). Plant seed is more tolerant of methyl bromide at lower moisture contents (Cobb 1958).

5.1.4 RESPIRATION RATE

Respiration rate can affect the toxicity of methyl bromide to insects. High levels of CO_2 which increase the respiration rate result in a similar increase in the toxicity of methyl bromide (Jones 1938). A significant relationship between the susceptibility of individual insects to methyl bromide and their respiration rates has also been shown (Bond 1956). Artificially raising the level of CO_2, may therefore be a useful strategy for increasing the effectiveness of a given dose of methyl bromide.

5.1.5 PRIOR TREATMENT

The manner in which insects are treated before exposure to methyl bromide can affect their response. The diet of insects has been suggested as a factor affecting the response of a number of species to fumigation. Small differences in the responses of the rice moth *Corcyra cephalonica* (Staint) reared on a range of diets (El-Buzz *et al* 1974) and of *T. casteneum* reared on either rice germ or bran have been noted (Halfawy *et al* 1978). However, in neither case was the significance of the differences tested. Small but significant differences in the responses on the dipteran leafminer, *L. trifolii*, have been found when they were reared on different cultivars of chrysanthemum (Mortimer and Powell 1984). Breeding *Anagasta* (= *Ephestia*) *kuehniella* (Zeller) and *P. interpunctella* on either corn meal or one of a range of diets caused variations in the LD_{50}s of up to four times, with insects raised on corn meal being more tolerant at all life stages (Amin *et al* 1979). Again, however, neither the variation nor the significance of the results was reported.

Rearing temperature had a significant effect on the susceptibility of khapra beetle, *T. granarium*, to methyl bromide. With a dose of $12\,\text{mg}\,\text{l}^{-1}$ the LT_{50} and LT_{99} were 6 and 20 hours respectively for larvae reared at 28 °C but these were significantly increased to 12 and 33 hours for larvae raised at 15 °C. Transferring larvae raised at 28 °C to lower temperatures for 1–2 months before fumigation also increased their susceptibility (El-Lakwah 1977a; b). A period of two days cold storage at 1–2 °C significantly increased the LD_{99} and $LD_{99.9}$ of methyl bromide to mature larvae of *L. trifolii*, though no significant effect was found with pupae (Mortimer and Powell 1984).

Prior exposure to methyl bromide will also affect the susceptibility of insects. *E. elutella* larvae that have been fumigated with $8\,\text{mg}\,\text{l}^{-1}$ for 14.5 h or $12\,\text{mg}\,\text{l}^{-1}$ for 3.5 or 7.5 hours, sufficient to kill a substantial proportion of individuals, showed a depression in the minimum effective concentration when immediately fumigated a second time at a low concentration. This has practical importance in that it indicates that low concentrations of methyl bromide that may persist at the end of a fumigation, either because of the time it takes to vent the gas or in conditions where there is a gradual decline in gas concentration, may contribute significantly to the success of treatment (Bell 1981, 1982).

5.1.6   DEVELOPMENTAL STAGE

The developmental stage of an organism will affect its response to methyl bromide. Plants and micro-organisms possess resting stages as seeds or spores when they are relatively insensitive to fumigation (Kolb and Schneiter 1950; Munnecke et al 1978). Other pests also vary in susceptibility throughout their life cycle. Figure 4.1 illustrates the different susceptibilities of the developmental stages of Sitophilus granarius to methyl bromide. The dose required to control young larvae of S. granarius amounted to less than 30% of the value required for the pupae. Adult beetles were also far less tolerant than pupae.

To obtain reliable data for the control of any insect it is important to include in any investigation, the most tolerant stage likely to be present in the commodity to be treated. The changes in susceptibility of the different developmental stages of S. granarius have been reported for phosphine (Reichmuth 1986) and carbon dioxide (Reichmuth 1990). When comparing the susceptibility of different developmental stages of different insect species to methyl bromide at 25 °C, pupae or diapausing larvae were found to be most tolerant (see Table 4.6). A comparison of the susceptibility of diapausing and non-diapausing larvae of E. kuehniella (Zeller) at 10 °C found an LD_{50} for diapausing larvae of 35.3 mg l^{-1} h with 95% fiducial limits of 29.4 and 39.8 mg l^{-1} h, and an LD_{99} of 80.9 mg l^{-1} h (65.6–26.5 95% FD). The corresponding figures for non-diapausing larvae were 25.1 mg l^{-1} h. (21.0–28.5 95% FD) for the LD_{50} and 65.2 mg l^{-1} h (52.9–94.5 95% FD) for the LD_{99} (Cox et al 1984).

5.1.7   RESISTANCE TO METHYL BROMIDE

The possibility of insects developing resistance to methyl bromide has concerned workers since the 1940s. Yust and Busbey (1942) showed that there were small differences in the response to methyl bromide of populations that were resistant to hydrogen cyanide when compared to a non-resistant strain. An FAO report (Champ and Dyte 1976) found that of seven stored products pests investigated, strains resistant to methyl bromide had been reported for six of them. Out of a total of 894 populations of all seven species, 42 (4.7%) showed resistance. However, in comparison to other pesticides this was a relatively small proportion. Nearly twice as many (9.7%) of the populations showed resistance to phosphine and nearly three quarters of all strains tested showed resistance to lindane.

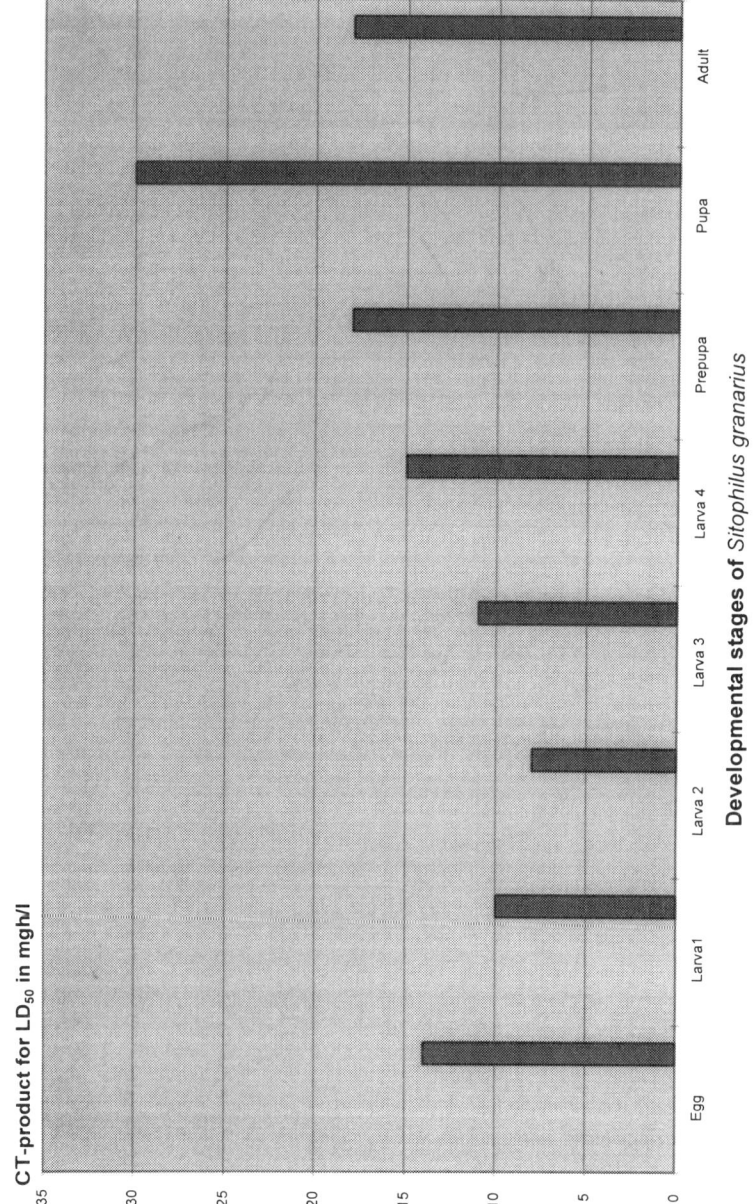

Figure 4.1. The change with age in the susceptibility to methyl bromide of the different developmental stages of *Sitophilus granarius* at 25 °C derived from data of Howe and Hole (1966)

Figure 4.2 illustrates the differences in the LD_{50} and $LD_{99.9}$ of different strains of *Prostephanus truncatus* (Detmers 1993). The biggest difference was between strain A from Togo with $LD_{99.9} = 40.1 \, \mathrm{mg \, l^{-1} \, h}$ and strain C from Tanzania with $LD_{99.9} = 23.6 \, \mathrm{mg \, l^{-1} \, h}$, a 1.7 fold difference in tolerance. The variation of the CTP increases pronouncedly from LD_{50} to $LD_{99.9}$ as do the corresponding fiducial limits. This increase in the fiducial limits as doses approach the LD_{100} is typical for results with probit analysis and can be explained by the normal distribution of the mortality data around LD_{50}, where the fiducial limits are smallest. It is thus debatable as to how far the significant differences between the CTPs of the different strains may be related to resistance. At least the results give a good impression of the variations of susceptibilities to methyl bromide of stored product pest insects that may be encountered in the field.

Upitis *et al* (1973) were able to stimulate the development of a low level of resistance (not more than a factor of 3) in laboratory experiments with *S. granarius* and related resistance to increased body weight, expanded developmental period and reduced respiratory rate. This change in weight and respiratory rate differs from the behaviour of phosphine resistant insects, which tend to stay active longer in the gas than the susceptible ones (Reichmuth 1990, Reichmuth 1991a). However, no relationship could be found between the uptake of methyl bromide and its toxicity. Bond and Upitis (1976) in later work found a roughly two-fold difference between both the LD_{50} and LD_{99} of absorbed fumigant between susceptible and resistant strains of *S. granarius*. Bell (1988) found a significant difference in the efficacy threshold, i.e. the lowest concentration of methyl bromide at which Haber's rule still applies, at $25 \, °\mathrm{C}$ between stock cultures of several species and strains that showed higher than normal tolerance to methyl bromide. This difference was not apparent at a lower temperature. The low level of resistance towards methyl bromide is in accordance with findings in field strains reported by Champ (1985) and Champ and Dyte (1976). This is different from the situation with phosphine, where a pronounced resistance is built up after repeated successive fumigations.

6.1 PRACTICAL METHYL BROMIDE TREATMENTS

The mortality data presented in Tables 4.6 and 4.7 as well as many of the reports in Tables 4.4 and 4.5 relate to laboratory experiments. The practical object of most fumigations is to eliminate the target organism, so that the real interest of laboratory experiments with a fumigant is to determine the LD_{99} or an even higher dosage. Experiments are seldom performed with sufficient test insects in order to determine these levels of mortality directly.

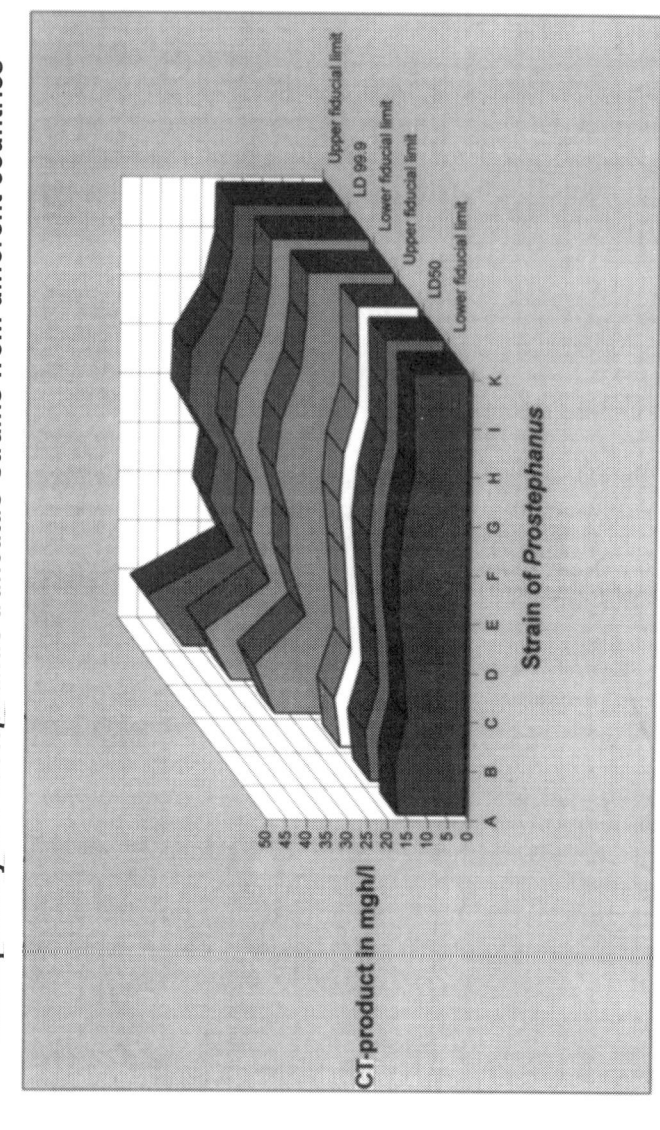

A/B : Togo ; C/D/E : Tanzania ; F : Mexico ; G : Guatemala ; H/I/K : Costa Rica

Figure 4.2. Comparison of LD_{50} and $LD_{99.9}$ of different strains of adult *Prostephanus truncatus* to methyl bromide (10 mg l^{-1}, four replications of the series) at 30 °C and 70% r.h. from different countries (data from Detmers 1993)

Theoretically, it is impossible to determine the LD_{100} because an infinite number of test insects would be required (Bliss 1935). Therefore, the probit lines are often extrapolated to obtain the dosage of fumigant needed to give some desired level of mortality (Brown 1954). Although there are theoretical objections to this practice, it appears to give acceptable values for adult insects. Obviously the standard error of such a predicted value is very high and its accuracy depends on the linearity of the probit line. Howe and Hole (1967) dealt with this problem in detail and demonstrated that for *Sitophilus granarius* the probit line was curved at the upper extreme. The calculated $LD_{99.9}$ was $75\,mg\,l^{-1}\,h$, whereas a dose of $52.5\,mg\,l^{-1}\,h$ caused complete mortality of 3600 insects and $50\,mg\,l^{-1}\,h$ gave 99.95% kill. Furthermore, at a dose of $45.7\,mg\,l^{-1}\,h$ only about 1% of insects survived and these had only 1% of the fecundity of a control group so that the number of insects in the next generation was only about 0.01% of the control. Thus, even if the proportion of individuals killed can be accurately determined, sublethal effects may result in further additional reductions in the size of a population. Similar curving of the probit line has been reported for the western flower thrips *Frankliniella occidentalis* (Pergande) (Macdonald 1993). Nevertheless, this fallacious use of probit analysis is widespread and is probably inevitable in the selection of a dose level to ensure the complete mortality of pest insects. Fortunately the errors in calculating responses seem to result in an overestimate of the dose required to achieve high levels of control.

In a few cases sufficient insects have been tested, either to extend the dose response line well beyond the LD_{99} or to reduce the uncertainty that a given dose will achieve 100% mortality to extremely low levels. Since 1939, on the basis of a United States Department of Agriculture circular (Baker 1939), a number of Plant Health authorities have required 'probit 9 security' for quarantine controls, particularly for fruit flies (Greany 1994). Probit 9 represents a mortality of 99.99683% and Probit 9 security is usually defined as the dose at which there is a 95% chance of this level of kill being achieved, i.e., the upper 95% fiducial limit of the $LD_{99.99683}$. Workers have therefore been required to test sufficient insects to verify that treatments meet this standard. In some cases these tests have involved over 750 000 individuals being subjected to a given dose. Data for a number of fruit fly species, *Anastrepha ludens* (Loew.) on stone fruits, *A. suspensa* (Loew.) on Grapefruit, *Dacus cucurbitae* (Coquillett) on cucumbers, and *Ceratitis capitata* (Wiedemann) and *D. dorsalis* (Hendel) on a range of fruits are shown in Table 4.5. Breeding large numbers of fruit flies is relatively easy and although the assessment of such large numbers of insects as used in these tests is undoubtedly labour intensive, such large scale tests are relatively easy on these species. Many other pests, however, cannot be reared or handled so easily in such large numbers. Nonetheless, where they are of quarantine significance, some species from other groups have also been subjected to large scale tests. For example eggs of the codling moth, *Cydia pomonella* (L.), a quarantine organism in Japan, has been the subject of a number of studies that approach in scale some of those carried out on fruit flies (Table 4.5).

The use of Probit 9 as a measure of quarantine security has recently been subject to criticism. There does not appear to have ever been any explained reasoning for the use of probit 9 mortality for quarantine purposes and there appears little justification for its use, except possibly for fruit flies, where many individuals may be found in one fruit. Probit 9 does not take into account the level of infestation (Vail *et al* 1993) nor the problems that survivors might face on introduction to a new area (Landolt *et al* 1984). More systems based approaches are now being proposed for quarantine purposes (Vail *et al* 1993, Moffitt 1989, Hara 1994). Using the binomial distribution law it is possible to calculate the confidence limits for the true unknown survival proportion, based on the number of survivors in a given test. Such confidence limits can even be combined to calculate the overall confidence limit of a sequence of operations or treatments (Couey and Chew 1986). By looking at a whole production system the probability of quarantine pests being transported with produce can thus be determined. When this is linked to knowledge of the pests biology and behaviour, an overall assessment of the risk posed by the pest can be determined and the necessary level of additional controls, to reduce this to an acceptable level, calculated on a firm scientific basis.

For quarantine treatments on live plants, fruit or vegetables, accurate dosing is critical as there is often only a narrow window between acceptable levels of control, which are frequently defined by Plant Health authorities, and unacceptable phytotoxicity. Treatments should therefore usually be carried out in purpose-built chambers allowing conditions to be closely controlled and minimising gas leakage (Anon 1982). Treatment schedules are based on laboratory determined dose response parameters. Most recent quarantine schedules either specify the CTP to be obtained (Anon 1994a,b,c) or the concentration of gas that must be present at the end of the treatment period (USDA 1992). Some governments such as the US (USDA 1992), Australia (Australian Quarantine and Inspection Service 1985, 1992) and New Zealand (New Zealand Ministry of Agriculture and Fisheries) publish their own statutory treatment schedules; in other cases international organisations such as the FAO (Anon 1977) and European and Mediterranean Plant Protection Organisation (EPPO) take on the task of proposing schedules which may then be accepted by member nations as effective treatments for plant quarantine purposes.

Quarantine treatments however only account for a very small proportion of the use of methyl bromide (Chakrabarti and Bell 1993). In most other areas there is a tremendous gap between the dosage required to achieve a given mortality from laboratory data (Haber 1924, Peters and Ganter 1935, Reichmuth 1986, Reichmuth 1990, Reichmuth 1991b, Unger *et al* 1992, Winks 1982, 1984, 1985, Bell 1988, Bond 1984) and the recommended fumigation schedules (Anon 1982, Howe and Hole, 1967, Bond 1984, Pflanzenschutzmittel-Verzeichnis 1994, Monro 1969). In practice significant quantities of the gas may be lost during treatment due to leakage from structures (Merkblatt 71 der Biologische Bundesanstalt für Land und Forstwirtschaft, 1993), through or round sheeting (Van Wambeke 1989, Klein 1989) and by sorption of the gas on commodities and surfaces of the treated structure (Banks and Pinkerton 1987, Banks 1985, Franz *et al* 1992). For fumigated flour mills and warehouses for example, the loss of methyl bromide within two days comprises 50–90% of the initial dose (Reichmuth, unpublished data, Figure 3).

As shown in Figure 4.4, this loss can roughly be described by the equation

$$c = c_0 \exp\left((-n/24)t\right) \tag{2}$$

where:

c_0 is the theoretical concentration derived from the initial dose of methyl bromide per volume of the treated object, and the factor 24 corrects the formula when t is calculated in hours

n is an approximate factor for the leakiness of the structure. This factor is normally determined by pressure testing of structures prior to fumigation by applying a constant pressure difference p and determining the required air flow to hold this pressure difference, or by determining the half-life, $t_{1/2}$, of the pressure difference and deducing n from this half-life by use of the function: (Merkblatt 71 der Biologische Bundesanstalt für Land und Forstwirtschaft, 1993).

$$n = \Delta V/V = (p_0 \ln(2)\Delta t)/(10^5 t_{1/2}) \tag{3}$$

where:

ΔV and V are the exchanged volume and the total volume of the treated structure,

p_0 is the initial pressure in Pascals above atmospheric.

The model gives a good fit in practical situations, where often less than $2\,\mathrm{mg\,l^{-1}}$ of methyl bromide are left after only 24 hours of fumigation. In such cases, extending the treatment period beyond 48 hours would not increase the effectiveness of the fumigation. The goal has to be the improvement of gas tightness to avoid the indicated losses.

In Germany, structures to be fumigated have to pass a gas tightness check in advance (Merkblatt 71 der Biologische Bundesanstalt für Land und Forstwirtschaft, 1993). The maximum accepted loss rate, n, from a pressure test at 10 Pascal is 2.4 volume changes per day. This demand can be fulfilled for most buildings but often requires some improvement in sealing. The

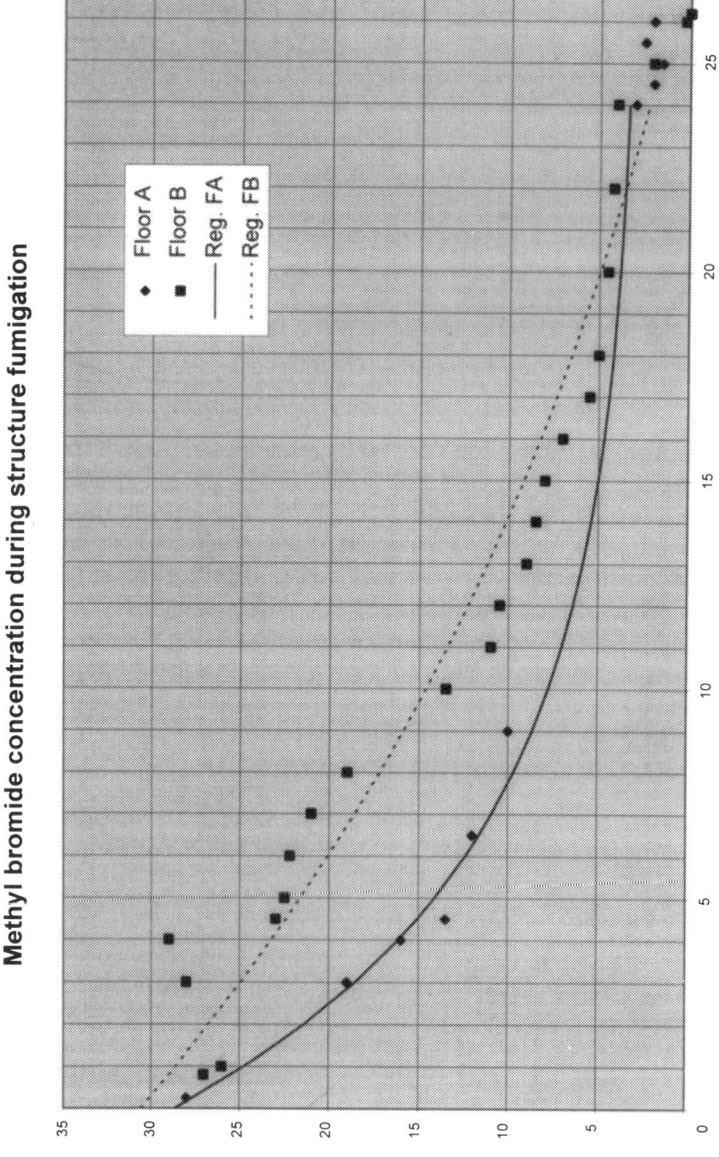

Figure 4.3. Methyl bromide concentrations during a methyl bromide fumigation of two floors A and B in a big warehouse; the graph contains also the regression of these concentrations with an e-function of the form of formula (2)

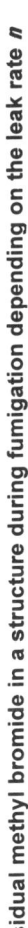

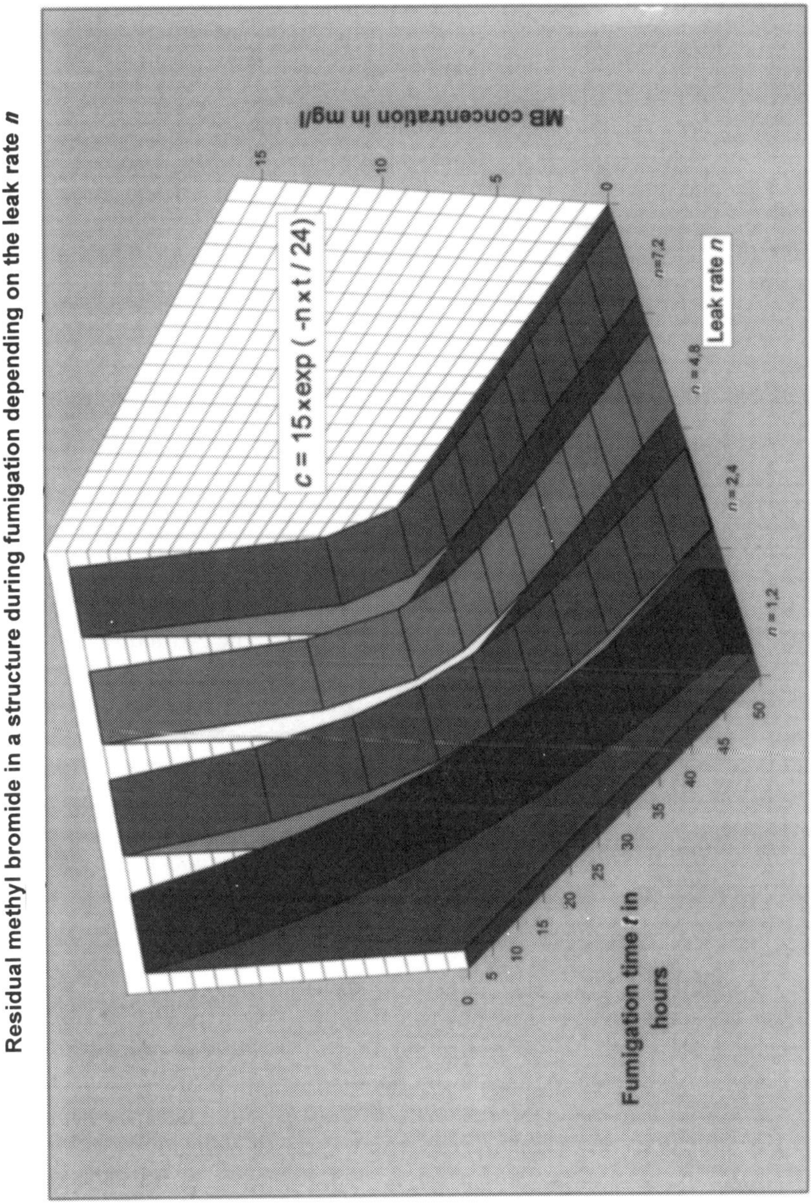
Residual methyl bromide in a structure during fumigation depending on the leak rate n
c = 15×exp (-n×t / 24)
MB concentration in mg/l
15
10
5
0
n=7.2
n=4.8
n=2.4
n=1.2
Leak rate n
0
5
10
15
20
25
30
35
40
45
50
Fumigation time t in hours

Figure 4.4. Methyl bromide concentration in a structure at different leak rates (n)

necessary procedures for this have been described in detail by Wohlgemuth (1990). Reichmuth (1993) proposed a method of calculating the necessary initial dose after correcting for different degrees of leakage of the treated structure. This leak rate was determined either by measuring the half-life of the pressure difference or the necessary throughput of air to obtain a pressure difference of 10 Pascal.

To obtain, for example, a CTP of $140\,\text{mg}\,l^{-1}\,h$ within two days, a sufficient dose to control all stages, apart from diapausing larvae, of insects in Table 4.6 at 25 °C, the maximum leak rate that can be accepted with an initial dose of $15\,\text{mg}\,l^{-1}$ is 2.4 volume changes per day (Figure 4.5). As the integral of function (2) between the lower, $t_1 = 0$ and upper limits, $t_u = t$, the CTP corresponds for each time t to the area under the respective curve of the gas concentration inside a fumigated structure (Figure 4.4), thus:

$$ct = \int (2)\,dt = \{c_0/(n/24)\}\{1 - \exp((-n/24)t)\} \qquad (4)$$

Figure 4.5 illustrates the different CTPs obtained in objects with different leak rates. The graph shows clearly that for leak rates above $n = 4.8$ the final CTP is achieved largely within the first 24 hours of exposure. In practice, many other factors such as wind and temperature will influence the rate at which gas is lost from a fumigated building. Nevertheless, the functions and graphs demonstrate that the greater the loss of gas during treatment, the smaller the CTPs attained inside the structure. In the example, with an initial dose of methyl bromide of $15\,\text{mg}\,l^{-1}$, an exposure period of 48 hours would give a CTP of $720\,\text{mg}\,l^{-1}\,h$, more than five times the required lethal dose of $140\,\text{mg}\,l^{-1}\,h$. The division of $140\,\text{mg}\,l^{-1}\,h$ as the effective dosage for complete control by $48\,h$ gives a concentration of $2.92\,\text{mg}\,l^{-1}$ as the theoretical constant lethal concentration. Only 20% of the initial $15\,\text{mg}\,l^{-1}$ in the example is therefore actually used to kill the insects. The rest (80% of the dose) is needed to compensate for leaks and is emitted into the atmosphere. A better standard of gas tightness is desirable to avoid these tremendous gas losses. However, the achievement of such is often quite costly.

One possibility to reduce these losses without the cost of better sealing is to change the application technique (Reichmuth 1991a). Instead of applying all the gas at the beginning of the fumigation, it is advisable to inject only about $7\,\text{mg}\,l^{-1}$. Recirculation inside the building will help to quickly achieve an even distribution. As the gas concentration drops, the losses are replaced by further injections to maintain a concentration of $7\,\text{mg}\,l^{-1}$. Thereby, a reduction of the total dose from $15\,\text{mg}\,l^{-1}$ to $10\,\text{mg}\,l^{-1}$ or less is possible without a loss of efficacy. On the other hand, a little more than the theoretical concentration for complete control is needed to ensure sufficient diffusion of the gas inside the structure, into cracks and crevices in concrete

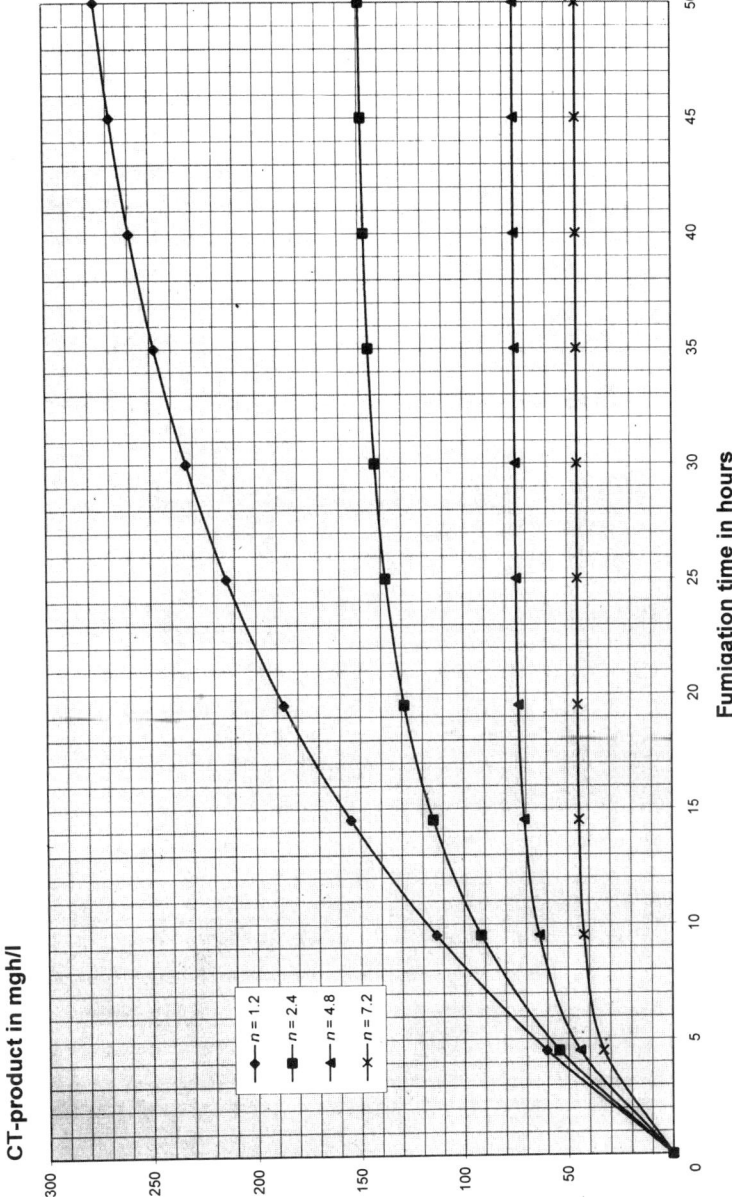

CT-product during a fumigation with 15 mg l⁻¹ depending on the leak rate n

CT-product in mgh/l

Legend:
- $n = 1.2$
- $n = 2.4$
- $n = 4.8$
- $n = 7.2$

Fumigation time in hours

Figure 4.5. CTPs required for complete control of stored product insects during fumigation with methyl bromide at 25 °C at different leak rates (n)

and wood to achieve complete disinfestation everywhere in the treated object.

The final formula to determine the right initial dose c_0 for a fumigation with methyl bromide depending on the achieved gas tightness expressed as leak rate n and the required CTP in a given exposure time can be derived from formula (4):

$$c_0 = 24(ct)/n\{1 - \exp(-nt/24)\} \qquad (5)$$

Figure 4.6 illustrates the values of c_0 required for three different CTPs at different leak rates (n). Halving the leak rate by improving the degree of sealing of a building leads to nearly 50% reduction in the initial dose required for values of $n > 1$. Of course, depending on the species and stages to be controlled, the dose should not be reduced below the effectiveness limit of about $3 \, mg \, l^{-1}$ at which Haber's rule ceases to apply (Section 3.1).

For soil fumigations, the method in which methyl bromide is applied can affect the depth to which the gas penetrates into the soil and how long it is

Initial dose c_0 as function of the leak rate n to achieve a certain CT-product

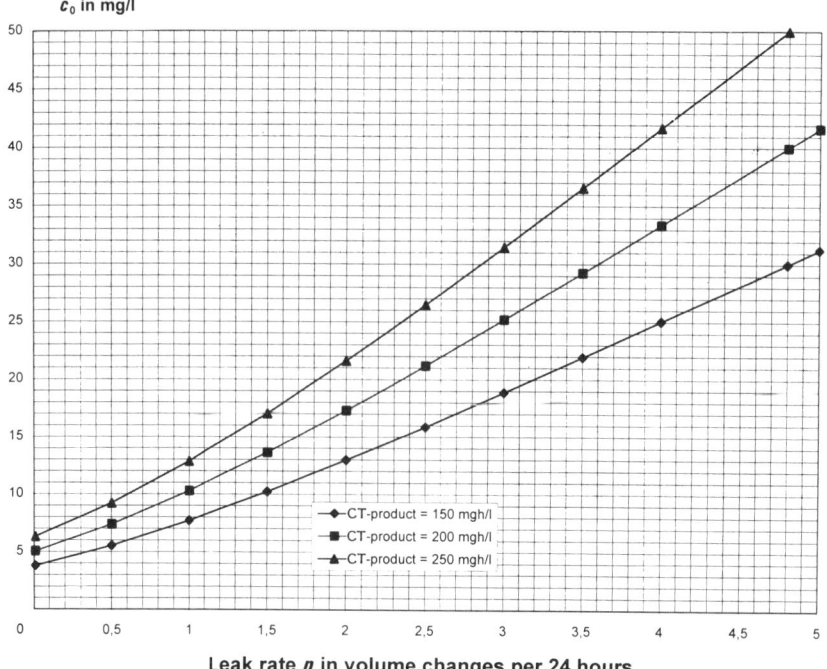

Figure 4.6. The initial dose of methyl bromide for fumigation of a structure depending on the degree of gas-tightness expressed as leak rate n per day at a pressure difference of 10 Pascal for different CTPs and 48 hours of exposure period

retained. The type of sheeting used to cover soil will also affect the retention of gas in the soil. Basile *et al* (1986) investigated the effect on the nematicidal properties and bromide residues after treatment of applying the gas cold or hot and when two different type of sheeting, 45 μm polythene or impervious tarping, were used. *Xiphinema* spp. were eradicated to a depth of 90 cm under the impervious tarping using 20 g m^2 of hot methyl bromide. When the gas was injected cold, 40 g m^2 were required to achieve the same results. Under polythene sheeting 60–80 g m^2 were required using hot gas for the same result. Inorganic bromide levels were also increased under the impervious sheeting. Injecting liquid methyl bromide six inches deep gave better control of the burrowing nematode *Radolphus similis*, than did surface application of cold gas, but still not as good as that obtained from injecting hot gas at the surface (O'Bannon and Tomerlin 1968).

Although advances are being made in the development and use of less permeable sheeting [see Chapter 6], soil fumigation still remains as much of an art as science. Most research only reports initial dosage and the effect on yield of the subsequent crop. Only in a few cases have CTPs been determined, for example by Klein (1989) while investigating ways of reducing the rate at which methyl bromide was lost from the soil, and by Henry *et al* (1992) while monitoring a large statutory fumigation carried out in the UK to sterilise a field infested with the newly introduced rhizomania disease of sugar beet.

The future use of methyl bromide overall is the subject of much debate. The Montreal Protocol has listed it as a potent ozone depleting chemical and thus there is a possibility of a worldwide ban on its production for all but a few limited purposes (Anon 1992). If methyl bromide is to continue to be used for soil fumigation, dosages are likely to have to be reduced to a minimum efficacious level. This will require much closer matching of field dosage rates to those actually required to control the target organism and a greater understanding of the behaviour of the gas in the soil. Some progress has already been made, Mignard and Benet (1989), for example, developed a model for predicting the diffusion of methyl bromide through the soil and other authors in this book report on some of the more recent developments. Much more work needs to be carried out, however. For any continued use in storage and quarantine, methyl bromide emissions would need to be closely controlled by applying reduced dosages and by recapture plus possible recycling of gas.

REFERENCES

Abdalla N and Lear B, 1975, *Plant Disease Reporter*, **59**, 224–8.
Adu OO and Mithu M, 1985, *Insect Sci. Applic.*, **6**, 75–8.

Amin ME, Kamel AH, Ismail II and El-Nahal AKM, 1979, *Bull. Ent. Soc. Egypt, Econ. Ser.*, **11**, 57–63.

An Z-Q, Grove JH, Hendrix JW, Hersham DE and Henson GT, 1990, *Soil Biol. Biochem.*, **22**, 715–20.

Anonymous, 1975, *FAO Plant Protect. Bull.*, **23**, 12–25.

Anonymous, 1977, *FAO Plant Protect. Bull.*, **25**, 49–67.

Anonymous, 1982, EPPO Recommendations on fumigation standards (2nd ed), 31 pp

Anonymous, 1992, United Nations Environment Programme/OzL.Pro4/15.

Anonymous, 1994a, *EPPO Bulletin*, **24**, 315.

Anonymous, 1994b, *EPPO Bulletin*, **24**, 316.

Anonymous, 1994c, *EPPO Bulletin*, **24**, 317.

Armstrong JW, Harvey JM, Garcia DL, Menezes TD and Brown SA, 1988, *J. Econ. Entomol.*, **81**, 1120–3.

Armstrong JW, Schneider EL, Garcia DL and Couey HM, 1984, *J. Econ. Entomol.*, **77**, 680–2.

Armstrong JW and Couey HM, 1984, *J. Econ. Entomol.*, **77**, 1229–32.

Armstrong JW and Garcia DL, 1985, *J. Econ. Entomol.*, **78**, 1308–10.

Australian Quarantine and Inspection Service, 1985, Plant quarantine treatment schedule, Canberra, Australia.

Australian Quarantine and Inspection Service, 1992, Plant Quarantine Manual, Canberra, Australia.

Baker AG, 1939, The basis for treatments of products where fruitflies are involved as a condition for entry into the United States, USDA Circular 551, Washington, DC.

Banks HJ, 1985, in: Champ BR and Highley E (eds.), Pesticides and Humid Tropical Grain Storage Systems, Proceedings of an International Seminar, Manila, Philippines, May, ACIAR Proceedings no 14, 179–193.

Banks HJ and Pinkerton A, 1987, *J. Stored Prod. Res.*, **23**, 105–3.

Barker PS, 1967, *J. Stored Prod. Res.*, **2**, 247–9.

Basile M, Lamberti F, Melillo VA and Basile AC, 1986, Rivista Ortoflorofrutt. Ital, no. 3.

Bell CH, 1976, *J. Stored Prod. Res.*, **12**, 1–10.

Bell CH, 1977, *J. Stored Prod. Res.*, **13**, 119–27.

Bell CH, 1978, *Pestic. Sci.*, **8**, 529–34.

Bell CH, 1981, *Pestic. Sci.*, **12**, 59–64.

Bell CH, 1982, *Pestic. Sci.*, **13**, 442–6.

Bell CH, 1988, *Pestic. Sci.*, **24**, 97–109.

Bell and Glanville 1973, *J. Stored Prod. Res.*, **9**, 165–70.

Bendheim U and Shoshan AE, 1979, *Refuah Veterinarith*, **22**, 154–8.

Bennet RG, 1969, *J. Stored Prod. Res.* **5**, 119–26.

Benson NR, Covery RP Jr and Haglund W, 1978, *J. Am. Soc. Horticult. Sci.*, **103**, 156–8.

Bingefors S and Hahlin M, 1955, *Svenske Frotiding*, **5**.

Blackburn SE, Consden R and Phillips H, 1944, *Biochemical Journal*, **38**, 25–9.

Blanton FS, 1942, *J. Econ. Entomol.*, **35**, 31–4.

Bliss CI, 1935, *Ann. Appl. Biol.*, **22**, 134–67.

Bogs D, 1976, *Nachrichtenblatt Pflanzenschutz in der DDR*, **30**, 221–2.

Bond EJ, 1984, Manual of fumigation for insect control, *FAO Plant Prod. Paper* **54**. FAO, Rome, 432 pp.

Bond EJ, 1956, *Canadian J. Zool*, **34**, 405–15.

Bond EJ and Buckland CT, 1976, *J. Econ. Entomol.*, **69**, 725–7.

Bond EJ and Buckland CT, 1978, *J. Econ. Entomol.*, **71**, 307–9.
Bond EJ and Munro HAU, 1961, *J. Econ. Entomol.*, **54**, 451–4.
Bond EJ and Upitis E, 1976, *J. Stored Prod. Res.*, **12**, 261–7.
Bowley CR and Bell CH, 1981, *J. Stored Prod. Res.*, **17**, 83–7.
Brown WB, 1954, *Pest Infest. Res. Bull*: No. 1.
Brown WB, 1959, *Pest Infest. Res. Bull*: No. 1, 2nd Ed.
Brust GE and Scott WD, 1994, *Hort. Science*, **5**, 471.
Calderon M and Leesch JG, 1983, *J. Econ. Entomol.*, **76**, 1125–8.
Carmi Y, 1985, *EPPO Bulletin*, **15**, 15–6.
Chakrabarti B and Bell CH, 1993, *Chemistry and Industry*, **22**, 992–5.
Champ BR, 1985, in, Champ BR, Highley E (eds), Pesticides and Humid Tropical Grain Storage, Proceedings of an International Conference, Manila, Phillipines, ACIAR Proceedings no. 14. 229–255.
Champ BR and Dyte CE, 1976, Report of the FAO global survey of pesticide susceptibility of stored grain pests. FAO Plant Prod. and Prot. Ser. No. 5, 297 p.
Champ BR and Dyte CE, 1977, *FAO Plant Protect. Bull.*, **25**, 49–67.
Cheetham T, 1990, *Ann. Entomol. Soc. Am.*, **83**, 59–67.
Cherif R, Leesch J and Davis R, 1985, *J. Econ. Entomol.*, **78**, 660–5.
Cintra AF, Almeida PR, Myazaki I and Neves HS, 1978, *Biologico*, **44**, 3–10.
Clerjeau M, Dauple P, Ginoux G, Leroux J-P and D'Oleon F, 1973, *Pepinieristes Horticulteurs Maraichers*, **138**, 35–42.
Cobb RD, 1958, *Bull. Calif. Dept. Agr.*, **47**, 1–19.
Coolen EA, Hendrickx GJ and D'Herde CJ, 1972, Rijksstn. voor Nematologie en Entomologie, Merelbeke, Belgium, publikatie 11, 22 p.
Couey HM and Chew V, 1986, *J. Econ. Entomol.*, **79**, 887–90.
Cowley JM, Baker RT, Engleberger KG and Lang TG, 1991, *J. Econ. Entomol.*, **84**, 1763–7.
Cox PD, Bell CH, Pearson J and Beirne MA, 1984, *J. Stored Prod. Res.*, **20**, 215–9.
Dead Sea Bromine Group, 1991, Methyl Bromide for Soil Fumigation.
Detmers HB, 1993, Empfindlichkeit der Entwicklungsstadien von *Prostephanus truncatus* (Horn) (Coleoptera: Bostrichidae) und *Teretriosoma nigrescens* Lewis (Coleoptera: Histeridae) gegenüber Methylbromid. [The Susceptibility of the developmental stages of *Prostephanus truncatus* (Horn) (Coleoptera: Bostrichidae), and *Teretriosoma nigrescens* Lewis (Coleoptera: Histeridae) to methyl bromide] PhD Thesis, Technical University, Berlin, 189 p.
Dickson DW and Hewlett TE, 1988, *Annals Appl. Nematology*, **2**, 95–101.
Dixon M and Needham DM, 1946, *Nature*, **158**, 432–8.
El-Buzz HK, Kamel AH, El-Nahal AKM and El-Borollosy FM, 1974, *Agricultural Research Review*, **52**, 21–9.
El-Lakwah Von F, 1977a, *Anz. Schädlingskde., Pflanzenschutz, Umweltschutz*, **50**, 118–22.
El-Lakwah Von F, 1977b, *Anz. Schädlingshutz, Pflanzenshutz, Umweltschutz*, **50**, 180–3.
El-Lakwah, Von F, 1977c, *Anz. Schädlingskde., Pflanzenshutz, Umweltschutz*, **50**, 68–73.
El-Nahal AKM, Barakat AA, El-Halafawy MA, and Hassan HI, 1986, *Bull. Ent. Soc. Egypt, Econ. Ser.*, **14**, 141–54.
Faucon JP, Arvieu JC and Colin ME, 1982, *Revue de Medecine Veterinaire*, **133**, 207–10.
Fisk FW and Shephard HH, 1938, *J. Econ. Entomol.*, **31**, 79–84.
Franz A, Reichmuth Ch and Wohlgemuth R, 1992, *Z. Lebensm. Unters. Forsch.*, **194**, 148–151.

Godden E and Howe RW, 1965, *Tribolium Inf. Bull.*, **8**, 1–76.
Goodey JB, 1949, *J. Helminth.*, **23**, 171–4.
Goodey T, 1945, *J. Helminth.*, **21**, 45–9.
Gostick KG, 1963, *Pl. Path.*, **12**, 62–4.
Greany PD, 1994, in: Paull RE and Armstrong JW (eds). *Insect Pests, and Fresh Horticultural Products: Treatments, and Responses*, (Wallingford, UK: CAB International), pp 69–84.
Haas HV, 1976, *Phytoparasitica*, **4**, 201–5.
Haber F, 1924, *Fünf Vortäge aus den Jahren 1920–1923*, **5**, 76–92.
Hague NGM and Clark WC, 1959, *Meded. LandbHogesh. Gent.*, **24**, 628–36.
Hah JK, Lee CK, and Yu KY, 1982, *Korean Journal of Plant Pathology*, **24**, 133–7.
Halfawy MA, Abdel Wahab AE, Heykal A and Asran AM, 1978, *Agricultural Research Review*, **56**, 155–9.
Hallman GJ and King JR, 1992, *J. Econ. Entomol.*, **85**, 1231–4.
Hara AH, 1994, in: Paull RE and Armstrong JW (eds). *Insect Pests, and Fresh Horticultural Products: Treatments, and Responses*. (Wallingford, UK: CAB International), pp 69–84.
Hartsell PL, Nelson HD, Tebbets JC and Vail PV, 1986, *J. Econ. Entomol.*, **79**, 1299–1302.
Hartsell PL, Vail PV, Tebbets JC and Nelson HD, 1991, *J. Econ. Entomol.*, **84**, 1289–93.
Henry CM, Bell GJ and Hill SA, 1992, *Pl. Path.*, **41**, 483–9.
Howe RW and Hole BD, 1966, *J. Stored Prod. Res.*, **2**, 13–26.
Howe RW and Hole BD, 1967, *J. Appl. Ecol.*, **4**, 337–51.
Hussaini SS, 1985, *Indian Journal of Nematology*, **15**, 88–92.
Hussaini SS, 1989, *Indian Journal of Nematology*, **19**, 14–17.
Irish DD, Adams EM, Spencer HC and Rowe VK, 1940, *Journal of Industrial Hygiene, and Toxicology*, **22**, 218–30.
Ironside DA, Swaine G and Corocoran RJ, 1978, *Queensland Journal of Agricultural, and Animal Sciences*, **35**, 29–33.
Jones RM, 1938, *J. Econ. Entomol.*, **31**, 398–9.
Kennedy J, 1961, *J. Sci. Food Agric.*, **12**, 96–103.
Khrone HE and Lindgren DL, 1958, *J. Econ. Entomol.*, **51**, 157–8.
Klein L, 1989, *Acta Horticulturae*, **255**, 213–25.
Kolb RW and Schneiter R, 1950, *J. Bacteriol.*, **59**, 401–12.
Landolt PJ, Chambers DL and Chew V, 1984, *J. Hort. Sci.*, **59**, 285–7.
Le Goupil M, 1932, *Revue de Pathologie Végétale et d'Entomologie Agricole de France*, **19**, 169–72.
Lewis SE, 1948, *Nature*, **161**, 692–3.
Liese W and Rütze M, 1985, *EPPO Bulletin*, **15**, 29–36.
Lindgren DL, 1936, *J. Econ. Entomol.*, **29**, 1174–5.
Lindgren DL and Vincent LE, 1965, *J. Econ. Entomol.*, **58**, 551–5.
Loveday PN and Winteringham FPW, 1951, Fate of insecticides in insects. In: Pest Infestation Research 1948, HMSO, London, pp 29–31.
Macdonald OC, 1993, *Ann. Appl. Biol.*, **123**, 531–7.
Macdonald OC and Mills KA, 1994, Brighton Crop Protection Conference – Pests, and Diseases, **3**, 183–190.
Macdonald OC, 1996, *Ann. Appl. Biol*, in Press.
Mackie DB, 1938, *J. Econ. Entomol.*, **31**, 70–9.
MAFF/PSD Advisory Committee on Pesticides, (1992) Evaluation of Fully Approved or Provisionally Approved Products: Evaluation of Methyl Bromide.
Maindonald JH, Wadell BC, and Birtles DB, 1992, *J. Econ. Entomol.*, **85**, 1222–30.

Majumder SK, 1954, *Mysore Cent. Food Tech. Res. Inst. Bull.*, 269–271.
Mignard E and Benet JC, 1989, *J. Soil Sci.*, **40**, 151–65.
Miller DP and Haggard HW, 1943, *Journal of Industrial Hygiene and Toxicology*, **25**, 423–33.
Moffitt HR, 1989, Proc. Washington State Horticultural Assn. 223–5.
Moffitt HR, Drake SR, Toba HH and Hartsell PL, 1992, *J. Econ. Entomol.*, **85**, 1855–8.
Monro HAU, 1969, *Manual of Fumigation for Insect Control. 2nd ed.*, FAO Agricultural Studies, **79**, FAO, Rome.
Mortimer EA and Powell DF, 1984, *Ann. Appl. Biol.*, **105**, 443–54.
Mortimer EA and Powell DF, 1988, *Ann. Appl. Biol.*, **112**, 33–9.
Munnecke DE, Ludwig RA and Sampson RE, 1959, *Can J. Bot.*, **37**, 51–8.
Munnecke DE, Moore BJ and Abu-el-Haj F, 1971, *Phytopathology*, **61**, 194–7.
Munnecke DEJ, Bricker JL and Kolbezen MJ, 1978, *Phytopathology*, **68**, 1210–16.
New Zealand, Ministry of Agriculture, and Fisheries, 1990, *Sentinel*, **3**.
New Zealand Ministry of Agriculture, and Fisheries. Treatment requirements, NASS:152.02, Border protection specifications: Clearance of fresh produce, Section 2.0. Wellington, New Zealand.
Nishimura MM, Umeda S, Ishizo S and Sato M, 1980, *Journal of Toxicological Science*, **5**, 321–30.
Noe JP, 1980, *Journal of Nematology*, **22**, 39–44.
O'Bannon JH and Tomerlin AT, 1968, Proceedings of the Soil, and Crop Science Society of Florida, **28**, 299–306.
Page ABP, Hague NGM, Jakabsons V and Goldsmith RE, 1959, *J. Sci. Food Agric.*, **10**, 461–67.
Parameswaran N and Rütze M, 1984, *Material und Organismen*, **19**, 133–40.
Peters G, 1936, *Neue Folge*, **21**, 1–120.
Peters G, 1942, *Neue Folge*, **47a**, 143 pp.
Peters G and Ganter W, 1935, *Z. Angew. Entomol*, **21**, 547–59.
Pflanzenschutzmittel-Verzeichnis, 1994, Teil 5, Vorratsschutz. Biologische Bundesanstalt für Land und Forstwirtschaft ed., Saphir Verlag, Ribbesbüttel, Germany, 42, Aufl, ISSN 0178–0638, 54 p.
Pike V, 1994 *Crop Protection*, **13**, 141–145.
Piper WR and Davidson RH, 1938, *J. Econ. Entomol.*, **31**, 460–1.
Powell DF and Gostick KG, 1971, *Bull. Ent. Res.*, **61**, 235–40.
Powell DF and Mills KA, 1981, *EPPO Bull.*, **11**, 371–6.
Powell DF, 1974, *Pl. Path.*, **23**, 110–3.
Powell DF, 1975, *Ann. Appl. Biol.*, **81**, 425–31.
Powell DF, 1981, *Pl. Path.*, **30**, 159–65.
Rajendran S, 1990, *Pestic. Sci.*, **29**, 75–83.
Raju CA, 1984, *Phillipine Journal of Coconut Studies*, **9**, 23–5.
Reichmuth Ch, 1986, Proceedings of the GASGA Seminar on Fumigation Technology in Developing Countries, Slough, UK. Overseas Development, and Administration, Tropical Development, and Research Institute, 88–98.
Reichmuth Ch, 1990, in, Champ BR, Highley E and Banks HJ (eds) Fumigation, and Controlled Atmosphere Storage of Grain, Proceedings of an International Conference, Singapore. ACIAR Proceedings no. **25**, 56–69.
Reichmuth Ch, 1991a, *Gasga Newsletter*, **15**, 14–15.
Reichmuth Ch, 1991b, New techniques in fumigation research today. in: F Fleurat-Lessard, and P Ducom, (eds) Proceedings of the 5th International Working Conference of Stored Product Protection, Bordeaux, France, September, 709–725.
Reichmuth Ch, 1993, Merkblatt 71 der Biologische Bundesanstalt für Land und

Forstwirtschaft ed., Saphir Verlag, Ribbesbüttel, Germany, **42**, 38 pp.

Reidel RM and Ellis MA (1983) Annual Report of the Ohio Agric. Res. Centre, 41–42.

Richardson LT and Monro HAU, 1962, *Applied and Environmental Microbiology*, **10**, 448–51.

Rohde WA, Johnson AW, Dowler CC and Glaze NC, 1980, *Journal of Nematology*, **12**, 33–9.

Roth H and Kennedy JW, 1973, *J. Econ. Entomol.*, **66**, 935–6.

Roth H and Kennedy JW, 1972, *J. Econ. Entomol.*, **65**, 1650–1.

Rütze M and Liese W, 1983, *Holz-Zentrablatt*, **113**, 1533–5.

Sadoshima T, Ushimaki A, Ohara A and Ushio S, 1993, *Research Bulletin of the Plant Protection Service, Japan*, **29**, 77–9.

Sanchez-Riviello M and Rhode RH, 1976, *Southwestern Entomologist*, **1**, 178–80.

Schacher A and Knülle W, 1980, *Anz. Schädlingskunde, Pflanzenschutz, Umweltschutz*, **53**, 166–9.

Scheffrahn RH and Su J, 1992, *J. Econ. Entomol.*, **85**, 845–7.

Shepard HH, Lindgren DL and Thomas EL, 1937, *Univ. Minn. Ag. Exp. Sta. Tech. Bul.*, **120**, 23.

Shivanandappa T and Rajendran S, 1987, *Pesticide Biochemistry and Physiology*, **28**, 121–6.

Soderstrom EL, Brandl DG, Hartsell PL and Mackey B, 1991, *J. Econ. Entomol.*, **84**, 936–41.

Stark JD, 1994, Chemical fumigants, in: *Insect Pests and Fresh Horticultural Products: Treatments and Responses*. Paull RE and Armstrong JW (eds). (Wallingford, UK: CAB International) pp 69–84.

Stephan ZA, Al-Maamoury IK and Michbass AH, 1991, *FAO Plant Production and Protection Paper*, **109**, 343–50.

Sumner DR, Dowler CC, Johnson AW, Glaze NC, Phatak SC, Chalfant RB and Epperson JE, 1983, *Plant Disease*, **67**, 1071–5.

Tanaka K, Sunnagaw K, Oda Y and Hokama T, 1986, *Research Bulletin of the Plant Protection Service, Japan*, **22**, 67–78.

Thompson RH, 1966, *J. Stored Prod. Res.*, **1**, 353–76.

Unger W, Reichmuth Ch, Unger A and Detmers H-B, 1992, *Zeitschrift für Kunsttechnologie und Konservierung*, **6**, 244–59.

Upitis E, Monro HAU and Bond EJ, 1973, *J. Stored Prod Res.*, **9**, 13–7.

USDA Animal, and Plant Health Inspection Service, 1992, Plant Protection, and Quarantine Manual, United States Department of Agriculture, Hyattsville, Maryland.

Vail PV, Tebbets BE, Mackey BE and Curtis CE, 1993, *J. Econ. Entomol.*, **86**, 70–3.

Van Wambeke E, 1989, *Acta Horticulturae*, **255**, 243–54.

Verma BR, 1986, *Journal of Entomological Research*, **10**, 138–143.

Vincent LE, Rust MK and Lindgren DL, 1980, *J. Econ. Entomol.*, **73**, 313–7.

Vincent LE and Lindgren DL, 1975, *J. Econ. Entomol.*, **68**, 53–6.

Vincent LE and Lindgren DL, 1977, *J. Econ. Entomol.*, **70**, 497–500.

Von Oettingen NF, 1955, The halogenated hydrocarbons, their toxicity, and potential dangers. US Public Health Service, Washington, DC, Publication No. 414.

Williamson DL, Summy KR, Hart WG, Sanchez-Riviello M, Wolfenbarger DA and Bruton BD, 1986, *J. Econ. Entomol.*, **79**, 172–5.

Winks RG, 1982, *J. Stored Prod. Res.*, **18**, 159–69.

Winks RG, 1984, *J. Stored Prod. Res.*, **20**, 45–56.

Winks RG, 1985, The Biological efficacy of fumigants: time/dose response pheno-mena, in: BR Champ, E Highley (eds), *Pesticides, and Humid Tropical Grain Storage Systems*, Proceedings of an International Seminar, Manila, Philippines, May, ACIAR Proceedings no 14, 211–221.

Winteringham FPW, Helyer GC and McKay MA, 1958, *Biochemical Journal*, **69**, 640–8.

Winteringham FPW and Barnes JM, 1955, *Physiology Review*, **35**, 701–39.

Wit AKH and van der Vrie M, 1985, *Med. Fac. Landbouww. Rijksuniv. Gent*, **50/2b**, 705–12.

Witherell PC, 1984, *Florida Entomologist*, **67**, 254–62.

Wohlgemuth R, 1990, *Merkblatt 66 der Biologischen Bundesanstalt für Land und Forstwirtschaft ed., Saphir Verlag, Ribbesbüttel, Germany*, **42**, 20 pp.

Yokoyama VY, Miller GT and Hartsell PL, 1987, *J. Econ. Entomol.*, **80**, 1226–1228.

Yokoyama VY, Miller GT and Hartsell PL, 1992, *J. Econ. Entomol.*, **85**, 150–156.

Yokoyama VY, Miller GT, Hartsell PL, 1994, *J. Econ. Entomol.*, **87**, 730–735.

Yust HR and Busbey RL, 1942, *J. Econ. Entomol.*, **35**, 343–345.

5

Methyl Bromide as a Soil Fumigant

L KLEIN

Dead Sea Bromine Group, Beer Sheva, Israel

C.H. Bell, N. Price and B. Chakrabarti: The Methyl Bromide Issue
© 1996 John Wiley & Sons Ltd

1.1 INTRODUCTION

In soils cropped frequently with the same plant, or closely related crops, there is a build-up of detrimental soil-borne plant pathogens which adversely affect the yield, quality and stability of production.

The optimal growing conditions provided for repeated cropping in intensive and irrigated agriculture (greenhouse and outdoor high-value crops) also provides optimal conditions for the occurrence and development of many soil-borne pests and pathogens. If unchecked, these can build up to such levels that outbreaks may entirely destroy the crop or render it uneconomical. This, of course, is not acceptable especially in greenhouses where a lot of money has been invested and a high return is expected.

The difficulty in controlling soil-borne diseases is that the pathogens exist deep in the soil which serves to shield them against most pesticides. The tremendous volume of soil encompassed in one acre of land to a depth of only 30 cm is sufficient to illustrate the inherent problems in getting all of this volume of soil treated to kill the pathogens. The inoculum has to be reached and effectively controlled to a considerable depth.

Experience has shown that pathogens can be controlled by fumigants that penetrate deeply into soil spaces and pores. During the past 40 years several compounds have been used as soil fumigants, each with its particular advantages and limitations. Of these, it is generally accepted that methyl bromide has proved itself as giving the most efficient, reliable and economical control of a wide range of soil-borne pests and diseases. Indeed, fumigation with methyl bromide is the most widely-used method of soil disinfestation especially in protected cropping. A list of its characteristics is given below.

Specific gravity, liquid at 0 °C	1.732 (Water at 4 °C = 1)
Relative density of the gas at 20 °C	3.27 (air = 1)
Boiling range	3.5°–4 °C
Vapour pressure	1420 mm Hg at 20 °C
Solubility in water	~1.6% at 20 °C

2.1 THE CASE FOR SOIL FUMIGATION WITH METHYL BROMIDE

Certain characteristics, which are the result of methyl bromide's physical and chemical properties give it the unequivocal status of the No. 1 fumigant against soil-borne pathogens.

Methyl bromide is a clear colourless liquid, packed in pressurized steel cylinders of various sizes and under natural pressure in cans. It boils at

3.5°–4 °C so that when in air at normal temperatures the compound becomes a gas, odourless at low concentrations. For safety reasons, for all methyl bromide used on soil, 2% chloropicrin (in specific cases amyl acetate) is added as a warning agent, because it is a potent tear gas and so reveals leaks that might expose applicators to the dangerous gas.

The uniqueness of methyl bromide lies in the following factors.

2.1.1 BROAD SPECTRUM ACTIVITY

No other treatment matches methyl bromide fumigation for its broad-spectrum efficacy. Methyl bromide controls weeds, all species of nematodes, soil-borne insects, fungi and some soil-borne bacteria and parasitic plants. When methyl bromide is mixed with chloropicrin to a ratio of approximately 70:30 (a common mixture in many countries), the rate of control of certain fungi is increased because the mixture has a synergistic effect. As an 'all purpose' fumigant, methyl bromide is a single treatment which is effective against most soil-borne pathogens, and thus eliminates the need to use a number of alternative pesticides, which could pose additional problems for the environment. Moreover, it releases the grower from the need to identify the different pathogens which may attack the crop.

2.1.2 QUICK AND DEEP PENETRATION INTO THE SOIL

Most fumigants are limited in the depth of their penetration into the soil. Due to its physical properties (high vapour pressure, low boiling point, heavier than air) methyl bromide is outstanding in its rapid and extensive penetration to considerable depths, and in its spread throughout the soil atmosphere in sufficiently high concentrations to control the pathogens. It penetrates soil to a greater depth than chloropicrin or methyl isothiocyanate (MITC) compounds (Reber 1967).

2.1.3 VERY SHORT EXPOSURE PERIOD

Because of the high toxicity of methyl bromide to pathogens and its rapid effect, control can be completed in 24–48 h, depending on soil temperature. Other fumigants need a much longer exposure to work.

2.1.4 QUICK DISSIPATION FROM THE SOIL AFTER FUMIGATION

The aeration period following fumigation takes only a few days. This means that planting can begin very soon after the treatment as compared with other fumigants for which planting takes place after weeks of aeration. This is an important economical advantage in intensive cultivation.

2.1.5 PENETRATION INTO UNDECOMPOSED MATERIAL

Plant residues, e.g. stems, roots, tubers etc., which are left in the field from previous crops may be infested with pathogens. Such infested organic material is a primary source of infestation for the coming crop. Methyl bromide is the only fumigant which is able to penetrate the undecomposed material and control both the pathogen and the host, thus eliminating the carry-over of pests to successive crops.

2.1.6 INCREASED GROWTH RESPONSE EFFECT

A beneficial side-effect in methyl bromide fumigated soil is the subsequent improved growth and yields of plants, beyond that attributable to disease control. This growth stimulation was noticed even in soils where known pathogens were absent (Chen *et al* 1991). This phenomenon is also known with other disinfestation methods but is most noticeable with methyl bromide.

The unexpected beneficial effect on plant growth in the absence of a target pest can be explained in several ways:

(a) Following the fumigation, minerals are released into the soil and they become more available to the plants causing a more vigorous growth. Chemical and physical analyses of fumigated soils have confirmed this (Chen *et al* 1991).

(b) Control of minor pathogens. These are secondary non-specific parasites which grow and multiply on the plant roots, but are not considered as obvious soil-borne pathogens. The activity of such microorganisms does not affect the plants in a major way, but their cumulative influence during the lifespan of the plant may be detrimental (Chen *et al* 1991).

(c) Also the stimulus of re-establishment of beneficial micro-organisms, freed from other competition, could provide a reason for better plant growth.

No less important is the occasional situation where a decreased growth response is observed. This subject will be dealt within Section 8.1 'Special problems associated with soil fumigation'.

2.1.7 GOOD EFFICACY IN A WIDE RANGE OF TEMPERATURES

It is widely accepted that methyl bromide remains effective at lower temperatures, conditions in which other fumigants are often ineffective.

2.1.8 EFFICACY WITH MORE THAN ONE CROP

Frequently, a single application of methyl bromide has been found effective for more than one season and for more than one crop. This is especially so for outdoor crops, though it depends on the severity of inoculum, the sensitivity of the crops to the diseases, and post-fumigation practices (Klein, unpublished).

2.1.9 DISADVANTAGES

The above-mentioned unique properties made this compound the fumigant of choice for most control situations on soil. However, using methyl bromide has certain drawbacks. Its high human toxicity necessitates considerable care in its use. In soils it can break down to inorganic bromides which are toxic to some plants or can be taken up by others and accumulated in the edible parts of certain species. Its broad spectrum of activity means that some beneficial organisms such as mycorrhizae may be adversely affected. Finally, because of the skill and care needed to apply methyl bromide, it is relatively expensive to use.

3.1 SOIL-BORNE PESTS CONTROLLED BY METHYL BROMIDE

Methyl bromide is applied before sowing or planting or replanting a whole range of cultivated crops:

- In seedbeds and nurseries of vegetables, tobacco, fruit trees, forestry and ornamental plants.
- For greenhouse and other protected crops.
- For flowers and other high-value vegetable field crops.
- Prior to planting and replanting in deciduous orchards, vineyards, citrus groves and tropical and subtropical plantations.
- In restoration or resodding of lawns and golf courses.
- In treating bins or stacks of potting soil.

3.1.1 PLANT-PARASITIC NEMATODES

Almost all plants can be attacked by plant parasitic nematodes (eelworms), tiny worms that feed and may breed on and in plant roots. Damage by nematodes can depress crop yields and degrade crop quality, opening the way to invasion by soil-borne fungi which may result in the death of the host plant. Nematodes can also transmit plant viruses, and even break down resistance of crops bred for resistance to soil-borne pathogens. The sensitivity of crops susceptible to soil diseases is further increased by nematode attack.

Methyl bromide is the most efficient fumigant for penetrating and killing nematodes within their cysts and galls, and nematodes sheltering within non-decomposed plant residues, e.g. cyst nematodes, root-knot and root-lesion nematodes, and pin, dagger and stem nematodes.

3.1.2 PLANT DISEASES

Plant diseases of varying severity may be caused by soil-borne fungi and bacteria. Outbreaks of damping-off, wilting, and root and crown rots may affect plants so severely that no crop is left to harvest.

Fungi are the primary target of methyl bromide fumigation which controls all life stages (e.g. spores, mycelium and resting structures) of the soil-borne pathogens that cause:

damping-off: *Rhizoctonia solani, Pythium* spp., *Sclerotium bataticola (Macrophomina phaseolina), Phytophthora* spp., and *Thielaviopsis basicola.*

crown rot: *Sclerotium rolfsii, Sclerotinia* spp., *and Phytophthora cactorum.*

root rot: *Pythium* spp., *Stromatinia* spp., *Fusarium* spp., *Sclerotium bataticola (Macrophomina phaseolina), Monosporascus eutypoides, Rhizoctonia solani, Pyrenochaeta* spp., (corky root of tomatoes, pink rot of onions), *Armillaria* spp., and *Phytophthora fragariae.*

wilt: *Fusarium* spp., *Verticillium* spp., and *Curvularia* spp.

Methyl bromide also controls many bacteria including bacterial canker (*Corynebacterium michiganense*) bacterial wilt (*Pseudomonas solanacearum*), common scab (*Streptomyces scabies*), in addition to some soil-borne viruses and their vectors, e.g.: Big-vein virus in lettuce.

3.1.3 WEEDS

Weeds, both annual and perennial, compete with plants for light, nutrients and water, also serving as carriers of pests that attack crop plants. At best, crop yield and quality are curtailed by weed infestation—at worst, a crop may be so overrun as to render harvesting pointless.

Methyl bromide effectively controls the seeds of annual weeds with the exception of certain weeds such as mallow (*Malva*), legumes (*Leguminosae*) and horseweed (*Conyza* or *Erigeron* spp.). It controls perennial broad-leaf weeds like sorrel, blackberry and bindweeds (*Convolvulus* spp.); perennial grasses like bermudagrass (*Cynondon dactylon*), johnsongrass (*Sorghum halepense*) and quackgrass (*Agropyron repens*); and other monocotyledons like wild onions and garlic (*Allium* spp.). Perennial weeds will not be completely controlled if deep-rooted in insufficiently cultivated medium and heavy soils. Purple nutsedge or nutgrass (*Cyperus rotundus*), for example, is controlled only in light soil and under optimal conditions.

3.1.4 PARASITIC PLANTS

Parasitic plants, such as broomrape (*Orobanche* spp.), attack the roots of a very wide range of host plants, drawing off water and nutrients and causing deterioration and a marked reduction in yield. Heavily parasitized plants may die as a result.

Methyl bromide is the only treatment that controls the seeds of all broomrape species.

3.1.5 SOIL-BORNE INSECTS

Some soil insects live in the soil, especially mulched soil, for part or all of their life cycles. There, they are protected, and feed on plant roots. The injuries they cause stunt and even kill crop plants and also open the way to secondary infections.

Methyl bromide controls all the life stages of mole crickets; wire, cut and other worms; various grubs; ants and termites.

Details on the effect of methyl bromide on target organisms are discussed in Chapter 4.

4.1 FACTORS PERTINENT TO SUCCESSFUL FUMIGATION

In order that the many advantages of methyl bromide can be fully exploited, it is important that it be used with proper attention to the efficiency of the fumigation procedure, soil preparation, conditions at the time of fumigation, post fumigation practices and general management. All these factors are inter-related in soil fumigation (Figure 5.1).

Failure to take these factors into account can mean that problems will be encountered during subsequent cropping.

The following detailed descriptions of soil preparation, pest activation and application conditions are designed to ensure a commercially successful fumigation.

4.1.1 SOIL PREPARATION

Proper soil preparation is the key and the most important single factor for successful pest control by methyl bromide.

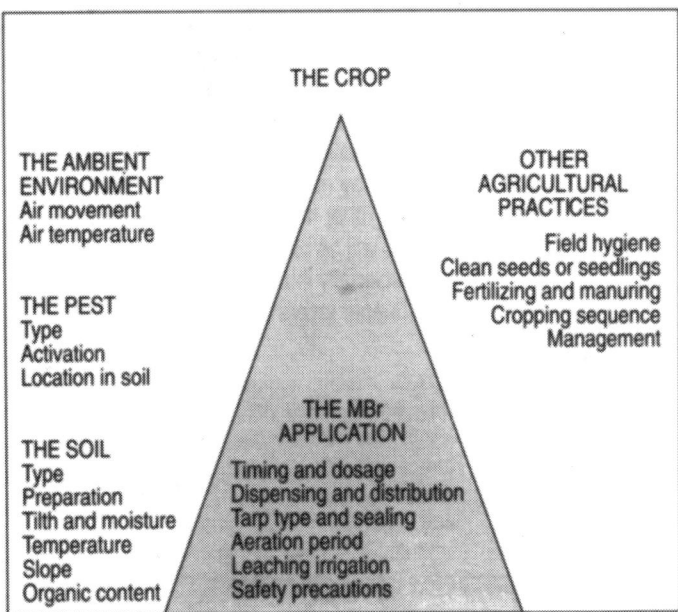

Figure 5.1. Factors ensuring successful fumigation

4.1.1.1 Removal of Crop Trash

The previous crop's trash may be infected by pathogens. Often diseased plants are left in place until the end of the crop. Diseased plants harbour great numbers of resting structures of pathogens, sometimes thousands of propagules in one gram of tissue. If these are incorporated into the soil, which is often the case, they enrich the pathogenic reservoir considerably. Any control method is less effective when a high amount of inoculum is involved. It is good hygienic sense to remove and destroy this trash immediately after the harvest. It is bad practice to plough it in and hope that the methyl bromide treatment will sterilize it. Firstly, part of the applied methyl bromide will combine chemically with the organic matter, necessitating a far higher dose of methyl bromide than normal to be effective against the pests inside the undecomposed trash. Secondly, the trash causes large air spaces in the soil that result in an uneven distribution of the fumigant. Thirdly, when mechanical application is used, the injection tines could be blocked by trash causing furrows to remain open during injection, rendering the entire operation inefficient.

4.1.1.2 Cultivation to Seed-bed Tilth

Deep cultivation aids gas penetration—a plough-sole (a layer of compacted soil immediately below the depth of cultivation) will act as a gas barrier and pests beneath it will probably survive the fumigation. Large unbroken clods offer sheltered lodging to the pests inside them and because of channelling, the gas will not reach them. The soil has to be worked so as to get a good porosity. Only if the soil is friable and porous, will the gas diffuse laterally and vertically to great distances in a uniform manner.

4.1.1.3 Soil Moisture Content

Soil moisture is a very important factor governing success or failure of soil fumigation. Moisture is needed to activate dormant weed seeds and disease agents, rendering them more susceptible to treatment with methyl bromide. Too little moisture in the soil may result in no weakening of pest dormancy. However, soils which are too wet or have waterlogged patches are not fumigable, since all pores are blocked with water and diffusion of the fumigant is hindered. It is especially important to avoid excess moisture in the deeper layers of heavy soils. For optimum results, soils should be kept moist for at least 10 days prior to the treatment. The only exception is when methyl bromide is aimed at killing *Armillaria mellea*, which is found very deep in the soil, by deep placement of methyl bromide. The mellea-infested roots remain moist even in fairly dry soils and thus methyl bromide can diffuse through to the site of action.

In summary—soils correctly prepared have to be in seedbed condition: cultivated to at least 40 cm, loose, moist, with no lumps, without undecomposed material and as level as possible. If required, manure should be applied before fumigation and be in a good physical state. Failure in preparation means that the pest inoculum may survive and the expense of the treatment will have been wasted.

4.1.2 TEMPERATURE

The effectiveness of methyl bromide increases with an increase in temperature. Distribution of the gas is enhanced and susceptibility of pathogens are increased at higher temperatures (Ebben *et al* 1983, Munnecke and Bricker 1978). Methyl bromide is less effective at low temperatures and should not be used at temperatures below 8 °C in the soil measured at a depth of 15–20 cm, the optimum being about 25 °C. On cool but sunny days, it is advisable to delay treatment for a few hours so that the soil under the tarp warms up. Sometimes, depending on the pest involved, it is worthwhile to compensate for marginal conditions with higher dosages for better efficacy. On the other hand, in very hot conditions, the permeability of the plastic covering to methyl bromide is increased, which may cause loss of methyl bromide gas through the film. It is advisable to delay treatment until after a particularly hot or a very windy day.

4.1.3 DOSAGE

Dosage rates recommended for commercial applications range from 50–100 g/m^2 (1–2 lbs/100 ft^2). Under optimal conditions and against most target organisms mentioned earlier, a dosage of 50 g/m^2 is sufficient. The correct dosage is often a function of the degree of pathogen infestation of the soil, a fact often overlooked in commercial soil fumigation. Therefore, in greenhouses, where soil might be highly infested with pathogens which are difficult to control and where the fumigation is aimed at achieving maximum results, higher dosages, e.g., 75–100 g/m^2 are used. Other factors like soil type, temperature, organic material and no less important—knowing the disease incidence of previous crops—are factors to consider when determining the dosage.

Typical instructions for soil fumigation using Metabrom 980 (trade mark of methyl bromide 98% manufacture by Dead Sea Bromine Group, Israel) are given in Table 5.1 in section 7.1

4.1.4 EXPOSURE TIME

In order to be effective, methyl bromide must remain in contact with the target organism for sufficient time and in sufficient concentration to kill the pathogen (5.1). In most cases and under favourable conditions, the control is achieved in 24–48 h and even less in many situations. Longer sheeting times would not necessarily achieve better results. However, at lower temperatures or for safety and environmental reasons, local recommendations dictate exposure times of 4–10 days. (U.K., Belgium)

4.1.5 SOIL TYPE

Heavier types of soil are more difficult to fumigate than lighter types. It is harder to achieve optimal soil preparations, the organic content and the clay fraction are higher, causing a quicker breakdown of methyl bromide, and absorption of methyl bromide to the soil is greater. This usually necessitates increased rates to compensate for these factors and a higher level of inorganic bromide residues are formed (see 8.1.1). Deeper layers of soil may be too wet at the time of fumigation, especially if they are fine-textured and therefore less penetrable to the gas. However, if approached with a thorough knowledge and comprehension of all its aspects, fumigation of heavier types of soil can be equally effective.

4.1.6 ORGANIC MATERIALS

Organic manures or any other organic materials have the same character-istics as heavy soils. Methyl bromide disintegrates more quickly in the presence of organic material, and greater absorption of methyl bromide may cause high levels of inorganic bromides. Organic material should be added prior to the fumigation, preferably before the first cultivation, and the dressing must not exceed 200 m³/ha. If a larger dressing is required, or if the manure is lumpy, it must be applied at least three months before fumigation so that it can decompose and mix with the soil.

4.1.7 POST FUMIGATION HYGIENE

Methyl bromide has no residual effect in soil and therefore any contaminant will flourish when introduced into the fumigated area. Every effort should be made to prevent re-infestation of the treated area via diseased planting material or by unhygienic actions. Only clean, healthy planting material and certified seeds should be used. Removal of diseased plants and their destruction outside the field is very important. Agricultural equipment which has been used in infested soil can also carry pathogenic micro-organ-isms and infest newly fumigated areas. All equipment should therefore be washed and/or fumigated separately prior to being introduced into the fumigated areas. Likewise, run-off water either from excess irrigation or after heavy rainfall may carry soil-borne pathogens from infested to fumigated areas, especially when sloped. A small trench, dug around the fumigated area, is a good preventative measure to direct the contaminated water away.

All the above-mentioned factors are inter-related and can be manipulated to a greater or lesser extent in order to guarantee the success of the fumigation.

5.1 BEHAVIOUR OF METHYL BROMIDE IN SOIL

The behaviour of the fumigant in the soil is subject to a number of factors determining distribution, absorption, degradation, dissipation and perme-ation through the film (Van Wambeke et al 1985).

A major principle of soil fumigation is to obtain the highest possible concentration of the fumigant, derived from the dosage, as quickly and evenly as possible, and to maintain this lethal concentration for the required exposure time. This principle is realized by the Concentration-Time Product equation.

5.1.1 CONCENTRATION × TIME PRODUCT (CTP)

In order to be effective, methyl bromide must remain in contact with the target organism for sufficient time and in sufficient concentration to kill the pathogen.

Treatment recommendations are given on the basis of a dosage of methyl bromide expressed in weight per given area over a length of time, e.g. 500 kg/ha for 48 h. Initially after the fumigation, the concentration of methyl bromide is uneven—it could be higher in the upper layer of 20–30 cm and low in the deeper layer. It might also differ in a horizontal plane. After some time the concentration reaches an equilibrium although it will still vary at different depths. Certain factors continue to influence and modify the concentration available in the soil to act against the soil pests, namely, leakage to the air and absorption by the soil. After 48 h the concentration will have fallen to very low levels. It is the combination of Concentration (c) and Time (t) of exposure which dictates whether the target organism will be controlled or not. The multiplication of these two factors gives a product known as CTP and is expressed as mg/h/l (milligram hours per litre) g h/m^3 or oz/h/1000 cu. ft. In many cases a certain CTP can be achieved either by exposure of the pathogen to high concentration of methyl bromide for a short duration or a lower concentration for longer periods. To a certain extent, it is possible to juggle with the two components of the formula to obtain the same CTP, and the same rate of control of many soil pathogens, although it is likely that, to be lethal, the gas concentration must remain above a minimum critical level for a period, the actual level depending upon the organism. At low concentration of the fumigant under laboratory conditions, prolonged exposure may achieve the necessary CTP but may not kill some important soil-borne pathogens (Ebben *et al* 1983).

The actual CTP needed to eradicate a pathogen can be determined under controlled laboratory conditions where little leakage or absorption occurs. In the laboratory, therefore, the theoretical and the actual CTP are similar. In the field, however, because of a constant decrease in concentration, it is impossible to calculate the CTP from a single measurement of gas concentration—it has to be measured several times throughout the fumigation and the CTP can then be calculated. Therefore, the actual value of the CTP is always smaller than the theoretical. Figure 5.2 illustrates a typical decrease in concentration of methyl bromide with time and the actual CTP would be equivalent to the area under the curve.

Pests differ considerably in their susceptibility to methyl bromide and in light of the above, it would be more accurate to relate the susceptibility of pests to the CTP and not to dosage or concentration. Also, there are differences between life stages of the same organism. Sclerotia of a fungus are always more tolerant than mycelia and a nematode cyst is more tolerant

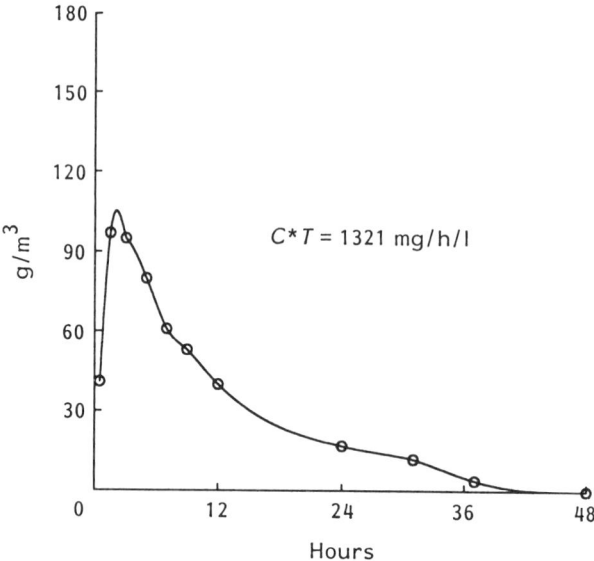

Figure 5.2. Concentration of methyl bromide in soil at a depth of 20 cm after application at a rate of 500 kg/ha

than a larva. In practice, several pathogens may be found in the area to be treated and therefore recommendations are aimed at killing the most tolerant stage or pathogen, making allowance for leaks and absorption. Generally speaking, the order of susceptibility is insects < weeds < nematodes < fungi < bacteria, the insects being the most susceptible and bacteria being the most tolerant to methyl bromide (Fletcher 1984). However, it is the author's opinion that control of weeds, insects and nematodes is achieved at a narrower range of CTP and that control of bacteria starts at a higher CTP than that shown by Fletcher (1984). Fungi are the primary objective for soil fumigation but there is a big variation in their susceptibility to methyl bromide (Ebben *et al* 1983, Munnecke *et al* 1978). Ebben (1983) has determined the susceptibility of selected soil-borne pathogens to methyl bromide in the laboratory and the following results were obtained: *Phytophtora* < *Rhizoctonia* < *Sclerotinia* < *Didimella* < *Pyrenochaeta* < *Verticillium* < *Phomopsis* < *Fusarium*. This was generally in agreement with the studies of Munnecke *et al* (1978) who also checked additional fungi, and divided the fungal sensitivities into three groups. The most sensitive were several species of *Phytophtora* and *Pythium*, moderately sensitive were *Armillaria*, *Sclerotium* and *Rhizoctonia* and least sensitive were *Whetzelinia*, *Fusarium* and *Verticillium*. Such relationships have been known for a long time from practical experiences; that is *Pythium* spp, *Rhizoctonia* spp and the Phycomycetes in general, are relatively easy to control with methyl bromide whereas *Fusarium* spp and sclerotia-forming fungi are more difficult to control.

To sum up—if the CTP of methyl bromide attained in a soil profile is sufficient to control the most tolerant species, it can be assumed that the fumigation is successful provided all the other factors necessary for fumigation are optimal.

A practical implementation of the CTP equation took place in Holland in the 1980's when impermeable films were used with reduced dosages of methyl bromide. In recent years, because of the ozone issue, extensive scientific work is taking place to optimize dosing and minimize emission (see Section 10.1).

5.1.2 MOVEMENT OF METHYL BROMIDE IN SOIL

Movement of methyl bromide after application is caused by mass flow and molecular diffusion. Because of the high vapour pressure of methyl bromide, the gaseous concentration of the chemical in the soil-pore spaces increases so rapidly that the movement of the chemical occurs first by mass flow, but the most important movement is by diffusion (Brown and Rolston 1980, Munnecke *et al* 1979). The behaviour of the gas in the soil is strongly influenced by simultaneously-occurring sink processes. The sink process can be reversible or irreversible. Reversible sink processes include physical absorption-desorption, chemical absorption-desorption and dissolution-distillation from the soil water. Irreversible sink processes include irreversible chemical bonding and decomposition (Brown and Rolston 1980). These processes determine to a large extent the movement of methyl bromide in its gaseous phase and govern its persistence in the soil. The irreversible sink process will be dealt with in Section 8.1.1 under inorganic bromide residues.

Factors affecting movement of methyl bromide include also temperature, soil type, (texture and composition), moisture content, porosity and physical barriers (Munnecke and Bricker 1978). The painstaking preparation of soil to seed-bed tilth are all directed towards the single objective of ensuring a uniform diffusion of methyl bromide. Low soil temperatures (below 8 °C at 15–20 cm depth) are unfavourable to diffusion and sterilization (Drosihn *et al* 1968) and give rise to uneven penetration (Galley and Hague 1967). For this reason, it is sometimes preferable in cool countries that soil fumigation for control of soil-borne diseases should be carried out at the end of the cropping period when the temperature is higher, rather than just before the planting of the next crop.

Soil porosity is of prime importance and good porosity will enable methyl bromide diffusion throughout the profile. Light soils are very porous and generally are not compact. Where compacted horizontal layers occur in soil, they act as barriers and prevent penetration of methyl bromide into the deeper layers. Moderate moisture content is desirable but, in wet soil penetration is markedly reduced and efficiency of fumigation impaired (Munnecke *et al* 1971). In fact, soil porosity and moisture content are inversely related to each other and the correct ratio of moisture/porosity should be maintained in the different soil types for good distribution of methyl bromide. Moisture is also important in the sorption of methyl bromide; more gas being absorbed by dry soils as opposed to moist soils. The composition and texture of the soil plays an additional role in the penetration of methyl bromide. Soil texture is often directly related to the success or failure of a field fumigation. It is difficult to study the effect of soil texture on fumigant behaviour in the field because other conditions, such as moisture content, variations in soil profile, and organic matter content, all vary so much that it is virtually impossible to get two fields for comparison that differ only in their soil texture. Different fractions of the soil differ in their sorptive properties to methyl bromide, the sorption decreasing in the sequence peat > clay > sand. However, the majority of the sorption that occurs in fumigated soils is attributed to organic matter. Generally, coarse-textured soils are more easy to fumigate successfully, than fine-textured soils such as clay loams and clays.

Depth of penetration of gas is influenced by all the factors mentioned above (Van Wambeke *et al* 1985). In a surface application there is usually rapid penetration of methyl bromide to the top 20–30 cm of soil, i.e.: into the depth of cultivation, where the highest concentrations can usually be measured within 1 h of the application. Gas concentration decreases with increasing depth and this profile is maintained throughout the exposure time (Klein, unpublished). Concentrations of methyl bromide decrease with depth, due to leakage and absorption. Most of the pathogens are found in the top 30–40 cm of soil and generally satifactory CTP's are achieved in this root zone. Drosihn (1968) found that diffusion of methyl bromide into undisturbed glasshouse soil occurred to a depth of more than 1 m within a few hours.

With injection methods of application (scc Scction 6.1), initial concentrations are highest at 20–30 cm either side of the injection. Upward transport causes the concentrations at 10, 20, and 30 cm from the soil surface to equalize after about a day. At depths below 30 cm, transport is slower. In deep injection, methyl bromide can penetrate up to 8 ft (2.4 m) in dry soil (Rackhom *et al* 1968, Kolbezene *et al* 1974).

The fact that methyl bromide is heavier than air is important only in the first stage after application in its downward movement, but the gas tends to move down when the fumigation takes place in sloping fields and in such cases, it is recommended to fumigate according to the contours. Its movement can sometimes pose problems if it enters greenhouse drainage systems and finds its way into neighbouring houses.

After removal of the sheeting, the remaining methyl bromide dissipates into the atmosphere. In the first hours there is a rapid decrease in concentration in the upper 30 cm soil layer but residues of methyl bromide have been detected in deeper layers of heavy soils even after 10 days, hence the range of aeration periods recommended by the manufacturers. Loss of the compound from the upper soil layers continues for a further 2–3 days, virtually disappearing from the top layers of medium and heavy soils and completely from sandy soils.

However, lower layers of heavier types of soil may retain some methyl bromide which can persist at such depths as 1 m or more for as long as 15 days (Drosihn *et al* 1968, Maw and Kempton 1973).

In light of the above, the degree of success of soil fumigation is not a matter of chance. It is based on a clear and definable set of rules and on factors whose influence can be analyzed.

5.1.3 EFFECT ON SOIL MINERALIZATION

Soil fumigation is a potent agricultural tool that may cause chemical and biological changes even in situations where there are no root diseases. Increased or decreased crop growth in fumigated soils can result from these changes.

As mentioned earlier, unexpected improved plant growth has been observed following soil fumigation beyond that attributable to pest control (Chen *et al* 1991, Altman 1970, Rovira 1975, Millhouse and Munnecke 1979). One of the reasons for this phenomenon is that the fumigation causes an increased concentration of nutrient ions in the soil solution. Mineralization is common to most disinfestation techniques, especially steaming. Following methyl bromide fumigation of the soil, there is an increase in extractable nitrogen, copper, manganese, zinc, magnesium, calcium, potassium, phosphorous, iron and chlorine in the soil and a resultant increase of most of the elements to be found in leaves (Millhouse and Munnecke 1979, Smith 1963). Although there is a small increase in available soil phosphorus, this does not result in an increase in the leaves of the crops, this being attributable to suppression or elimination of mycorrhiza needed primarily for phosphorous uptake.

Perhaps the most significant effect of methyl bromide is on the soil nitrogen. Some researchers attribute enhanced crop growth to the effect of methyl bromide on soil nitrogen levels. Fumigated soils are uniformly higher on NH_4-N content which accumulates and tends to persist in the soil; it is electrostatically bound to the clay soil constituents and does not leach. Methyl bromide is highly toxic to nitrifying bacteria (*Nitrosomonas* and *Nitrobacter*) and therefore the biological change of ammoniacal nitrogen to nitrate nitrogen is temporarily inhibited. This can cause the well recognized nitrification lag, manifested by high soil NH_4-N levels and a deficiency of NO_3-N. Nitrates are leachable and therefore in short supply. This factor may be a disadvantage in cool damp soils causing ammonia toxicity to some crops. So, depending on the season of the year, this effect can lead to increased or decreased plant responses, causing an adverse effect on some plants including excess of vegetative growth.

In practice, it is recommended to reduce nitrogen fertilization in the pre-planting basic dressing and only to add nitrate-N as a post-planting top dressing if needed. Nitrifying bacteria are not completely eliminated and eventually after 7–10 weeks, nitrification is resumed in fumigated soils. In most fumigations carried out under accepted conditions, adverse effects will not occur.

6.1   APPLICATION METHODS

Methyl bromide is applied in one of the two ways; basic manual application or mechanized injection. Whichever method is used, it is standard procedure to use clear plastic tarpaulins, currently polyethylene, to confine this volatile gas in the soil.

6.1.1   MANUAL APPLICATION

This method is a surface application in which methyl bromide is applied to the soil, which has been pre-tarped with plastic sheets either with a thick film of 0.1–0.15 mm for multiple use or with a thin film of 0.03–0.05 mm for a single use. The application involves the use of pressurized cylinders of different sizes, or cans.

6.1.1.1   The Hot Gas Manual Method

The main method of manual application is the so-called 'hot gas' method, where liquid methyl bromide from cylinders is vaporized in a vaporizing device (heat exchanger) prior to its introduction under the plastic sheeting. The gaseous methyl bromide is transported to the area to be fumigated by a delivery system, either flat, perforated polyethylene tubing (layflat tubes) or drip-irrigation lines pre-placed on top of the soil before spreading the plastic film over them.

Worldwide (outside the USA) this is the main method of application and almost the only one used in greenhouses. This method is especially suitable for treating greenhouse soils for operator safety because the fumigant is

Figure 5.3. Sheeting over layflat dosing strips prior to fumigation of a green-house by the hot gas method

delivered from outside the structure. In some countries, this method is also widely used for outdoor fumigations (Israel, South Africa, Kenya).

6.1.1.2   Cold Gas Manual Method (Raised Tarp Fumigation)

Methyl bromide in pressurized cylinders or under natural pressure in cans is applied to soil under tarps raised above the soil surface. The liquid methyl bromide is discharged into a series of small vessels (evaporating jars) from which it volatilizes beneath the sheeting. This method is most suited to small plots.

In the past this method of fumigation was very common. Nowadays, its usage with cylinders is limited but, fumigation using cans is common in a number of countries. Methyl bromide in cans of 1 lb (454 g) and of 1.5 lb (681 g) is used to fumigate small areas such as tobacco seedbeds (Zimbabwe, Brazil, Yugoslavia, Greece) but also bigger areas like greenhouses in Greece and Turkey. The cans are spaced on the soil before placing the sheeting. the cans are subsequently pierced with a special can opener to release the gas. The vapour pressure of methyl bromide forces the fumigant into a trough or any receptacle from where it evaporates, spreads over and penetrates into the soil.

The cold gas method is less efficient under marginal conditions than the hot-gas method (Galley and Hague 1967, Haas and Klein 1976). Heated fumigant penetrates the soil more quickly, to greater depth and spreads more evenly, especially at low soil temperatures. In addition, less residual bromide (see Section 8.1.1) occurs in the soil (Haas and Klein 1976).

Figure 5.4. Dosing from a can under raised sheeting to treat a small seed bed

6.1.1.3 Fumigation of High (Walk-through) Tunnels

The term 'high tunnels' refers to semicircular plastic houses, much smaller than conventional plastic greenhouses, a few metres wide, about 2 m high and 30–50 m long. Such tunnels are numerous worldwide.

High tunnels can be fumigated either by conventional soil fumigation as described earlier or by introducing the gaseous (vaporized) methyl bromide into the space of the structure, allowing the gas to settle and penetrate into the soil that has been prepared for fumigation. This 'space fumigation' of the tunnel structure can only be used in tunnels covered with one piece of new sheet which has no holes or tears and if the doors can be sealed effectively.

In hot climates, liquid methyl bromide can be discharged through a polyethylene pipe placed inside the tunnel and the liquid, sprayed through nozzles or holes, volatilizes immediately. The dosage given is the same as for conventional soil fumigation and the same effect is achieved.

There are some advantages of the high tunnel fumigation method:

(a) A saving in plastic sheeting as compared with the conventional soil fumigation method.
(b) The convenience of the method—no labour is required to cover the soil and to remove the tarp after the fumigation.
(c) A more effective fumigation is achieved along the edges of the plastic walls because there is no need to remove the plastic sheet. Thus re-infestation is reduced.

6.1.1.4 Single Tree Site Application

Another method is the 'single tree site' fumigation through a single point deep injection using a special probe (Anon 1983a, Munnecke 1977). This entails using a perforated probe that is pushed into the ground to the desired depth. A metering device attached to the cylinder or to the probe is used to measure one pound of methyl bromide, and this liquid is forced down through the probe into the ground at a marked planting site of each tree (Anon 1983b). This single-point application is useful when a single or small number of trees become diseased and have to be replanted. It is sometimes used when replanting an entire orchard. It provides excellent distribution of methyl bromide throughout the soil profile. For maximum benefit, efficacy can be increased by covering the area with a tarp and applying a small amount of methyl bromide beneath the tarp as a surface application.

6.1.2 MECHANIZED INJECTION

In this method, methyl bromide from cylinders is applied by injecting the fumigant through chisels set 30 cm apart to a depth of 20–25 cm into the soil, the treated area being simultaneously sealed by polyethylene sheeting approx. 30 micron (0.03 mm) thick (shallow injection). Tarp-laying fumigation rigs allow much quicker application of the fumigant than by manual means. Sophisticated injector/tarp-laying machines enable large areas to be fumigated accurately, efficiently and safely by a mechanized glued-strip system. Successive tarps are glued together at each pass of the machine, one edge being buried during application, the other glued to the adjoining sheet, until the entire field is fumigated in one continuous operation with the outer edges buried in soil. The thin plastic sheets can be used only once.

Another method of mechanized injection is deep placement of the fumigant without covering the treated area with plastic sheets (non-tarp chisel application). The fumigant is injected to about 0.6–0.9 m deep with up to three shanks 1.5–1.8 m apart and the shank channels tightly compacted. The deep placement of methyl bromide is aimed at killing soil-borne pathogens which may be found in the deeper layers of the soil, such as *Armillaria mellea* (oak root fungus). This is the only practical way to bring the gas to where it is required in sufficiently high concentration and to control the deeply buried inoculum (Kolbezene *et al* 1974). The treatment is carried out mainly prior to replanting deep-rooted perennial crops such as in deciduous orchards, vineyards and other plantations. In the USA it is carried out on a broad-acre basis by injecting the fumigant to approx. 0.9 m deep and through 2–3 chisels spaced 1.5–1.8 m apart in a continuous line pattern, regardless of the future rows. In South Africa the same technique is used through a single shank which injects the chemical in a line pattern exactly where the rows will be placed.

Although all methods of methyl bromide application are in use for replanting fruit trees, the most common is mechanized injection. This can be applied with polyethylene strips of approx. 2 m wide, if spacing between the rows permits, thus giving protection to the young trees until they are well established. Savory (1966) concluded that it is characteristic of trees affected by replant disease that the more severe retardation of growth occurs in the first two years after being planted in the affected site. More often, however, the deep placement technique without tarp is used.

This does not produce adequate lethal concentrations of methyl bromide in the top layer of the soil. For complete control of pathogens in this layer (when deemed necessary) the treated area should be tarped and the tarp left in place for at least a week. Sometimes a complementary treatment with a nematicide is carried out.

Mechanized injection is carried out in outdoor fumigation only, and is the predominant method in the USA. It is also used in some European countries (Italy, France, Spain, UK), and in Israel, Australia and South Africa. Mechanized application is generally restricted to qualified contractors/ custom applicators.

Field fumigation, both by manual methods or by mechanized injection, may be done in two ways:

(a)   As a single overall operation, the tarps being joined or glued to each other.

(b)   Alternate-strip fumigation: alternate strips treated and the remaining strips then treated after 48–96 hours. In the manual method one edge of the tarps is freed, pulled over the adjacent strip and sealed on the opposite side. In mechanized injection, the existing tarps are removed

Figure 5.5. Mechanized injection for broad-acre fumigation

from the field and the alternate strips are then treated ensuring good overlap on the previously-treated strips.

6.1.3 NARROW-STRIP FUMIGATION

Most fumigations, whether indoors or outdoors, are carried out on a broad acre basis. In many situations however, the fumigation is carried out in mulched strips (strip fumigation) of 0.80–1.20 m where the film is retained as a mulch for the lifespan of the crop. Fumigation of such strips is for row-crops, such as Cucurbits and Solanaceae and for strawberries at times, and is carried out by mechanized injection with 2–3 tines or by the hot-gas method, using a pre-placed drip irrigation system.

While narrow-strip fumigation is economical in terms of saving considerable amounts of methyl bromide as compared to broad-acre fumigation and in eliminating the need for special plastic, there is a potential problem in using this method. Under the same conditions and dosages the concentration of methyl bromide under the plastic strip decreases faster than under the wide tarp of broad-acre fumigation. This is due to quicker diffusion and escape of the gas sideways through the edges of the tarp (Klein 1989). This is known as the 'edge effect' and stems from the high ratio of circumference to the area of the treated strip. The narrower the strip, the more evident is this 'edge effect'. Consequently, while the more susceptible soil-borne pests such as weeds, nematodes and certain diseases, will be eliminated, difficult-to-control fungi requiring higher concentrations may not be completely controlled. Therefore, being somewhat less effective than broad-acre fumigation, the strip fumigation method is not recommended for greenhouses or for nurseries nor when growing outdoor high-cash crops such as vegetables and flowers. Nor is it recommended in cases of high infestation, or when the anticipated benefit of blanket fumigation will justify its cost. Nevertheless, soil fumigation in strips will provide satisfactory results if approached with knowledge and comprehension of all its aspects.

7.1 FORMULATIONS FOR SOIL FUMIGATION AND DOSAGE RATES

7.1.1 METHYL BROMIDE/CHLOROPICRIN 98/2

This formulation contains 98% of methyl bromide by weight and is the formulation of choice for soil fumigation. The added 2% chloropicrin (tear gas), serves only as an efficient warning agent. This formulation is applied either manually for cold or hot-gas methods or mechanically for injection.

In France the use of chloropicrin is forbidden by law and the fruity-smelling amyl acetate is used as a warning agent.

Figure 5.6. A field under narrow-strip fumigation

7.1.2 VARIOUS MIXTURES OF METHYL BROMIDE AND CHLOROPICRIN

For mechanized application only, various mixtures are used. The percentage of chloropicrin is increased in some formulations above 2%, because its own fungicidal properties render the mixtures more synergistic. The most common mixtures are 70/30 or 66/33 methyl bromide/chloropicrin w/w and they are especially effective in controlling *Verticillium* wilt (Wilhelm *et al* 1974). The proportions of the fumigant components may be tailored to individual field requirements and the chloropicrin component can be increased up to 50%. When using these mixtures, the instructions are somewhat different from those of the standard 98/2 formulation.

Dosage rates for the formulation Metabrom are set out in Table 5.1. 100% methyl bromide is never used for soil fumigation.

8.1 SPECIAL PROBLEMS ASSOCIATED WITH SOIL FUMIGATION

8.1.1 INORGANIC BROMIDE RESIDUES IN THE SOIL

8.1.1.1 Accumulation of Bromides in Soil

Naturally-occurring bromides in soils are generally present at less than 10 ppm, the level depending on soil type and geographical situation.

Following fumigation, some of the methyl bromide dissipates into the atmosphere and part of the remaining fumigant undergoes a variety of processes including decomposition in the soil, resulting in the production of inorganic bromides which may reach levels of 20–40 ppm (Maw and Kempton 1971, Anon 1985).

Clay particles and soil organic matter play the most significant part in this decomposition in the soil during fumigation (Brown and Rolston 1980, Haas and Klein 1976, Malkomes 1971). Therefore, rates of bromide production are greatest with clay soils, intermediate with loam and least with sand. Daelmans (1978) calculated that the rate of degradation of methyl bromide in soil was about 6–14% per day at 20 °C.

The inorganic bromide formed in soil in this way can persist for considerable periods of time, even when normal cultivation and watering regimes are in operation (Kempton and Maw 1970), and small but significant increments above the untreated level are detectable more than one year after the application. These soluable inorganic bromides are available for uptake by plants and may accumulate in edible crops, but even high bromide levels in soil are not phytotoxic to vegetable crops, nor to most other plants (Maw and Kempton 1971).

The level of inorganic bromides which results in soil and plants from fumigation is related to the following factors (Hoffmann and Malkomes 1979, Kempton and Maw 1973):

1. Soil type, its structure and texture
2. Percentage of organic matter in the soil
3. Rates of methyl bromide applied
4. Exposure time
5. Soil temperature
6. Interval beteen fumigation and planting
7. Moisture content of the soil
8. Type of plastic sheeting
9. Soil compaction
10. Irrigation practice and/or rainfall
11. Method of application
12. Fumigation frequency.

Most of these factors can be manipulated to a greater or lesser extent (Van Wambeke *et al* 1985, Van Wambeke 1974, Gums 1989) without affecting fumigation efficacy, in order to reduce the level of bromide ions to a minimum.

It is worth noting that it is not only methyl bromide that increases bromide levels in soil. Other fumigants such as methyl isothiocyanate releasing compounds and dichloropropene were found to do so to some

Table 5.1. Pest-controlling dosage rates of Metabrom 980 in kg/ha according to soil type

Crops	Dosage in KG/HA by soil type[1]			Aeration	Remarks
	To control nematodes, annual and perennial weeds[2] and broomrape	To control damping-off fungi, e.g. *Rhizoctonia, Pythium* spp., *Thielaviopsis basicola, Phytophtora* spp.	To control fungi causing rots and wilts[3] e.g. *Sclerotium* spp., *Pythium* spp., *Phytophtora* spp., *Sclerotinia* spp., *Pyrenochaeta* spp., *Armillaria* spp., *Verticillium* spp., *Fusarium* spp., *Curvularia* spp.	In days, by soil type and temperature[4]	
Nurseries of vegetables, tobacco and flowers	350–500	500		7–14	Do not fumigate soils to be used for celery nurseries.
Vegetables: cucumbers, melons, tomatoes, eggplants, peppers. courgettes (squash)	350–500	500	750–1000	7–14	

Crop				Aeration period (days)	Remarks
Strawberries (nursery and field)	350–500	500	500–750	7	
Leafy vegetables (lettuce, celery, etc.)	350–500	500	500–750	14	Maximum dosage for leafy vegetable crops is 750 kg/ha. Big-vein disease is also controlled.
Flowers: annual and perennial, cut flowers, ornamentals	350–500	500	750–1000	7–14	Even light soils must be leached before planting carnations.
Bulbs and corms (on light soils only)	350–500	500	750	7	Do not plant bulblets in fumigated soil.
Citrus replanting			500	14	
Deciduous replanting			750–1000	14	

[1] When a range of doses is given, the smaller dose relates to light soils, the larger to medium and heavy soils.

[2] Purple nutsedge (nutgrass, Cyperus spp.) corms and seeds of horseweed (Erigeron, Conyza), mallow (Malva) and legumes are not efficiently controlled.

[3] The dose rate to control Fusarium in all soil types is 1000 kg/ha.

[4] For light soils and/or high temperatures, a shorter aeration period is sufficient. For medium and heavy soils, and all soil types at low temperatures, a longer aeration period is required. The long aeration is also desirable for direct seeded crops. If rain is expected during the aeration period, do not remove the plastic sheets entirely but allow for aeration while protecting the soil from direct rain by opening at the sides only.

exent, and soil sterilization by steaming was also found to raise bromide levels in soil and crops by as much as six times that present before steaming (Roorda Van Eysinga and DeBes 1984).

8.1.1.2 Uptake of Inorganic Bromides by Plants

Inorganic bromides present in soil can be taken up by plants along with other nutrients. The amount of bromide in the soil directly influences the bromide content of the plant growing on it, resulting in the so called 'bromide residue' problem. Plant species differ in their ability to take up bromide from the soil and the distribution to different plant organs is not uniform. The accumulation of the bromide ion is greater by far in the leaves than in the fruits or roots; older leaves have higher bromide content than younger leaves (Maw and Kempton 1973, Hoffman and Malkomes 1979, Kempton and Maw 1973, Van Wambeke 1974). In lettuce, for example, the outer leaves accounted for the bulk of the bromide present in the plant (Kempton and Maw 1972). Therefore, concern has been expressed specific-ally for leafy vegetable crops, e.g. lettuce, celery etc., the edible part of which may have a high bromide level at harvest (Roorda Van Eysinga and DeBes 1984). Vegetable fruits (e.g. tomatoes) seldom accumulate bromides above the accepted level (Maw and Kempton 1971, Hoffman and Malkomes 1979, Vanachter 1979), and much lower bromide levels are found in the fruit than in the leaf (Maw and Kempton, 1973). The residue in the tomato fruit decreases with time. The accumulation of bromide in the plant organs is related to the amount of transpiration and therefore the content in the fruit is lower than that in the leaves (Hoffman and Malkomes, 1979). In a survey carried out in Belgium it was demonstrated that pre-planting waiting periods had a clear influence on decreasing levels of bromides in lettuce (Vanachter *et al* 1975).

To minimize residues of inorganic bromide derived from soil fumigation, in order that stipulated tolerance levels are not exceeded, the following recommendations should be adopted, especially when growing leafy vege-table crops:

- Fumigation should take place only when essential and in no case should there be more than one treatment per annum, per individual area.
- A minimum recommended dosage to achieve control of the target organisms should be applied.
- The recommended exposure time should not be exceeded.
- As long an interval as possible between fumigation and planting time should be allowed (Maw and Kempton 1973, Vanachter 1979, Van Assche and Van Wambeke 1981).
- Leafy vegetables should be avoided as the first crop (Butters and

Fletcher 1985). Crops such as tomatoes as the first crop after methyl bromide fumigation are preferable (Vanachter 1979).

● Soil should be leached after fumigation (see Section 8.1.1.3). Leaching highly organic soils is difficult so leafy vegetable crops should not be grown in such soils within two years of treatment with methyl bromide.

When implemented, the above recommendations have in practice proved to be very effective in reducing bromide residues in treated soils and in crops.

It should be noted that fumigation with mixtures of methyl bromide and chloropicrin reduces the uptake of inorganic bromides because of the competition between the bromide and chloride ions (Vanachter *et al* 1978).

The following is a list of selected crops with their Maximum Residue Limits [MRL] of inorganic bromides expressed in mg/kg as accepted by the Codex Alimentarius Commission:

Celery	300
Cucumber	50
Garden pea	500
Lettuce	100
Squash	200
Strawberry	30
Tomato	75

8.1.1.3 Leaching Bromide Residues

Soil leaching after methyl bromide fumigation is needed for two reasons:

(a) To eliminate toxic residues which cause phytotoxicity to sensitive crops.

(b) to prevent accumulation of bromides in the edible parts of vegetables, especially in leafy crops.

While most plant species display a considerable tolerance to bromide ions present in the soil after treatment with methyl bromide at commercial rates, carnations in particular, but also garlic and onions, are susceptible to bromides in the soil following fumigation, and display phytotoxic symptoms (Drosihn *et al* 1968, Haas and Klein 1976, Kempton and Maw 1971, Kempton and Maw 1974). Leaching the soil can reduce the levels of soil bromides available to plants to below phytotoxic concentrations (Drosihn *et al* 1968, Haas and Klein 1976, Kempton and Maw 1970). Other plant species sometimes show growth retardation in fumigated soil, but the reasons are not related to inorganic bromide residues in the soil.

Inorganic bromides in the soil are very mobile and tend to move as water moves in the soil (Brown *et al* 1979). Leaching is therefore an effective means of removing bromides which can be washed away from the plant-root danger zone, thus reducing the levels of inorganic bromides available to the plants. Leaching can in some cases lower the water-soluble bromide content to a value close to that existing prior to the fumigation, and in other cases, by more than 50% in the top 50 cm of the soil (Anon 1985).

Leaching must not be done too soon after fumigation and about a week should be allowed before starting. This will ensure the complete escape of the gas from deeper layers of the soil. The amount of water required to leach inorganic bromides successfully will vary according to soil type. The irrigation rate must correspond to each soil's water intake capacity—too high a rate would cause undesirable run-off.

Experiments with different soil types showed that in light soils there were no difficulties in significantly reducing the bromide concentration. Results suggested that 75 mm of water in one operation was enough to cause a reduction of 52% and 152 mm reduced the bromide by 85% (Maw and Kempton 1973, 1971). However, the amount of water required for a very efficient leaching of bromide is 200–400 mm according to soil type (Vanachter *et al* 1981a). Efficient leaching for heavy soils requiring higher rates, is only possible if the soil is permeable or well drained. The presence of a good drainage system helped to lower the residual bromide concentration of the ground water considerably (Gums 1989). It is interesting to note that the use of drip irrigation in tomatoes significantly reduced the uptake of soil bromine ions. The constant flow of water together with the nutrition solution, prevents the plant roots taking up the extractable bromide from the soil solution and thus the bromide accumulation is lowered.

In Holland, leaching of soil following fumigation was very common until the use of methyl bromide was discontinued in 1991. The leaching resulted in the pollution of surface water in the canals with bromides. In Holland, the conditions are very unusual with a soil level lower than sea level and the water from canals not flowing to the sea or rivers. The glasshouse growing area is very dense in the Westland and the leaching of greenhouse soil caused bromide pollution of the surface water, which was to be recycled. Research carried out in the Antwerp region of Belgium showed that, where the drainage water is directed to rivers, there is a fast dilution effect and no build-up of bromide residue was recorded in the surface water. It was therefore concluded that, under Belgian conditions, the bromide concentration of surface water was not a problem to the environment (Vanachter *et al* 1981b).

8.1.1.4   Inorganic Bromides and Human Health

There are stipulated tolerance levels for inorganic bromides in many vegetables and fruits expressed as the Maximum Residues Level (MRL).

MRL's are expert assessments of the maximum concentration of pesticide residues likely to occur in or on a commodity, either resulting from the approved use of a pesticide or arising from environmental sources. They usually apply to the whole product, not just the part normally consumed. The potential exposure of consumers to residues in food is assessed by comparison with the Acceptable Daily Intake (ADI). This is defined as the amount of pesticide which can be consumed daily in the diet over a whole life-time in the practical certainty, on the basis of all the known facts, that no harm will result. The ADI incorporates a large safety margin and is expressed on a body-weight basis.

An important question is whether inorganic bromides are prejudicial to public health.

The official (FAO/WHO) ADI for inorganic bromides for humans is 1 mg/kg of body weight per day. For a 60 kg person, this would correspond to an ADI of 60 mg bromide/day. In studies carried out in the Netherlands and Belgium, it was shown that an average intake of bromides was 7.6–8.5 mg per day (Coosemans and Van Assche 1983, Poulson 1983) which falls well below the FAO/WHO standards.

In the UK an assessment was made in 1982 of likely intake of bromide ion from all sources in the diet and it was concluded that the use of methyl bromide as a soil fumigant should not present a concern to the consumer (Ministry of Agriculture, Fisheries and Food, 1982). This was confirmed again in 1989 by the Advisory Committee on Pesticides (Anon 1990).

It is imperative to point out that the US Environmental Protection Agency (EPA) in their memo of July 1989 on their review of inorganic bromide tolerances, 'has determined that because of its long history in use as a human drug, its existing toxicological data base and its environmental ubiquity, inorganic bromide is not of toxicological concern'.

It has also been reported elsewhere (Anon 1991) that, in 1990, the EPA rescinded the need for residue limits for inorganic bromide arising from fumigation, since these were no longer considered to be of toxicological significance.

In the report of the Scientific Committee of the EEC for Pesticides on the use of methyl bromide as a soil fumigant, it was mentioned that bromide ion residues resulting from soil fumigation, may cause some local water pollution, e.g. in polder areas of the Netherlands, but, in most areas this is likely to be of minor importance (Anon 1985).

8.1.2 DECREASED GROWTH RESPONSE

Occasionally, a decreased growth response is observed in fumigated soils. This is defined as a decrease in growth and yield resulting from soil fumigation and can be regarded as a phytotoxic side effect of the fumigation (Chen *et al* 1991).

Phytotoxic effects resulting from the use of methyl bromide in soil fumigation may be attributed to the following:

(i)   The action on plants of the compound *per se*.
(ii)   The action of inorganic bromide formed by the breakdown of methyl bromide in soil.
(iii)   Indirect action through effects of methyl bromide on soil microflora.
(iv)   Effect on abiotic factors.

Many cases have, in fact, been cited in the literature of plant injury or growth retardation following planting in soil previously treated with methyl bromide. It was not possible in all cases to conclusively attribute such effects to methyl bromide fumigation and in some cases we know now that the reasons for the phytotoxicity were different from those specified in early publications.

Phytotoxicity due to methyl bromide *per se*, is very rare and may be caused if aeration is not sufficient. A very low concentration of methyl bromide may be left in the soil, especially in the deep layers of heavy type at low temperatures. When the recommendation for the aeration period is not followed, the methyl bromide moves upwards or, alternatively, the roots reach that zone and the injury occurs.

The issue of inorganic bromides has already been discussed earlier. Therefore the most common cause of decreased growth is the indirect action through the effects of methyl bromide on soil beneficial microflora especially mycorrhizae (Menge 1982).

8.1.2.1 Effect of Methyl Bromide on Mycorrhizae

Occasionally, plants growing in fumigated soils grow poorly, stunted and off-colour. The problem is acute when susceptible plants are started as seeds in fumigated (or steamed) soils. This is apparently due to the high phosphorous absorption capacity of some soils, which can be attributed to their high calcium carbonate and/or specific clay fractions. This phenomenon has been observed on a number of crops such as pepper, celery, carrots, citrus seedbeds, unrooted vineyard cuttings, peanuts, cotton, onions (Munnecke and Van Gundy 1979, Menge 1982, Krikun *et al* 1983), although some of these crops did not manifest the same symptoms under different soil conditions.

This syndrome is usually due to the elimination by fumigation of a benefical group of micro-organisms called mycorrhizae. These micro-organisms live in symbiotic association with certain plants and form an essential link between those plants and the soil environment. These mycorrhizal fungi assist the plant in nutrient uptake, particularly of phosphorous but also zinc and manganese. Soils with a high calcium carbonate content (and therefore with low available phosphorous due to its high phosphorus-absorbing capacity) are those in which the presence of mycorrhizae are most beneficial for crop growth. Therefore disturbing these symbiotic relationships by fumigation, adversely affects plant growth. However, when rooted seedlings of the same sensitive crops are transplanted into the fumigated field, these symptoms are not evident and often enhanced growth will occur presumably because the mycorrhizal fungi have been established on the roots prior to transplanting (Munnecke and Van Gundy 1979). Mycorrhizal fungi are extremely susceptible to methyl bromide. Menge *et al* (1979) have reported that *Glomus* spp. was more sensitive to methyl bromide than all of the soil-borne plant pathogenic fungi previously mentioned—twice as sensitive as *Phytophtora* spp., about four times more sensitive than *Verticillium*, and about nine times more sensitive than *Sclerotium rolfsii*. They concluded that it is unlikely that methyl bromide dosage could be reduced sufficiently to allow survival of mycorrhizal fungi without at the same time allowing survival of pathogenic fungi.

Frequently these disorders can be overcome by large additions of phosphates or by inoculating the soil with mycorrhizal fungi (Ross and Harper 1970, Ross 1971). Especially effective is fertilization with phosphoric acid. It was proved (Hass *et al* 1987) that the development of stunted peppers was greatly alleviated by introduction of phosphoric acid into the trickle irrigation system. It has become a standard practice in the Arava Valley, in the south of Israel, to provide the pepper crop with phosphoric acid following pre-plant fumigation with methyl bromide.

8.1.2.2 Influence on Other Beneficial Micro-organisms

While methyl bromide exhibits high toxicity to a broad spectrum of pathogens including some beneficial micro-organisms, it is selective to some other beneficial micro-organisms in the soil. Therefore, it is not right to call methyl bromide a sterilant as it is sometimes described in the literature—it is actually a partial sterilant. Because of this property, methyl bromide has been used in fumigation of mushroom compost as a substitute to 'peak heating' by steam. In the composting process, different groups of micro-organisms take a part in preparing the compost as a substrate for growing the mushroom. In this process methyl bromide is used and it is very important that the gas does not affect beneficial micro-organisms but controls the pathogens (Hayes 1969).

Actinomycetes is one group of micro-organisms that is less sensitive to fumigation and not affected by commercial rates of methyl bromide (Chen *et al* 1991, Van Assche 1971). Also, many bacteria are not controlled by fumigation and many saprophytic nematodes are able to repopulate the soil a short time after fumigation. A possible explanation is that these beneficial organisms are found in the soil in much greater numbers compared with the pathogenic populations. Following fumigation they survive in greater numbers than the pathogenic forms. Since they can live and multiply in the absence of living plants unlike the pathogenic forms, they serve in many cases as antagonists to reinfestating agents, provided that the infestation is not massive and that the disinfestation dosage is not excessive.

Perhaps the most known case of tolerance to methyl bromide is that of *Trichoderma* spp. This beneficial fungus is extremely tolerant to methyl bromide as well as to other fumigants (Chen *et al* 1991). Commercially recommended dosages do not affect *Trichoderma* and even 3–4 times the recommended dosages do not reduce its population in the soil significantly (Klein, unpublished).

When beneficial micro-organisms are suppressed following fumigation, it takes 30–60 days for these organisms to become re-established and the 'biological buffering' is presumably re-established.

9.1 SOIL FUMIGATION WITH METHYL BROMIDE IN REPRESENTATIVE CROPS AND SITUATIONS

Methyl bromide is applied on a very wide range of crops in greenhouses and in the open field. In the following, the rationale behind the use of methyl bromide on four representative crops and situations will be described.

9.1.1 METHYL BROMIDE IN SEEDBEDS AND NURSERIES

Propagation material must be disease-free in order to guarantee the future commercial crop. Therefore seedbeds and nurseries must get the best soil treatment in order to prevent large crop areas from being infested through diseased seedlings. Such infested fields are potential areas for unnecessary use of pesticides and the resulting negative implications on the environment. Seedbeds are by nature relatively very small areas, compared to the commercially-grown field and the cost of fumigation with methyl bromide per single plant at the nursery stage is negligible.

Tobacco seedbeds are a good representative example of the value of methyl bromide in seedbeds. The first noticeable advantage of fumigation is the weed control achieved, which is especially important at the seedbed stage. Weeding is a slow, tedious, laborious, costly job which is likely to injure the young plants, reducing the stand.

Seedlings in a fumigated seedbed are vigorous, disease free and uniform. There is a better stand, the yield of seedlings is increased, and therefore less area of seedbed is needed to produce the same amount of seedlings, which on planting out in the field provide a good basis for a successful crop.

In fumigated tobacco seedbeds, it has been shown that, due to a stimulation of the seedlings, there is a shortening of the growing period by 10–14 days. This has also been observed in vegetable and other seedbeds. This phenomenon results because of the absence of damage by soil-borne pests and because of a better availability of nutrients in the soil. Transplanting such high quality seedlings to the field, results in a better take and stand. Such healthy seedlings will survive transplanting shock and will be less vulnerable to nematodes and disease. Their roots function better in absorbing water and nutrients, resulting in rapid growth and eventually higher yield.

Indeed, there is a total dependence on methyl bromide in many countries where seedlings are raised on soil.

9.1.2 METHYL BROMIDE IN OUTDOOR, INTENSIVE, HIGH-VALUE CROPS

Infestations of injurious root disease organisms occur in all agricultural lands and often prevent successful cultivation of preferred crops. Today, with the growing trend of specialization in one or a few economically interrelated crops, only dependable year-after-year production of high quality produce will justify the high investments in such crops. Soil-borne diseases constitute a major hazard to this stability.

Strawberries are a typical high-cash intensive crop. The introduction and continued use of methyl bromide or methyl bromide-chloropicrin mixture for strawberry cultivation typifies the procedure used in other high-value outdoor crops, like out-of-season vegetables, flowers, etc. In strawberries, which suffer from many soil-borne pathogens, preplant soil fumigation has removed the major risk to successful production by controlling the serious and widespread pathogens. What was considered a normal yield in California a few decades ago—3–5 tons/acre (7.5–12.5 tons/ha) represented only a fraction of the potential yield. Today, 25–30 tons/acre (62.5–75 tons/ha) and even more are obtained after fumigation and the grower now usually continues to plant on the same land year after year.

To quote an article by Wilhelm and Paulus (1980): 'The impact of soil fumigation in California is reflected in the common reference to distinct eras "before fumigation" and "after fumigation" among growers who remember the difficulties, and frustrations of strawberry cultivation before the introduction of the soil treatment'. The features most distinctive of the period before fumigation were the speculative nature of strawberry cultivation, the uncertainty of achieving economic yields, and a constant search for new land. The yield potential was far from being realized and it can be said with certainty that soil-borne diseases claimed the greater part of the harvest (Wilhelm and Paulus 1980). Soil fumigation has been an integral part of strawberry cultivation in California since about 1960. In addition to achieving high, dependable yields through the control of *Verticillium* wilt, other soil-borne diseases and weeds, the practice has made possible far-reaching changes in crop management e.g. clear mulching, drip irrigation, fertilization etc., and this has been of immense value to the industry.

In many agricultural regions, especially those which export their produce to foreign markets, there is a full dependence on soil fumigation with methyl bromide. In California, Israel and other countries where strawberries are grown as an annual crop, both the bearing and nursery fields are fumigated before each crop is planted. This tendency is gaining momentum in more countries. Many fields have been fumigated and replanted with strawberries even 15 times without build up of toxic residues or destruction of desirable soil microflora (Wilhelm and Paulus 1980).

In strawberries, the excellent results of soil fumigation with methyl bromide greatly contributed to a realization of the genetic yield potential. In fact, it made redundant the need to breed cultivars resistant to soil-borne pathogens, and allowed breeding efforts to concentrate on factors of fruit production and quality. In other crops such as tomatoes, the demand for resistance has over-ridden all other considerations and resistant cultivars are being bred, without eliminating the need for and the importance of soil fumigation with methyl bromide.

Another important point is whether and how a single soil fumigation benefits more than one crop. The carry-over effect to successive crops depends on the proper sequence of the crops, their susceptibility to soil pathogens, correct management, degree of success of the fumigation before the first crop and post fumigation practices. Little mention is made in the literature of this aspect, but Wilhelm and Paulus (1980) mention several crops that were planted subsequent to strawberries, that had benefitted from the fumigation. In Holland it was recommended that a main crop could be followed by a leafy vegetable crop in the same season. This practice is applicable to countries like the UK and Belgium today. The residual effect of methyl bromide is kept while the bromide residue problem in leafy vegetables diminishes or disappears. In Israel a lot of research has been carried out in this area, with the intention of incorporating methyl bromide in crop rotation. Market vegetables like tomatoes, potatoes, peanuts, carrots, radishes, sweet potatoes and wheat have all benefitted even as the third crop after fumigation. In one particular case four successive potato crops over a four-year period, enjoyed increased yields and quality, albeit in decreasing order (Nachmias *et al* 1995). Much of the economic strategy in fumigation of outdoor crops in Israel, in certain regions and conditions, is based on growing several crops, taking advantage of the residual effect of a single treatment. The expense of the fumigation is more than recovered by the first crop, while the others benefit from the results (Nachmias *et al*

1994). Alternative fumigants to methyl bromide do not have this ability to enable residual protection to more than one crop.

9.1.3   METHYL BROMIDE IN GREENHOUSES

Greenhouses constitute perhaps the most intensive area of agriculture—a great deal of money is invested in structures, equipment and other inputs to ensure favourable conditions for plants to give maxiumum results. At the same time optimal conditions are provided for pathogens, which, if not controlled, can build up to levels that completely destroy an entire crop or render it unprofitable. The monocultural nature of greenhouses with a limited number of crops, on one hand, and on the other, the many soil pathogens, often with a wide range of hosts, guarantee the existence and continuity of such problems. Another factor of relevance is the very short interval between crops. The grower has to choose a chemical which is quickly effective and rapidly dissipates from the soil after treatment. At present, the farmer's choice is weighted towards methyl bromide. It is thus possible to have an interval of no more than 7–10 days between crops, enabling a very rapid change over and maximum exploitation of the greenhouse resource. Steaming of the soil provides a similar solution though it is not as effective in deep soil as methyl bromide, and it has some other undesirable properties, especially the very high cost involved. Similarly, artificial growing media offer solutions to many of the problems of soil infestation, but again, this method is extremely expensive and not without problems. In Holland, soil fumigation with methyl bromide in greenhouses was replaced by artificial media and steaming for reasons connected with special conditions relevant only to Dutch agriculture. (e.g. pollution of surface water and air near the greenhouses).

9.1.4   METHYL BROMIDE TREATMENT BEFORE REPLANTING FRUIT TREES

Problems in replanting fruit trees are universal and often occur where trees are replanted in orchards previously planted with the same cultivar. The trees fail to grow satisfactorily even though recommended cultural practices are being followed (Parker *et al* 1966). The incidence and severity of replant problems, evidenced by stunting and/or tree mortality, may appear in a range of fruit tree orchards, such as in stone and pome trees, in tropical and subtropical orchards, and in vineyards.

The etiology of replant diseases of fruit trees appears to be complex, but it can be concluded from accumulated research results that the causal agents are soil-borne organisms including plant parasitic nematodes, such as *Pratylenchus* on apples, cherry and peach and parasitic fungi, such as

Rhizoctonia solani in apple and cherry, *Phytophthora* spp. and *Pythium* spp. on peach, cherry and apple, *Thielaviopsis basicola* on cherry. Several *Fusarium* spp. and many other soil-borne organisms have been shown to be associated with roots and soils of poorly grown apples (Mai and Abawi 1981). Other abiotic factors are also involved, but usually as secondary contributing factors.

Two general types of replant diseases have been recognized—specific and nonspecific (Hoestra 1968, Savory 1966). Nonspecific replant diseases affect several fruit tree crops, are present in most orchard soils, have a patchy distribution in the orchard and are associated with high numbers of plant parasitic nematodes. In contrast, specific replant disease does not correlate with plant parasitic nematodes, has as even distribution throughout the orchard and is limited in activity to one crop or to closely related fruit tree crops. For example: in Specific Apple Replanting Disease, the soil will grow anything except apple and probably pears. In some countries, the syndrome is called 'soil tiredness' or 'soil fatigue'. Where replant problems occur, good results have been obatined with control measures including the use of wide-spectrum soil fumigants, such as chloropicrin, saturation with formalin and methyl bromide; this last fumigant yields extremely satisfactory and reliable results, where both types of the replant problem exist. Trees in methyl bromide fumigated soil are more vigorous, have greener leaves, larger shoots and greater trunk circumference. They also have a deeper and more extensively branched frame root system with healthier feeder roots. Such root systems take up water and minerals from the soil more effectively and thus stimulate growth and yield.

Root pieces from deep rooted tree and vine crops can survive in soil. These roots die very slowly and often remain alive for several years. These residual roots are 8–16 times more resistant to fumigants than the pests or pathogens residing in them (Munnecke and Van Gundy 1979). When deep penetration is required, the 'seedbed condition' required for ordinary fumigation is usually too moist, because the soil at the lower depths will be too wet. Where preplant problems include the fungus *Armillaria mellea*, the nematodes *Tylenchulus semipenetrans* and *Meloidogyne* spp. and grape fan leaf virus, in deeply buried roots, an entirely different strategy is necessary for their control. The soil is dried during the summer, to enable deep penetration of the fumigant. This may result in the top 0.5 m of the soil being too dry and an additional treatment especially for this layer, may be necessary if target organisms are found there.

Many of the characteristics of methyl bromide, namely its broad spectrum efficacy, its ability to penetrate deep into soil and to reach root pieces left deep in the soil from the previous planting, and in which pests may be sheltered, and its rapid action, makes methyl bromide the preferred fumigant.

10.1 REDUCING EMISSIONS AND DOSAGES OF METHYL BROMIDE DURING SOIL FUMIGATION

The commonly-used films for methyl bromide fumigation e.g. Low Density Polyethylene (LDPE) are permeable to the gas and actually provide an inadequate barrier. By not preventing the emission of methyl bromide to the atmosphere during fumigation, the use of these films therefore requires relatively high dosages of the fumigant. The magnitude of this emission, 70–80% under extreme conditions, depends on factors such as temperature, thickness of the film and dosage. Worldwide, LDPE is the only film which is currently commercially available for this use.

On the one hand, methyl bromide is universally accepted as being vital to the viability of many crops under conditions of intensive agriculture and to the economy of many developed and developing countries; on the other hand, it is seen as being potentially harmful to the ozone layer. Thus it was requested in the Montreal Protocol of 1992 that, as a matter of urgent and utmost importance, research into reducing emissions of methyl bromide into the atmosphere during its use be examined to find ways of reducing fumigant dosage without affecting its efficacy in pest control.

A study on the permeability of different types of films was carried out in the US (Munnecke *et al* 1978, Kolbezene and Abu-El-Hay 1978), but the commercial use of impermeable films never materialised. More research into this question was followed in Holland (Van Wambeke 1984, 1989, De-Heer *et al* 1984) in the 80's resulting in the enforcement of the use of virtually impermeable films (VIF), with which the dosage of methyl bromide was reduced from $100 \, g/m^2$ to $40 \, g/m^2$ and the exposure time increased from 5 to 10 days. This application was in practice for a number of years until the use of methyl bromide in Holland was phased out in 1991.

Following the decisions of the Montreal Protocol, a new, much more detailed research programme has been started through a multi-national project carried out in several European countries and in Israel. In this project, new types of VIF are being tested for soil fumigation.

The practical importace of the CTP equation, (discussed in Section 5.1.1) is reflected in the usage of VIF. Under such films, the time (T) during which high concentrations of methyl bromide are retained in the soil is increased. In this way, using VIF enables the use of optimised dosage (C) resulting in an effective CTP and the same efficacy on pathogen control as compared with the standard treatment. Thus, the use of VIF for methyl bromide fumigation minimizes emission to the atmosphere during the fumigation.

Plastic sheeting can be divided into three categories according to their permeability to methyl bromide (Figure 5.7). Permeable sheets are characterized by a certain rate of permeation, until a certain point in time which is called the 'break through' point, when the rate of permeation changes from

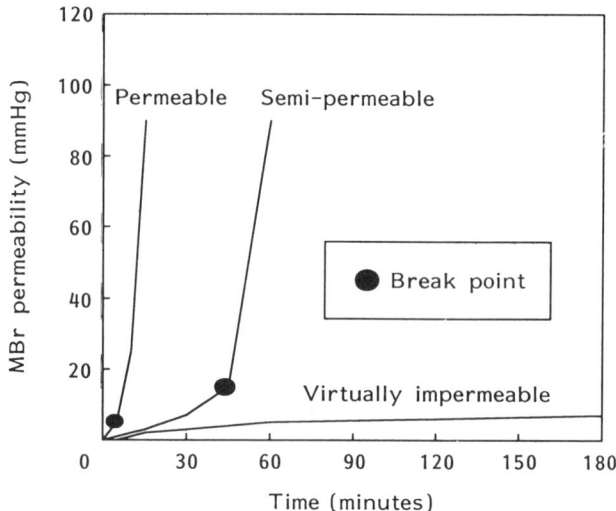

Figure 5.7. Permeability of plastic films to methyl bromide in a permeability cell

a linear to an exponential rate. Gas tight films are virtually non-permeable. The rate of permeation is extremely low and the 'break-through' point is not reached, whereas it is reached after a few minutes with permeable films such as the commonly-used LDPE. There is an intermediate category with films whose rate of permeation is slow and the 'break-through' point is achieved after a relatively long time. These films are defined as partially permeable to methyl bromide, e.g. polypropylene.

The multi-national research involves two VIF, co-extruded multi-layer films, 30 micron (0.03 mm) thick, which contain a barrier layer in the middle, 8–10 microns thick. The outer layers are LDPE. The barrier layer is made either of polyamide (nylon) or of ethylene vinyl alchohol (EVOH) both of which are virtually impermeable to methyl bromide. This extremely low permeation did not change even at 70 °C unlike with LDPE or HDPE. Permeability of poor-barrier films, such as LDPE and HDPE, increased exponentially upon temperature increase from 30–70 °C. In practice, temperatures of more than 50 °C can prevail under plastic mulch during sunny summer days in hot climates.

The efficacy of the VIF's was checked in Israel, both in broad-acre fumigation and in strip fumigation, under different soil types and climatic conditions in greenhouse and outdoors on the following crops: tomatoes (against *Fusarium* crown rot), muskmelons (sudden wilt), carnations (*Fusarium* wilt), basil (*Fusarium* wilt), cucumbers (yield decline), eggplants

(*Verticillium* wilt, Orobanche) and potatoes (*Verticillium* wilt) (Klein 1994, Gamliel *et al* 1994a, b, c, 1995).

Full dosage varied between 50 and 70 g/m² according to recommendations for standard treatments, and these were compared to half-dosages applied under the VIF's, and also to half-dosages applied under LDPE. The impermeable films retained methyl bromide in the soil for longer periods as compared with the standard treatments where the concentrations quickly decreased and reached sub-lethal levels after two days. The resultant CTP's from the reduced dosages under VIF were similar or higher at all depths to those for the full dosage applied under LDPE film (Figure 5.8).

Highly effective control of the different pathogens mentioned above was achieved by methyl bromide at half the normal dosage applied under VIF to a depth of 40 cm. Treated crops showed better growth, increased yield and improvement of quality (Gamliel *et al* 1994a, c, 1995). Half-dosages applied to LDPE for comparison were generally less effective.

Combining physical and chemical methods of soil disinfestation is another possibility for reducing dosages of methyl bromide. A combination of methyl bromide at reduced dosages together with solarization (Gamliel *et al* 1994b, c) gives effective control of *Fusarium* crown rot of tomato and of sudden wilt of melons similar to that obtained with methyl bromide in reduced dosage in conjunction with VIF or full dosage under LDPE.

The commonly used mixtures of methyl bromide and chloropicrin, at

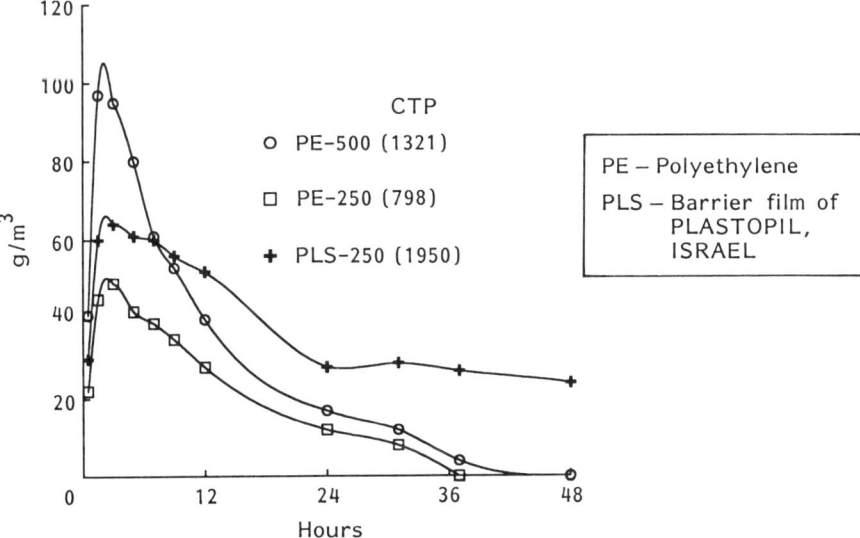

Figure 5.8. Concentration of methyl bromide in soil at a depth of 20 cm after application at a rate of 500 Kg/ha and 250 Kg/ha under different plastic sheeting

similar dosage rates to methyl bromide on its own, give excellent control in many situations, especially where *Verticillium* wilt is involved because of the synergistic effect of both components, thus reducing the quantity of the methyl bromide component per acre.

VIF's are much more expensive than LDPE. The insertion of the barrier layer increases sheeting cost by about 2–2.5 times. Under the commonly accepted dosage of methyl bromide e.g. $50 \, g/m^2$, the cost of fumigation per area of the new technology using half-dosages of methyl bromide will be more expensive. However, at a dosage of approximately $70 \, g/m^2$ and above, which is applied in many situations, the cost of fumigation of the new technology will be on a par with the standard treatment.

A potential problem with this type of plastic sheet is that a higher level of inorganic bromide residues may be formed in the soil, which can then be taken up by plants (Van Wambeke *et al* 1985, Van Wambeke 1989). However, as bromide ions can be removed easily in most situations, this fact is not of real concern as far as residues in food is concerned. On the other hand, by retaining high concentrations of methyl bromide for relatively long periods under VIF, a daily rate of degradation of 6–14% may be expected (Daelmans 1978), and so emission of methyl bromide to the atmosphere will reduce.

At the end of 1995, the parties to the Montreal Protocol will decide about the future use of methyl bromide, and might insist on mandatory introduction of VIF for soil fumigation. From the extensive multi-national research which has taken place in the last two years in eight countries, coupled with previous knowledge, it is apparent that the use of VIF for soil fumigation with optimised dosage of methyl bromide, can minimize the environmental hazards without affecting efficacy.

It can be seen from the foregoing that fumigation of soil with methyl bromide is a highly technical and complex operation, where current application results in highly effective pest and disease control in a wide range of agricultural and horticultural practices. Replacement of such a complex technology would not be easily achieved.

REFERENCES

Altman J, 1970, in: Toussoun, TA, Bega RV, and Nelson PE., (eds), *Root Diseases and Soil-borne Pathogens*, (Berkeley: University of Califoria Press) p 216.

Anon, 1983a, *Diciduous Fruit Grower*, July, p 242.

Anon, 1983b, *Farmer's Weekly*, August 12. Probing for profit.

Anon, 1985, Agriculture, Report of the Scientific Committee for Pesticides p. 15.

Anon, 1990, Report of the working party on pesticide residues: 1988–1989. Supplement to issue No.8 of the Pesticides Register. Ministry of Agriculture Fisheries and Food. Health and Safety Executive p 3.

Anon, 1991, Currently regulatory status of methyl bromide. Fumigants and phero-mones, 24, p. 2, Fumigation Service and Supply Inc. Indiannapolis.

Brown DB and Rolston DE, 1980, *Soil Science*, **130**, 68.

Brown AL, Burau RG, Meyer RD, Rasky DJ, Wilhelm S and Quick J, 1979, *Calif. Agric.* 11–13.

Butters RE and Fletcher J, 1985, Recommendations for ADAS on bromide residues in protected crops.

Chen Y, Gamliel A, Stapleton JJ and Aviad T, 1991, in: Katan, J and Devay JE (eds), *Soil Solarization* (Boca Raton: CRC Press) p 103. Commission of the European Communities EUR 10211 En. The use of Methyl Bromide as a fumigant of plant growing media.

Coosemans J and Van Assche C, 1983, *Acta Hort.*, **152**, 315.

Cuany and Lavergne, 1974, *Phytiat. Phytopharm.*, **23**, 183.

Daelmans A., 1978, Thesis Kath. University Leuven.

De-Heer H, Hamaker Ph, Tuinstra LGMth and Van Der Burg AHM, 1984, *Acta Hort.*, **152**, 109.

Drosihn UC, Stephan BR and Hoffman GM, 1968, *Z. Pflkrankh Pflpath. Pflscutz*, **75**, 272 (in German).

Drosihn UCZ, 1968, *Pflkrankh. Pflpath. Pflschutz*, **75**, 665 (in German).

Ebben MN, Gandy DG, Spencer DM, 1983, *Plant Pathology*, **32**, 429.

Fletcher JT, 1984, *Diseases of Greenhouse Plants*. (London and New York: Longman) p 351.

Galley DL and Hague NGM, 1967, Proc. 4th Br. Insectic. Fungic. Conf. Brighton, **1**, 56.

Gamliel A, Grinstein A, Klein L, Peretz I, Nachmias A and Katan J, 1994a, Proc. 9th Congress of Mediterranian Phytopathological Union, Kasadasi-Aydim Tur-kiye, p. 365.

Gamliel A, Grinstein A, Katan J, Klein L and Ucko O, 1994b, *Phytoparasitica*, **2**, 163.

Gamliel A, Grinstein A, Klein L, Katan J and Ucko O, Maduel A and Peretz I, 1994c, *Hassadeh*, **74**, 1011, (in Hebrew).

Gamliel A, Grinsten A, Klein L, Peretz I and Katan J, 1995, *Phytoparasitica*, **23**, 65.

Gums MF, 1989, *Acta Hort.*, **255**, 337.

Haas HV and Klein L, 1976, *Phytoparasitica*, **4**, 123.

Hass JH, Bar-Joseph B, Krikun J, Barak R, Markovitz T and Kramer S, 1987, *Agronomy Journal*, **79**, 905.

Hayes WA, 1969, *Span*, **12**, 3.

Hoestra H, 1968, *Meded. Landbouwhogesch. Wageningen*, p 99.

Hoffman GM and Malkomes HP, 1979, in: Mulder D, (ed), *Soil Disinfestation*, (Amsterdam: Elsevier Scientific), p 291.

Kempton RJ and Maw GA, 1970, *Rep. Glasshouse Crops Res. Inst., 1969*, 78.

Kempton RJ and Maw GA, 1973, *Ann. Appl. Biol.*, **74**, 91.

Kempton RJ and Maw GA, 1972, *Ann. Appl. Biol.*, **72**, 71.

Kempton RJ and Maw GA, 1971, *Rep. Glasshouse Crops Res. Inst. 1970*, 84.

Kempton RJ and Maw GA, 1974, *Ann. Appl. Biol.*, **76**, 217.

Klein L, 1989, *Acta Hort.*, **255**, 213.

Klein L, 1994, Industry Workshop on the practical implementation of the Montreal Protocol: The African Perspective. 29–31 Aug. 1994, Swaziland. Organized by the South African Department of Health. (Reduction of emission to the atmosphere during fumigation with Methyl Bromide).

Kolbezene MJ, Munnecke DE, Wilbur WD, Stolzy LH, Abu-El-Hay FJ and Szuszkiewicz TE, 1974, *Hilgardia*, **42**, 465.

Kolbezene MJ and Abu-El-Hay FJ, 1978, Proc. Int. Agric. Plastic Congress, San Diego, p 476.

Krikun J, Haas JH and Bar-Joseph B, 1983, *Phytoparasitica*, **11**, 228.

Mai WF and Abawi GS, 1981, *Plant Disease*, 859.

Malkomes HP, 1971, *Z. Pflkrankh. Pflpath. Pflschutz*, **78**, 464.

Maw GA and Kempton RJ, 1973, *Soils and Fertilizers*, **36**, 41.

Maw GA and Kempton RJ, 1971, Proc. 6th Br. Insectic. Fungic. Conf., p 231.

Menge JA, 1982, *Phytopathology*, **72**, 1125.

Menge JA, Munnecke DE, Johnson EL and Carnes DW, 1978, *Phytopathology*, **68**, 1368.

Millhouse DE and Munnecke DE, 1979, *Phytopathology*, **69**, 793.

Ministy of Agriculture, Fisheries and Food, 1982, Food Surveillance Paper No. 9, HMSO, p14.

Munnecke DE, 1977, *Proc. Int. Soc. Citriculture*, **3**, 853.

Munnecke DE and Bricker JL, 1978, *Plant Disease Reporter*, **62**, 628.

Munnecke DE, Bricker JL and Kolbezene MJ, 1978, *Phytopathology*, **68**, 1210.

Munnecke DE and Van Gundy SD, 1979, *Ann. Rev. Phytopath.*, **17**, 405.

Munnecke DE, Moore BJ and Abu-El-Hay F, 1971, *Phytopathology*, **61**, 194.

Munnecke DE, Kolbezene MJ and Wilbur WO, 1978, Proc. Int. Agric. Plastic. Congress, San Diego, p 482.

Nachmias A, Tsror L (Lahkim), Livescu L and Krikun J, 1995, *Acta Hort*, **382**, 145.

Nachmias A, Tsror L, Livescu L, Peretz I and Maharshak G, 1994, *Phytoparasitica*, **22**, 80.

Parker KG, Mai WF, Oberly GH, Brase KD and Hickey KD, 1966, *N. Y. Agric. Exp. Stn. Bull.*, **1169**, p 19.

Poulson E, 1983, *Food and Chemical Toxicology*, **21**, 421.

Rackhom RL, Wilbur WD, Szuszkiewicz TE and Hara J, 1968, *Calif. Agric.*, **22**, 16.

Reber H, 1967, *Gartenbawiss*, **32**, 291.

Roorda Van Eysinga JPNL and DeBes SS, 1984, *Acta Hort.*, **145**, 262.

Ross JP, 1971, *Phytopathology*, **61**, 1400.

Ross JP and Harper JA, 1970, *Phytopathology*, **60**, 1552.

Rovira AD, 1975, *Soil Biol. Biochem.*, **8**, 241.

Savory BM, 1966, *Res. Rev. Commonw. Bur. Hort. East-Malling*, **1**, p 64.

Smith DH, 1963, *Soil Sci. Soc. Proc.*, **27**, 538

Van Assche C and Van Wambeke E, 1981, *Acta Hort*, **122**, 175.

Van Assche C, 1971, Proc 6th Br. Insectic. Fungic. Conf., Vol. 3, 706.

Van Wambeke E, Vanachter A and Van Assche C, 1985, in: Comportment et effcts secondares des pesticides dan le sol. Versailles 4–8 Juin 1984. Ed INRA publ. (Les colloques de l'INRA No. 31). (Behaviour of the soil fumigant Methyl Bromide under different application conditions).

Van Wambeke E, 1974, *Agric. and Environ.*, **1**, 277.

Van Wambeke E, 1984, *Acta Hort.*, **152**, 137.

Van Wambeke E, 1989, *Acta Hort.*, **255**, 243.

Vanachter A, 1979, in: Mulder D, (ed), *Soil Disinfestation*, (Amsterdam: Elsevier Scientific), p 163.

Vanachter A, Van Wambeke E and Van Assche C, 1975, *Meded. Fac. Landbouw-wet. Rijksuniv. Gent.*, **40**, 1085.

Vanachter A, Van Wambeke E and Van Assche C, 1978, 3rd Int. Congress of Plant Pathology, München, August 1978, (Abstracts p. 380, p 435.).

Vanachter A, Van Pee G, Van Wambeke E and Van Assche C, 1981a, (ed), Fac. Landbouww. Rijksuniv. Gent, 46/1 343.

Vanachter A, Feyaerts J, Van Wambeke E and Van Assche C, 1981b, Med. Fac. Landbouww. Rijksuniv. Gent, 40/1 351.
Wilhelm S, Storkan RC and Wilhelm JM, 1974, *Agric. and Environ*, **1**, 227.
Wilhelm S and Paulus AO, 1980, *Plant Diseases*, **64**, 264.

6

Methyl Bromide in Storage Practice and Quarantine

B CHAKRABARTI
Central Science Laboratory, MAFF, Slough, UK

C.H. Bell, N. Price and B. Chakrabarti: The Methyl Bromide Issue
© 1996 John Wiley & Sons Ltd

1.1 INTRODUCTION

Methyl bromide (MeBr) has been in use for more than fifty years for controlling a wide spectrum of pests such as insects, mites, nematodes in soil and seeds, fungi in timber, vertebrate pests and also to comply with various quarantine and phytosanitary requirements. These may be in force for perishable and durable food commodities, other stored products, soil and timber.

The versatility of MeBr stems from its high toxicity to a wide range of pests, considerable powers of penetration and effectiveness at relatively low concentrations, even at low temperatures. Fumigation with MeBr is fast, typically lasting between 2 and 48 hours and the gas airs off rapidly with minimal disruption to normal commercial practices. Inorganic bromide residues produced as a result of fumigation are generally found to be within acceptable limits. In circumstances where fumigations were considered virtually impossible such as plant quarantine, the advent of MeBr has made it possible. Methyl bromide is considered as an ideal fumigant. Some of its advantages and disadvantages are set out in Table 6.1.

2.1 AREAS OF APPLICATION

Fresh products such as fruit, cut flowers and vegetables and other stored commodities such as grain, legumes and timber, which are kept for a longer period are usually fumigated with MeBr either to contain infestation within specific geographical boundaries or to reduce damage by pests, or both. Perishable commodities such as fresh fruit, vegetables and cut flowers are transported across the continents and there are stringent quarantine regulations to limit the spread of pest and diseases from one area to another. Disinfestation treatments are carried out under precise conditions according to international agreements to ensure total control of quarantine pests. Currently, there are comparatively few approved quarantine treatments which do not use MeBr.

Structures or an area within structures are fumigated to control termites (in dwellings), rats and insects (on board ships, warehouses, silos, trucks and railcars), rodents and cockroaches (in aircraft) and various pests (in mills,

Table 6.1. Factors in favour and against the use of MeBr in storage practices

Advantages	Disadvantages
• Effective against all insect species including the pre-adult stages	• No lasting residual effect
• Effect against mites	• Highly toxic
• Can ensure complete control even at as low a temperature as 5 °C	• Handling by skilled and certified personnel only
• Can be used for treating most commodities without causing any deleterious effect	• Environmental implication as a potential ozone depleter
• Residues are not usually a problem	
• Commodities can be treated *in situ*	• May cause fruit injury and off-flavour in certain circumstances
• Good penetration in most circumstances	
• Treatment can be completed between 2 and 48 hours, useful for quarantine purposes	
• Simple procedures for applications in different situations	
• No incidence of resistance has been reported	

food processing areas, warehouses and museums). Fumigation is used when the infestation is widespread and present in inaccessible areas (Figure 6.1) of a structure so that a localised treatment can only achieve partial control. Methyl bromide is used principally for its ease of application and the minimum disruption of production and transportation schedules.

3.1 METHODS OF APPLICATION OF METHYL BROMIDE

Methyl bromide is available under pressure as liquid in steel cylinders, in bulks of various quantities and in small cans of capacity 1 or 1.5 lb. It is applied either as atomised liquid by forcing it through spray nozzles which cause instantaneous vaporisation by absorbing heat from the atmosphere, or as gas delivered from a heated vaporiser. Although several devices have been developed for the latter method, a coil of copper tubing immersed in hot water is used almost universally for dosing large quantities of MeBr from cylinders. The distribution of the vaporised MeBr is accomplished by an

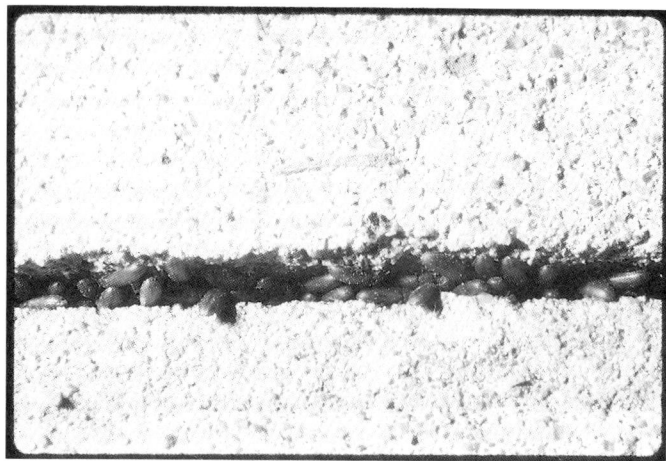

Figure 6.1. Heavy infestation inside the crevice of a wall

elaborate system of hose, piping and perforated lay-flat tubing to suit individual situations. When using disposable cans for small-scale fumigation, the fumigant is normally delivered directly into the enclosure from the cans under its own pressure via a proprietary puncturing device connected to a length of flexible tubing ending in a small orifice or other atomising device. During application it is imperative to avoid contact of liquid MeBr with surrounding materials, especially with food commodities, because painted surfaces are liable to damage, fresh fruits may get skin damage and flour, nuts etc. may acquire high residues and off-flavour.

4.1 MONITORING AND DETECTION

Any treatment with MeBr requires the fumigant to be present in adequate concentrations over the treatment period. This can only be established by monitoring the gas concentrations over that period. Also there are a limited number of plant species and varieties of fruit and vegetable which could be susceptible to injury by MeBr fumigation. Therefore preliminary tests are necessary to establish the parameters and to observe the effects on the marketability and shelf-life of the commodity. Chloropicrin is phytotoxic and to avoid injury, MeBr containing this compound as a warning agent is not used for treating perishable commodities.

To practise safe and effective use of MeBr, various methods of detection and analysis of the fumigant have been developed. Appropriately calibrated instruments such as gas chromatographs (GC), infrared, thermal conductivity meters TCM) and interference refractometers are available to monitor the gas at working levels. Portable TCMs and 'fumiscopes' (Anon 1976) which also work on the principle of thermal conductivity are generally used for monitoring MeBr under practical operating conditions. Halide and electronic leak detectors and detector tubes of several ranges are used to measure MeBr near the TLV* level (OES† in UK) of 5 ppm, while the former two are non-specific the latter gives the definitive concentration of the gas. In the Netherlands the TLV of MeBr is 0.3 ppm and a properly calibrated portable GC is the only instrument capable of measuring MeBr at such levels (Dumas and Bond 1985) apart from sensitive detector tubes.

In practice, success of a MeBr fumigation is demonstrated in two ways: (a) by attaining the correct CTP (5.1) required for the target pests; (b) by placing test insects of the same species or test insects with response similar to that of the most tolerant stage of the pest likely to be encountered. To ensure that the necessary CTP is achieved, fumigant concentrations are monitored over the fumigation period at regular intervals.

5.1 DOSAGES

Although there is a broad range of susceptibility amongst pests to MeBr, it is effective against most pests at relatively low concentrations. The pupal stages of insects and the egg and hypopal stages of mites are usually hardest to kill. To control mites a twice as much higher dose than that used for insects is required (Anon 1993a). Doses typically range from $10-150 \, \text{g m}^{-3}$ with exposure periods from 2–48 hours depending on the type of the enclosure, target pest and the nature and quantity of material being fumigated.

In Europe, for dosage calculations, dried food commodities are grouped together on the basis of their sorption capability (5.1.2.3) and temperature (Anon 1993a). For rodent control a concentration of $4 \, \text{g m}^{-3}$ needs to be maintained at least for four hours (Bond 1984). It is noticeable that the nominal concentration and time products [dosage (g m^{-3}) × time (hours)] are significantly higher than those necessary for the control of pests (Anon 1993a). This is because the correct amount of MeBr for the success of a treatment has to incorporate a sufficient excess to allow for leakage and sorption.

---

*The threshold limit value (TLV) of an air-borne substance has been defined as the average concentration of that substance over the exposure period i.e. a normal eight-hour working day, to which anyone may be repeatedly exposed, day after day, without experiencing any adverse effect.

†In the United Kingdom, the equivalent term for TLV is occupational exposure standard (OES) and is explained in a booklet (EH40) produced by the Health and Safety Executive, which is revised yearly (Anon 1994a).

5.1.1 CONCENTRATION-TIME PRODUCT

In dealing with fumigants, the dosage rate is considered as a combination of gas concentration ($mg\,l^{-1} = g\,m^{-3} = oz$ per 1000 ft^3) and time of exposure in hours (h) and is termed a Concentration-Time (c × t) Product (CTP). For MeBr the effect of a CTP at a specific temperature on a population of insects will be approximately the same, whatever the variation between concentration and time, provided the concentration is above a minimum threshold value (Bell 1978). For plant quarantine work, recommendations based on the c × t principle are particularly valuable because they promote uniformity in standards and permit reliable certification of goods so treated (Anon 1994b, d).

The CTP required to control different insects pests varies with temperature but for most species of stored product insects at temperatures above 15 °C a CTP of 180 gh m^{-3} is considered to be satisfactory (MAFF 1974). Higher values are required for lower temperatures and for tolerant species (Table 6.2).

5.1.2 FACTORS INFLUENCING GAS DISTRIBUTION AND CONCENTRATIONS IN A FUMIGATED AREA

The principal objective of any fumigation is to maintain an adequate concentration of the fumigant vapour within the treated area for a sufficient length of time for all the pests, including different stages of their life cycle, to receive a lethal dose.

5.1.2.1 Convection

Any enclosure is liable to feature temperature, moisture and pressure gradients. These give rise to local gas movements and air streams. Exposure to wind causes further pressure differentials, while sun shining on the roof or walls can cause pronounced local heating. Movement of MeBr is influenced by the much greater density of the vaporised gas than air (approx. 3.25 times heavier). There are also temperature differences due to a cooling effect by

Table 6.2. Pest species and CTPs required for their control at 15–20 °C

Pest	CTP of MeBr (gh m^{-3})
Mice	12
Cockroaches	65
Stored product insects	200
Mites (all stages)	600
Nematodes in seeds	800–1200

the volatilisation of the gas at the atomising nozzle. These factors can slow down the achieving of an even spread of the fumigant.

5.1.2.2 Gaseous Diffusion

The movement of molecules of a gas is influenced by its partial pressure, concentration gradients, relative densities of gas and air, temperature, size of molecules and sorption factors etc. Gases will always move down a concentration gradient upon release into an enclosure, at a rate determined by the combined effect of the factors listed above.

5.1.2.3 Sorption

Sorption removes fumigant gas from the atmosphere of the treated area. This could reduce the effective concentration of MeBr necessary for controlling the target pests. The term 'sorption' covers three types of phenomena, (a) *absorption* which occurs when the gas enters the commodity forming a non-attainable solution, e.g. with oil and fat contents of a product, (b) *chemisorption* which is a process in which MeBr reacts with the components of a commodity to form another compound and (c) *adsorption* which occurs when the gas molecules attach themselves by physical forces to the surface of the material. This last form of sorption is largely reversible and desorption occurs during ventilation after fumigation, usually increasing the time required for aeration.

5.1.2.4 Permeation

Materials are often fumigated under gas-proof sheetings with edges weighted down and sealed to the floor. Use of sheeting with high permeation may cause fumigation failure; it is therefore important to use sheets of known low permeabilities or to measure it (Wontner-Smith and Chakrabarti 1994) (Figure 6.2). Loss by permeation is higher at higher temperatures and rate characteristics are different for different gases. For MeBr, sheetings with permeabilities below $20\,\mathrm{mg\,m^{-2}\,h^{-1}}$ at 20 °C are arbitrarily considered as the highest allowable limit for a successful fumigation. Permeabilities of low density polyethylene (LDPE) show an inversely proportional relationship to the thickness. Under similar conditions polythene/nylon laminated sheet of 30 micron thickness shows about 10 times lower permeability than 125 micron LDPE film (Wontner-Smith and Chakrabarti 1994).

5.1.2.5 Leakage

Apart from the loss of fumigant gas by sorption and permeation through the fumigation sheet or enclosure, considerable leakage occurs due to poor

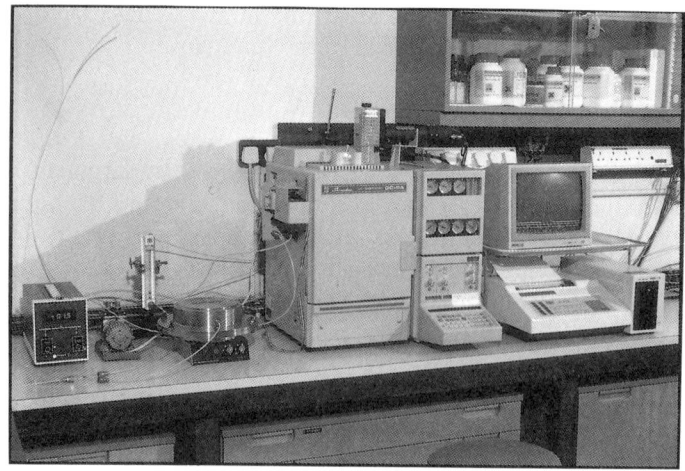

Figure 6.2. Gas chromatograph linked permeability apparatus used at CSL

sealing through cracks and gaps in an enclosure which may not have been sealed properly during pre-fumigation inspection. Leakage can be controlled to a large extent by the use of proper sealing materials, gas-proof sheeting, masking tape and sometimes by a coat of oil-based paint on walls and ceilings (Wainman and Chakrabarti 1973, Banks 1986).

5.1.2.6 Gas Tightness

When MeBr is vaporised, its volume increases many times. As a result, whenever it is applied to an enclosed space under atmospheric pressure, it generates a positive pressure. Depending on the air-tightness of the structure there will be a continuous leakage which will reduce the effectiveness of a treatment, and which may increase hazard in the surrounding areas.

Fumigation chambers and sealed structures can be routinely tested for gas-tightness using a pressure decay method (Banks and Annis 1977, Banks 1984) and if necessary, can be improved for better retention of the fumigant. The method involves pressurising a sealed structure using a blower. When the air input is cut off, the rate of pressure decay is monitored on a water gauge.

5.1.2.7 Vacuum Fumigation

A great many fumigations are carried out at atmospheric pressure in chambers, in sealed containers or in warehouses under gasproof sheets, but a limited number of fumigations are conducted in specially built sealed chambers under reduced pressure. The details of the technique vary widely but commonly the term 'Vacuum fumigation' is used to cover all such applications. In contrast with atmospheric fumigation where penetration into a packaged or bagged commodity can be slow and may take 16–24 hours or longer, vacuum fumigation may achieve similar or even better control in a much shorter time, typically 2–4 hours.

Where the circumstances demand rapid treatment, lower doses or quick turnover of goods such as in quarantine stations, vacuum fumigation methods have obvious advantages. In addition to 'sustained vacuum' treatments when the vacuum is maintained between 25 and 150 mm of Hg for the whole treatment period, many other vacuum regimes have been tried and practised with various commodities and circumstances (Monro 1958, Brown and Heuser 1953a, b, 1956, Monro and King 1954, Lepigre 1949). In general, vacuum fumigation techniques are considered intrinsically safer, since the treatment takes place in a chamber at reduced pressure from which no leakage can occur and the residual gas at the end of the treatment can be swept away by an 'air wash' procedure before the chamber is opened. The process involves restoring the pressure of the chamber to atmospheric with fresh air and reducing pressure alternatively. The cycle is usually repeated 5–6 times until the chamber is considered safe to enter. This also presents the possibility of recycling MeBr, instead of releasing it to the atmosphere.

5.1.3   RESIDUES IN FOOD COMMODITIES

During fumigation MeBr is sorbed by the treated commodities, some of which may desorb during ventilation when the greater part of the MeBr is removed in the gas phase from the enclosure. The quantity remaining depends on environmental factors, kind of packaging, stowage and the nature of the commodity. There is usually however, a small but variable amount of unchanged MeBr and a permanent residue of inorganic bromide, resulting from the chemical reaction between the fumigant and some constituents of the material, which remains behind.

Inadequately aired foodstuffs may sometimes retain MeBr or develop bromide residues to levels above the internationally accepted tolerances of 0.01 mg/kg and 50 mg/kg in bread and cereals respectively (FAO/WHO 1967, 1980). For spices the tolerance level for inorganic bromide residue is 200 mg/kg (Reeves et al 1985). Many food commodities contain naturally occurring inorganic bromides, sometimes in greater quantities than those derived from MeBr fumigation. Consumption of normal amounts of these foods with such residue levels is not considered hazardous to human health (USEPA 1989, Health and Welfare, Canada 1975, MAFF 1982).

Methods of analysis have been developed to determine free MeBr in foods using gas chromatography (Heuser and Scudamore 1968, 1969, Greve and Hodgendoorn 1979, Fairall and Scudamore 1980, Dumas 1982). Inorganic bromide may be determined using a selective ion electrode (Banks and Desmarchlier 1976) or by gas chromatography (Heuser and Scudamore 1970, Stijve 1977).

6.1 EXISTING USES

6.1.1 COMMODITIES

Methyl bromide is used to control various insect pests, mites, rodents and fungi (Bruch 1961, Richardson and Monro 1962, Bond 1984) and is the fumigant of choice for many commodities because of its ease of use. While some fruits are susceptible to injury by MeBr, many plants, vegetables and fruits remain unaffected at the concentrations necessary for the effective control of target pests. Because of marked differences in treating fresh products and stored products, these are divided under two broad-based headings; perishables for the former and durables for the latter group.

It is estimated (Anon 1995) that out of the 18–24% of MeBr used globally for the disinfestation of both durable and perishable goods, about a third is used for perishables.

It is estimated that the post-harvest losses due to stored-product pests in the USA is about 9% and up to 20% in the developing countries. Total food losses due to pests including pre-harvest is considered to be 48% (Pimental 1991). In tropical countries where insect populations are more active and the standards of storages, transport facilities and food processing plants and premises are not so sophisticated, post-harvest food loss could be between 35 and 70%, whereas in developed countries the losses could be 9–25% (Scott 1991).

In addition, the presence of stored product pests, their body fragments, excreta or secretary products, either in food when consumed or in contact with skin may cause allergenic reactions (Wirtz 1991). This emphasises the need for maintaining pest infestations at the lowest possible level and fumigation usually with MeBr is a major tool to achieve this objective.

6.1.1.1 Perishable Commodities

Perishables include a wide variety of commodities. These are grouped together according to their cultural similarity or by pest problems associated with them: (1) Apples and pears; (2) Berryfruit; (3) Citrus; (4) Cucurbits; (5) Grapes; (6) Root crops; (7) Tropical fruits; (8) Vegetables; (9) Cut flowers and ornamentals; and (10) Bulbs. Insect pests occurring in these commodities such as codling moth, fruit flies, colorado beetle and mango-seed weevil are considered as major destructive pests and regarded as of quarantine importance. This means that either these pests are absent from the importing country or that country has a policy of 'zero tolerance' for all kinds of live insects irrespective of any economic importance. Examples of perishables and their pests with countries where these are of economic importance are listed in Table 6.3.

6.1.1.1.1 Control of Infestation and Quarantine

Post harvest treatments on perishable commodities are mostly carried out for quarantine and phytosanitary purposes. An ideal treatment needs to be effective in controlling a wide range of pests and readily acceptable to the consumer and regulatory authorities. Other attributes should include rapid action against the target pest with low residues in the commodities, ease and safety of use and above all cost effectiveness. Taking all these into account MeBr fumigations of fresh fruits, vegetables and cut flowers are deemed to be the most reliable treatments available at the moment to plant health authorities around the world.

The treatments for disinfection are usually carried out at the country of origin or on arrival at its destination, as and when intercepted. Trading organisations in some countries treat perishable commodities immediately before export in order to facilitate their quick release into the retail outlets of the importing country and thus avoiding expensive fumigation costs on arrival, possible heavy spoilage due to delay and financial penalty.

6.1.1.1.2 Fumigation Procedures

The object of a fumigation in plant quarantine is to attain total kill of all stages of the target pests. Treatment schedules for this purpose have been developed based on the work of many research organisations around the world and selectively approved and introduced by regulatory agencies such as the Australian Quarantine and Inspection Service; Japan Ministry of Agriculture, Forestry and Fisheries; Ministry of Agriculture, Fisheries and Food, U.K.; USDA-Animal and Plant Health Inspection Service (APHIS). Each approved method has its schedule well defined to bring about the required degree of control. Work done by many individuals in plant quarantine for a wide range of commodities are listed in Table 6.4 with the dosage, exposure and temperature regimes. Generally, higher doses are needed for control at lower temperatures. In a situation where there is a possibility of product damage, alternative methods of control are to be sought.

Fumigation in chambers, freight containers, railway box-cars and fumigation under gas-proof sheeting are generally used for quarantine fumigation with MeBr. Most of the treatments are conducted under normal atmospheric pressure and fixed installations are tested for gas tightness (5.1.2.5) before

Table 6.3. Examples of perishables, their pests and countries where they are of economic importance

Commodity	Pests common names	Recommended treatment	Economically important to countries
Apples and pears	Codling moth, Apple maggot fly, Mites, Scale insects	MeBr	Australia, Chile, Europe, New Zealand, South Africa and United States
Stonefruits: apricots cherries nectarines peach plums	Codling moth fruit flies Oriental fruit moth Walnut husk fly Thrips	MeBr	Canada, Chile, Europe, New Zealand, United States
Citrus fruits: oranges lemons limes	Fruit flies Fullers rose weevil Scale insects	MeBr (limited use)	Australia, Brazil, Europe, Israel, Japan, South Africa and United States
Grapes	Fruit flies Vine moth Mealy bug Mites	MeBr	Australia, Brazil, Chile, Europe, Israel, South Africa and United States
Berryfruits: blackberry blueberry raspberry strawberry	Aphids Blueberry maggots Fruit flies Thrips Mites	MeBr	Australia, Brazil, Canada, Colombia, Israel, New Zealand, South Africa, United States and Zimbabwe

Commodity	Pests	Treatment	Countries
Vegetables: asparagus, beans, broccoli, cabbage, cauliflower, lettuce, peppers, tomato	Aphids, beetles, Fruit fly, Lepidoptera, Thrips, Weevils	MeBr some are sensitive to MB fumigation (Spitler and Couey 1983)	Most of the developing countries
Rootcrops: carrot, cassava, garlic, ginger, onion, potato, sweet potato, yam	Beetles, Mites, Nematodes, Scale, Thrips, Weevils	MeBr	Most of the developing countries
Cucurbits: cucumbers, melons, squash	Aphids, Fruit flies, Lepidoptera, Thrips	MeBr (very limited use)	Australia, Chile, Israel, Mexico, Netherlands, New Zealand, South Africa and United States
Tropical and Subtropical Fruits: avocado, banana, lychee, mango, papaya, pineapple	Fruit flies, Lepidoptera, Mites, Weevils, Internal and External feeders	MeBr used only selectively	Australia, Caribbean, India, Indonesia, Malaysia, Mexico, North and South America, Philippines, Thailand

Table 6.3. (*continued*)

Commodity	Pests common names	Recommended treatment	Economically important to countries
Ornamentals and flowers: carnations chrysanthemums orchids roses ornamental plants	Aphids External feeder Lepidoptera Mites Thrips Scale insects	Usually MeBr only	Australia, Colombia, Israel, Kenya, Malaysia, Netherlands, New Zealand, Singapore, South Africa, Thailand, United States and Zimbabwe
Bulbs: gladioli garlic lily narcissus tulip	Bulb aphid Bulb mite Bulb fly Thrips Internal and Surface feeders	Usually MeBr only	China, Israel, Netherlands, New Zealand, Taiwan and United States

Table 6.4. Examples of plant quarantine treatment with methyl bromide

Fruits, vegetables or plants	Pests and their life-stages	MeBr dosage (g m^{-3})	Period of exposure (hour)	Temperatures (°C)	References	Remarks
Apples and pears	Codling moth and larvae	32	2	23	Gaunce et al 1980 Tebbets et al 1986 Moffit et al 1983	100% pure MeBr is used for plant materials
	Diapausing larvae	56	2	6	Chapman 1940	
	Apple maggot	16	2	30	Richardson 1955	
	larvae and eggs	20	2	21	Johnson et al 1947	
	Oriental fruit moth – larvae	16	2	21	Phillips et al 1939	Fruit injury
	– eggs	25	2	21	Meheriuk et al 1990 Galetti and Berger 1987	May cause injury
Citrus species – grapefruits, lemon, oranges, tangerines etc.	Citrus blackfly, Mediterranean fruit fly	24–48	2	21	Anon 1992 Anon 1985, 1988 King and Benschoter 1991	
Grapes	Grape vine moth Mediterranean fruit fly,	24–48	2	10–30	New Zealand Ministry of Agriculture 1992b	No injury
	European red mites, Chilean spider mites	48	3	16	Soma et al 1991	May cause off-flavour
Stonefruits: apricots	Oriental fruit moth larvae	16	2	15–18	Johnson et al 1947	No fruit injury
	Mediterranean fruit fly, all stages	24	2.5	30	Armstrong and Couey 1984	
cherries	Codling moth, all stages	32	2	20–24	Anthon et al 1975	No fruit injury

continued overleaf

Table 6.4. (*continued*)

Fruits, vegetables or plants	Pests and their life-stages	MeBr dosage (g m^{-3})	Period of exposure (hour)	Temperatures (°C)	References	Remarks
plums} peaches} nectarines}	Codling moth, all stages	48	2	21	Armstrong et al 1988 Spitler and Couey 1983	
	Fruit flies, all stages	32	4	21	Hinsch and Harris 1992 Yokoyama et al 1987a, b, 1990a, b, 1994	No fruit injury
Berryfruits: blackberries} blueberries} raspberries} strawberries}	Codling moth, all stages	48	2	21	Anon 1994b	
	Fruit flies	48	3	21	Armstrong et al 1984	
		48	2	21	Anon 1994b	
Vegetables: avocado	Fruit flies	32	4	21	Anon 1994b	
beans	Seed beetles	48	2.5	21		
cabbages	External feeders – thrips, aphids	32	2	21		
peppers tomato	Fruit flies	40	1.5	21	Lipton et al 1982 Spalding et al 1978 Lipton et al 1982 Brecht et al 1986	Shows surface damage and decay. May delay ripenings
		48	2	21		
Rootcrops: carrot	Beetles, mites, nematodes, thrips	48	3	> 21	Anon 1994b	Under vacuum
ginger		48	2	> 21		
garlic	Beetles, mites, nematodes, thrips	48	2	> 21		
onion		48	2	> 21		
potato	European corn borer, tuberworm, beetles	40–48	2	> 21		
yam	European corn borer, tuberworm, beetles	64–40	4	21–34		

Commodity	Pests	Dose	Exposure	Temperature	References	Remarks
Cucurbits:						
cucumbers	Thrips, aphids and fruit flies	32–48	2–4	21	Armstrong and Garcia 1985 Stechmann et al 1988 Cowley et al 1991	May affect market ability
melons		32	4	21–26		No fruit damage
squashes						
Tropical fruits:						
lychee	Fruit flies, mites, weevils, moths and maggots	Wide range of doses and times			Anon 1994b Japan Ministry of Agriculture, Forestry and Fisheries 1987 New Zealand Ministry of Agriculture and Fisheries 1992a	(i) With vacuum fumigation, the exposure time can be reduced by 1 hour in each case.
mango						
pineapple						
papaya						
Bulbs:						(ii) For mites, treatments need to be repeated after 2 weeks.
gladioli	Aphids, mites, thrips, flies—all stages	48	3	> 21	Anon 1994c Bond 1984 Steinweden et al 1942	(iii) For treatments at lower temperatures, an extra hour of exposure would be necessary
lilies		48	4	> 21		
narcissus		48	3	21–27		
tulips						
Ornamental plants and cut flowers:						
bird of paradise	Aphids, mites, thrips, scale insects, leaf miner, mealy bugs	16	15	29–31	Carmi 1985 Junaid and Nasir 1956 Helyer and Ledieu 1989 Anon 1994c Gaur et al 1984 Bond 1984	Margin between exposure for effective control and phytotoxic effect on the product is very narrow. Sensitivity also varies with the species and temperature
carnations		12	2	> 21		
chrysanthemums		32	2	21–26		
orchids		48	2	21–26		
poinsettia		32	2	21–26		
roses						

fumigation. Sometimes regulatory authorities from the importing country specify the installation and supervise operation of the fumigation facilities of the exporter. Usually for fumigation of a commodity in a chamber, the amount to be loaded and how this will be arranged with regard to the category of the produce and the free space are specified by the importing country. Load factor or load limit which is the percentage of chamber volume which a commodity should occupy when loaded for fumigation, usually varies from 30–80% depending on the commodity.

As general rule a 30–60 cm gap is kept on the top of a load in a chamber or container to facilitate an even distribution of MeBr which is vaporised and circulated within the enclosure. However, excessive and continuous air movement may damage tender plants, fruits and vegetables. Therefore, with such materials, circulation of air is adjusted to be gentle and the period for mixing the gas is to be kept to a minimum.

Plant injury is also minimised by maintaining the humidity in the enclosure. Humidity is required to be maintained at a level high enough to avoid water loss, but not so high that water droplets are deposited on the commodity which also can affect plants adversely.

6.1.1.1.3   Methyl Bromide Fumigation Followed by Cold or Hot Treatment

Fumigation plus refrigeration of fruits has been approved in the Treatment Manual by USDA-APHIS (Anon 1994d). Methyl bromide fumigation followed by storage at −0.5 °C achieved more effective control of San-Jose scale *Quadraspidiotus perniciosus* (Comstock) and codling moth *Laspreyresia pomonella* (L.) than either of these treatments could have achieved independently (Moffit 1971, Morgan *et al* 1974, Waddell 1993). Combining MeBr and cold treatment also requires less MeBr or a shorter exposure time than MeBr alone would have needed (Couey 1983).

Injury to grapefruit was shown to be greater when fumigation was followed by cold storage than when the fruits were kept at ambient temperature (Benschoter 1979). To gain adequate control there is a possibility of product injury by either MeBr or cold treatment and a combined approach with reduced exposure to both may be successful.

By combining MeBr fumigation with heat, either the concentration required for control or the time of exposure can be reduced (Armstrong and Couey 1984, Sun 1946). Small scale trials are necessary to establish the precise parameters and also to monitor the scale of injury to a specific product before combination treatments are applied commercially. Also for acceptance by the regulatory authorities, extensive technical documentation will be necessary probably for each individual process with the target pest and commodity.

6.1.1.1.4 Conclusions

Following the withdrawal or discontinuation of other fumigants for a variety of reasons, MeBr has become the principal fumigant for quarantine security and phytosanitary purposes. It appears to be effective against most species of pests at 5–30 °C with doses of 24–48 g m^{-3} for 2 hours, without causing any appreciable damage to the fresh horticultural and floricultural commodities. For quarantine, MeBr is the fumigant of choice for its ease of use, degree of effectiveness and no known resistance problem in its 50 years of use. The operational procedures and the pitfalls are well documented. Even with a concerted research effort, it will be difficult to develop alternative treatments or procedures for such a wide area of application within a short time-scale.

6.1.1.2 Durable Commodities

The commodities classified as durables usually have a moisture content of less than 20% and can be stored for long periods if the goods are pest free and placed in a pest-free area. However, this is difficult to achieve in practice due to field infestation and/or cross infestation from other commodities from the usual movement of commodities in the trade. Food commodities are always stored post harvest for a period before consumption, processing or transportation to some other destination. Prolonged infestation of a food product in storage may render it unfit for human or animal consumption by producing conditions conducive to rapid mould growth and fungal decay. Therefore, in many situations MeBr is used as a quick remedial action to prevent disruption of the food supply chain. The durable commodities can be listed in broad terms in the following manner:

- Cereal grains—barley, wheat, rice, maize, rye and sorghum
- Pulses—peas, beans and lentils
- Grain and pulse products—flour, pasta, noodles and semolina
- Dried fruit and nuts—apricots, dates, prunes, peanut, walnut etc.
- Beverage crops—coffee, cocoa and tea
- Herbs and spices
- Oil-seeds—rape, ground nuts, linseed, sunflower etc.
- Seeds for planting
- Tobacco
- Unsawn timber and goods made of wood.

It is estimated that approximately 12–16% of the total annual agricultural uses of MeBr is for the disinfestation of durable commodities. Grain, timber and dried fruit and nut industries are heavily dependent on the use of MeBr as the principal means of controlling infestations (Anon 1995).

Fumigation is a widely used practice that has served an important role in controlling pests for many years in trade and industry. Fumigants are found particularly useful for the control of stored-product insects because of their capability to diffuse and penetrate into places and commodities where other forms of control are impractical or virtually impossible. Methyl bromide is one of the most used of a diminishing range of fumigants currently available (Banks 1994).

A broadly-based MeBr dosage schedule (Table 6.5) where durable commodities of similar nature are grouped together for the convenience of practical fumigation has been accepted by the European Plant Protection Organisation.

6.1.1.2.1 Fumigation of Stored Cereal Grain, Pulses and their Products

Fumigants are used for disinfesting a wide variety of cereals and pulses (Table 6.6). Choice of fumigant and application technique may depend on the nature of the crop and products made of it. Another important factor which is to be taken into consideration is the species of insects present and their life stages (Table 6.7).

Frequent transportation of cereals through international trade facilitates the spread of insect populations. Methyl bromide is used in many situations and success of a treatment will depend on various factors discussed in this section.

For fumigation purposes, the shape and size of a given storage are of importance. Unprocessed grain and pulses in bulk are usually stored in one of three different types of storage structures:

- Silo or vertical storage—in this the height of the structure is greater than the length or width; usually square, circular or rectangular cross-section.

In these types of storages, surface application of MeBr gives poor distribution and does not penetrate through the depth of the bulk. When applied with carbon dioxide as a carrier a better distribution of MeBr has been achieved (Calderon and Carmi 1973). If the structure is sufficiently gas tight a recirculation method is advocated for penetration and even distribution (Monro 1956, Storey 1967, 1971a, b). Fumigation usually takes 12–24 hours and the remaining gas at the end of the exposure period is blown out to avoid high residues.

- Flat or floor storage—usually rectangular in shape and the depth is less than either the length or width of the enclosure. These enclosures may

Table 6.5. Methyl Bromide dosage table*

Group Commodities	Dosage $g\,m^{-3}$			Exposure period	Remarks
	< 10 °C	10–20 °C	> 20 °C	(hrs)	
1. Rice, pea, beans cocoa and coffee beans, dried vine fruits, whole spices, museum artefacts	25	15	10	24	
2. Wheat, barley oat, maize, lentils	50	35	25	24	
3. Rice-bran, other cereals bran	70	45	30	48	
4. Sorghum, nuts, figs, timber	75	50	35	24	
5. Groundnuts, oilseeds, dates, soya, empty sacks	75	50	35	48	
6. Oilseed cakes and meals	120	85	60	48	Sorptive, poor penetration
7. Fishmeal, dried blood etc.	140	100	65	48	
8. Flour	50	50	40	48	Alternative fumigants are recommended

*Recommendation of European Plant Protection Organisation, Schedule 12 (Anon 1993a)

Table 6.6. Examples of cereal and legume crops

Common name	Scientific name
Barley	*Hordeum vulgare*
Beans	*Phaseolus* spp.
	Vigna spp.
Lentil	*Lens culinaries*
Maize	*Zea mais*
Oat	*Avena sativa*
Peas	*Pisum sativum*
Rice	*Oryza sativa*
Rye	*Secale cerale*
Sorghum	*Sorghum bicolor*
Wheat	*Triticum aestivum*

Table 6.7. Major pests in cereals and legumes

Common name	Scientific name
Saw-toothed grain beetle	*Oryzaephilus surinamensis*
Lesser grain borer	*Rhyzopertha dominica*
Granary weevil	*Sitophilus granarius*
Rice weevil	*Sitophilus oryzae*
Maize weevil	*Sitophilus zeamais*
Red flour beetle	*Tribolium castaneum*
Confused flour beetle	*Tribolium confusum*
Khapra beetle	*Trogoderma granarium*
Mediterranean flour moth	*Ephestia kuehniella*
Indian mill moth	*Plodia interpunctella*
Angoumois grain moth	*Sitotroga cerealella*
Dried bean beetle	*Acanthoscelides obtectus*
Cow-pea beetle	*Callosobruchus chinensis*

vary from trucks, freight containers railway box-cars to very large floor stores and ship's holds.
- Small farm bins and temporary storages—these are usually of very poor construction and very difficult to fumigate. Because of their high surface to volume ratio these types of structures require extra attention if a need for fumigation does arise.

Except for quarantine and phytosanitary purposes when some means of gas circulation is available, MeBr is not at present considered as a fumigant of choice of bulk grain due to its poor penetration.

The most efficient methods of fumigating bagged or cased cereal or pulse-based products is in specially designed gas-tight chambers equipped for applying the fumigant in a manner which will ensure its rapid and even distribution. The use of such chambers allows the minimum dose of fumigant to be used and gives the greatest assurance that the treatment will be effective. Fumigation may often be desirable when no suitable fumigation chamber is available. Moreover, when very large quantities are involved, fumigation in special chambers may be considered impracticable on account of the additional handling which would be necessary, and treatment *in situ* may be the remedy. Occasionally it is possible to seal a warehouse to render it gas-tight so that all the contents may be fumigated at one time. Many stores cannot be sealed in this way and is those where the volume of the free space may be very large in comparison with the volume occupied by the commodity, only part of the contents of a large store may require treatment.

In many of these cases the solution may be to cover completely the stack of commodity with gas-proof sheets and to introduce the fumigant under this covering. A number of fumigants have been used in this way but methyl bromide has proved particularly suitable and effectve. Early use of the method was confined to the treatment of individual small stacks each of which could be covered by a single sheet but subsequent experience has demonstrated the practicality of treating very large tonnages under a covering consisting of a number of large overlapping sheets. Methods have been developed for applying the methyl bromide so that adequate distribution is obtained and loss by leakage can be minimised by the use of proper sheeting and sealing techniques.

In finely-divided materials and also in commodities with high fat content such as wheat and soya flour and milk powder, a persistent odour or taint may be produced on fumigation with MeBr (Matthews *et al* 1970a, b). In these situations, penetration of the gas is also poor due to sorption and an alternative treatment may be of benefit. Otherwise, an exploratory test should be carried out before embarking on a full-scale treatment.

6.1.1.2.2 Fumigation of Dried Fruits and Nuts

The optimal storage of dried fruit and nuts is a critical need because most of these products are harvested over a relatively short period and are stored for a long time either prior to export or after importation depending on the local conditions and marketing strategies (Table 6.8). Unless controlled by some means, pests (Table 6.9) will deteriorate the quality of the product in storage and may cause substantial economic loss. (Table 6.10).

Currently most of the dried fruit and nuts are fumigated with MeBr at least once after harvest either to control insect populations or to meet phytosanitary requirements (Rhodes 1986) for export, import or even before releasing to the local retail outlets.

Under proper conditions, MeBr kills 100% of almost all target insects

Table 6.8. Examples of dried fruits, nuts

Common name	Scientific name
Apple	*Malus* spp.
Apricot	*Prunus armeniaca*
Banana	*Musa* spp.
Date	*Phoenix dactylifera*
Fig	*Ficus carica*
Peach	*Prunus persica*
Pear	*Pyrus* spp.
Prune	*Prunus domestica*
Sultanas, currants & raisins	*Vitis* spp.
Almond	*Prunus amygdalus*
Brazil nut	*Bertholletica excelsa*
Cashew	*Anacardium occidentale*
Chestnut	*Castanea* spp.
Coconut	*Cocos nucifera*
Colanut	*Cola acuminata*
Hazelnut	*Corylus* spp.
Pecan	*Carya illinoensis*
Pistachio	*Pistacca vera*
Walnut	*Juglans* spp.

Table 6.9. Major insect pests in dried fruits and nuts

Common name	Scientific name
Almond moth	*Ephestia cautella*
Curculio	*Curculio nasicus*
Codling moth	*Cydia pomonella*
Fruit flies	*Tephritidae*
Indian Meal Moth	*Plodia interpunctella*
Khapra beetle	*Trogoderma granarium*
Navel orange worm	*Amyelois transitella*
Saw-tooth grain beetle	*Oryzaephilus surinamensis*
Warehouse moth	*Ephestia elutella*

Table 6.10. World production of some major dried fruit and nut crops, 3 year average (90/91−92/93)*

Commodity	Quantity (tonnes)
Prunes	315 498
Raisins, sultanas and currants	910 098
Almonds	526 823
Hazelnuts	863 647
Pistachios	153 077
Walnuts	573 928

*Source: Horticultural Products Review, USDA/FAS, 1993

within 24–48 hours. In a specially built chamber where commodities can be treated under reduced pressure (vacuum fumigation 5.1.2.6), packaged dried fruits may need only 2–3 hours of exposure when fumigated with MeBr. As the oil contents of most nuts are high, fumigation under vacuum may cause taint problems.

With the exception of vacuum fumigation treatments, dried fruit and nuts are usually fumigated in purpose-built fumigation enclosures, in freight containers, in railway box-cars and under gas-proof sheets. Capacities of enclosures vary from a few cubic metres to 2–3000 cubic metres.

6.1.1.2.3 Fumigation of Herbs and Spices

Herbs and spices are usually dried flowers, fruits, leaves, barks and other parts of plants used mainly for culinary and medicinal purposes (Table 6.11).

These commodities are prone to attack by insects such as tropical warehouse moth (*E. cautella*), cigarette beetle (*L. serricorne*), grain beetle (*Oryzaephilus* spp.) drugstore beetle (*S. paniceum*) and flour beetle (*Tribolium* spp.).

These high value commodities are usually grown in tropical countries and due to the favourable weather conditions, the insect populations will be able to multiply very quickly and may damage the produce causing severe economic losses unless control procedures are in place at the time of harvest. Immediately after harvest, the commodities are dried and then either bagged or baled. Methyl bromide is the most favoured fumigant at the present time and the treatments are carried out in warehouses, in freight containers, in fumigation chambers and under gas-proof sheeting. A dose of 16 to 24 $g\,m^{-3}$ for 24 hours at or above 20 °C is recommended under normal

Table 6.11. List of some herbs and spices treated with MeBr

Herbs and Spices	
Common name	Scientific name
Bay	*Laurus nobilis*
Chillies	*Capsicum* spp.
Cinnamon	*Cinnamonum zeylanicum*
Cloves	*Syzygium aromaticum*
Corriander	*Coriandrum sativum*
Mint	*Mentha* spp.
Oregano	*Origanum vulgare*
Parsley	*Petroselinum crispum*
Rosemary	*Rosemaninus officinalis*
Saffron	*Crocus sativus*
Sage	*Salvia officinalis*
Turmeric	*Curcuma domestica*

atmospheric pressure and under sustained vacuum, at the same temperature a dose of $40 \, \mathrm{g \, m^{-3}}$ for 3 hours exposure (Bond 1984) is recommended.

6.1.1.2.4 Fumigation of Coffee, Cocoa Beans and Tea

After processing and packing, tea normally does not get infested unless it is in proximity with other infested commodities. Cocoa and coffee beans on the other hand may get infested with coffee bean weevil (*Araecerus fasciculatus*), tropical warehouse moth (*Ephestia cautella*) and warehouse moth (*Ephestia elutella*) amongst many other stored-product pests.

These commodities are produced in tropical countries where ambient temperature and humidities are conducive to insect multiplication and for coffee beans effective control measures are necessary to prevent severe losses. Usually there are agreements regarding nil tolerances of insects in beverage crops amongst countries or traders either for quarantine or phytosanitary reasons. Unlike tea which tends to be shipped in plywood chests, coffee and cocoa beans are transported in bags or in bulk. Due to high fat content in cocoa, multiple fumigations with MeBr are discouraged.

6.1.1.2.5 Fumigation of Tobacco and Related Products

Cigarette beetle, *Lasioderma serricorne* and tobacco moth, *Ephesita elutella* are the major pests in tobacco, another high value commodity grown in the tropical regions where the risk of infestation is high. Furthermore, it is transported around the world packed in various ways such as in large hessian sacks (hogsheads), in compressed bales and in large cardboard boxes. In the trade, even low level of infestation is usually unacceptable.

Methyl bromide used to be the main fumigant for treating raw tobacco or its finished products but currently its use is declining and alternative treatments are being favoured. When MeBr is used, loosely packed tobacco is treated in fumigation chambers or under gas-proof sheets with a dose ranging from $20\text{--}32 \, \mathrm{g \, m^{-3}}$ of MeBr for a period of 2–3 days, depending on the prevailing temperatures, higher doses being used at lower temperatures. But with compressed bales or densely packed tobacco it is difficult for MeBr to penetrate to the core of the pack and the treatments in such situations are carried out in a vacuum fumigation chamber, under a sustained vacuum of 25–150 mm of mercury with a dose of $64 \, \mathrm{g \, m^{-3}}$ or $84 \, \mathrm{g \, m^{-3}}$ for temperatures above or below 20 °C and for a 4-hour exposure period. In either of these methods, the main advantage of using MeBr is that it kills all developmental stages of the insects even at low temperatures (Benezet 1989).

6.1.1.2.6 Fumigation to Protect Objects of Arts and Heritage held in Museums Collections, Art Galleries, Books in Libraries

These are usually objects made from materials such as; bone, cotton, feather, horn, leather, paper, wood and wool etc. All these are attractive to insects and can become easily infested (Table 6.12). Many of these artefacts are collected from ethnic origins, traded and transported internationally and may harbour pests of quarantine importance.

Insects may damage or destroy (Nair 1986) some artefacts of historic, scientific or cultural significance and which may be irreplaceable. Therefore, many museums operate a control system by fumigating all incoming goods routinely which ensures that only insect-free objects are brought into the premises.

Museum objects are usually fumigated in purpose-built atmospheric or vacuum fumigation chambers (Figure 6.3), under gas-proof sheets and also, in a recently developed portable enclosure or 'bubble' (Smith 1988). This essentially consists of a ground sheet on which the materials to be fumigated are placed and a sheet covering this is sealed to the ground sheet with a special zip arrangement. Dosing and venting of the 'bubble' are also designed as integral parts of the system.

To hasten fumigation or to improve penetration the art objects are sometimes fumigated with MeBr under sustained vacuum of 25–150 mm of Hg. The dosage rates may vary from $40 \, \mathrm{g \, m^{-3}}$ to $64 \, \mathrm{g \, m^{-3}}$ for 3–4 hours exposure at 20 °C.

6.1.1.2.7 Fumigation of Logs, Timber and Wood Products

Wood commodities such as logs, bark, sawn timber; wooden products like timber dunnage, crates as well as toys, sports goods, plywood and packaging materials, are subjected to phytosanitary or quarantine regulations in the

Table 6.12. Examples of some common pests in museum artefacts (Pinniger 1994)

Common name	Scientific name
Book lice	*Liposcelis bostrychophila*
Carpet beetles	*Anthrenus* spp.
Clothes moth	*Tinea* spp./*Tineola bisselliella*
Drugstore beetle	*Stegobium paniceum*
Hide beetle	*Dermestes* spp.
Khapra beetle	*Trogoderma granarium*
Museum beetle	*Attagenus pellio*
Powder-post beetle	*Lyctus* spp.
Silverfish	*Lepisma saccharina*

Figure 6.3. Museum artefacts being fumigated in a vacuum fumigation chamber

international trade. The regulations may be specific for a country and designed to protect their own industries and forests from the invasion of any damaging pests (Table 6.13). As a result the criteria for allowing timber into each individual country may vary considerably. To prevent the spread of American oak-wilt disease (*Ceratocystis fagacearum*) into the member states of the EU, a control method has been developed using a dose of $240\,\mathrm{g\,m^{-3}}$ of MeBr with an exposure period of 3 days at or above $3\,°C$ (Liese and Ruetze 1985).

When pests are found during an inspection, the whole consignment is treated at the port of entry. Some countries stipulate a phytosanitary certificate to accompany any incoming goods packed in wooden frame or with wood-shavings or cotton wool.

Due to short exposure period required, the rate of penetration into a bulk and the degree of control attainable, MeBr enjoys commercial acceptability

Table 6.13. Examples of insect pests in logs, wood and bark products

Common name	Scientific name
Bark beetles	*Scolytidae* spp.
Book beetles	*Dentroctonus* spp.
Easter larch borer	*Tetropium cinnamopterum*
Longhorn beetle	*Cerambycidea scolytidea*
Powder-post beetles	*Lyctus* spp.
Sawyer beetles	*Monochamus* spp.
Woodwasp, sirex	*Urocerus gigas*

for treating large bulks of timber (Figure 6.4). Large bulks of logs, timber and bark materials are usually fumigated with MeBr under gas-proof sheets or on-board ships or barges. A dose of $32 \, \text{g m}^{-3}$ at temperatures of 15 °C and above is normally used when fumigation is carried out under gas-proof sheets (Burden and McMullen 1951) for a period of 24 hours while a dose of $48 \, \text{g m}^{-3}$ is recommended for in-ship fumigation.

To comply with the United Kingdom Forestry Commission Directive, all imported bark must be fumigated with MeBr in accordance with an approved method. This is to prevent importation of undesirable timber pests which could devastate UK forests and commercial woodlands. Moreover, if an infested consignment of timber is identified on inspection at the port of entry, the Forestry Commission depends on MeBr fumigation for an immediate cure (BPCA 1992). Countries such as Japan and Australia have made it a mandatory requirement that all shipments of wood and related materials used as wooden crates, pallets etc. in the load for imports, should be fumigated with MeBr.

Fumigation of wooden products are normally conducted in containers, fumigation chambers or special enclosures with a dose of $24–48 \, \text{g m}^{-3}$ of MeBr, depending on the temperature and gastightness of the structures.

6.1.1.2.8 Fumigation to Treat Seeds Before Transportation and Planting

Movement of seeds in international trade pose a problem due to the risk of spreading seed-borne pests. In many countries there are regulations regard-

Figure 6.4. Fumigation of sawn timber under gas-proof sheet

ing certification of seeds to control the spread of these pests, mainly nematodes (Table 6.14), inside and outside the country. Quarantine procedures for the treatment of seeds with MeBr to control stored-product insects in general has been compiled by the European and Mediterranean Plant Protection Organisation (Anon 1993b).

Methyl bromide is being used as a fumigant for seeds because of its ability to penetrate into large bulks of sacks or bags and is considered as a standard technique for controlling nematodes even when dormant. To control nematodes, CTPs higher than those needed for insect pests are necessary. Under certain circumstances, if the moisture or oil content of the seeds are high, fumigation with a high dosage of MeBr may result in loss of viability, delay in germination (Khanna and Yadar 1987) or sometimes retarded growth after germination (Hanson *et al* 1987). For lucerne seeds a range of CTPs are suggested (Goodey 1945, Lubatti and Smith 1948) for seeds of differing moisture content, when being fumigated at 12–15% for 24 hours as shown below.

% Moisture Content	Range of CTPs $gh\,m^{-3}$
Less than 10	1400–1500
10–11	1200–1300
11–12	1000–1100
12–14	800–900

Fumigation of seeds with over 14% moisture content is not recommended. Generally it has been found that if the moisture content is less than 12% and the recommended dosages are adhered to, there is less likelihood of any damage to the seeds (Blackith and Lubatti, 1965). Repeat fumigations with MeBr on a given consignment may reduce germination potential progressively (Strong and Lindgren 1961) and adversely affect growth and later yield from the plant.

6.2.1 DISINFESTATION OF STRUCTURES BY FUMIGATION

In many situations fumigation of a buildings are conducted at regular intervals. Structural fumigation is considered as an integral part of a pest management system. A structure may be empty or may contain commodities of various kinds such as agricultural products, either stored or being processed or it may contain materials, other than food which can be attractive to insects. When the infestation reaches a certain level which is unacceptable to the authorities of the establishment, a fumigation or other pest control treatment is contemplated. When the infestation is in inaccessible areas and needs to be controlled within a short time period, fumigation with MeBr is usually recommended. In many situations, irrespective of the

Table 6.14. Examples of some seed-borne nematodes which may be transmitted into the stems and bulbs

Seeds of:	Scientific names of nematodes
Aster	*Aphelenchoides ritzemabosi*
Beet	*Ditylenchus dipsaci*
Broad bean	*Heterodera schachtii*
Bentgrass	*Anguina agrostis*
Coconut	*Rhadinaphelenchus cocophilus*
Clover	*Ditylenchus dipsaci*
Carrot	
Field bean	*Ditylenchus dipsaci*
Grass	*Subanguina chrysopogoni*
Lucerne	*Ditylenchus dipsaci*
Millet	*Panagrolaimus* spp.
Oat	
Onion	*Ditylenchus dipsaci*
Plantain	
Pea	
Rice	*Aphelenchoides besseyi*
	Ditylenchus angustus
Rye grass	*Anguina funesta*
Rye	*Anguina tritici*
Wheat	

degree of infestation, buildings are fumigated regularly, sometimes as an easy option for controlling pests or as a clean up operation.

6.2.1.1 Fumigation Techniques and Procedures for:

● Buildings

There are two approaches when considering a fumigation of a structure. Either the building itself is used to contain the fumigant or the whole structure is covered with suitable sheeting material to contain the gas (Armitage 1956, Rasmussen 1967). In the first category flour mills, processing areas and museums where walls will be relatively gas-tight and other openings such as doors and windows can be sealed effectively. In the latter approach, dwelling houses, churches and similar structures with infestations such as drywood termites and wood boring insects inherent in the structure of the building, have to be treated effectively. Most flour mills and food processing areas are fumigated on a regular basis.

Methyl bromide has been found invaluable for structural fumigations

Table 6.15. Structures and their pests

Description	Range of Pests
Buildings used for the production, storage and distribution of food, e.g. flour mills, food processing plants, warehouses and structures engaged in similar activities.	Stored-product insects, mites, psocids, cockroaches and rodents.
Residential buildings, stately homes, art galleries, libraries, churches, castles and structures housing art and museum collections.	Wood boring beetles, dry wood termites, dermestid beetles, cigarette beetles, drugstore beetles, clothes moths, mites, cockroaches and rodents.
Transport vehicles such as trucks, railway box-cars, freight containers, aircraft, and ships, when empty (otherwise considered as a commodity fumigation).	Stored product insects, cockroaches and rodents.

and replaced the formerly used hydrogen cyanide largely because of an improved penetrative capability into the cracks and crevices of a structure. It can also penetrate into accumulated food debris in a flour mill or into the leaves of books in a library or into wooden beams etc. in a dwelling.

Fumigation with MeBr is a major undertaking and requires that a building be sealed extensively (flour mills warehouses etc.) or covered with tarpaulin (dwellings, truck and freight containers) to make it as gas-proof as possible (5.1.2.6). The advantage of a properly conducted MeBr fumigation is that within a reasonable length of time it renders the structure virtually free of living pests including the immature life stages of arthropods. The biology and control of insects in flour mills are described by Hill (1978). To fumigate, vaporised MeBr is dosed into a building with an elaborate system of pipes and perforated lay-flat tubing. Gas concentrations are monitored from areas considered vulnerable and may need an additional dose to maintain concentrations at the required level. Adverse weather condition will influence a fumigation treatment considerably and careful monitoring and further topping up with MeBr may be necessary. The duration of treatments for flour mills, dwellings etc. against insect pests would normally be 24 hours. The aeration period to release a structure as gas-free depends on the materials used for construction and also on the commodities it may contain. Aeration periods are typically 24 hours for flour mills and 48 hours or more for dwellings. At the moment there is no single method or material to replace MeBr for structural fumigation for situations where it is currently used (UNEP 1992).

- Transport vehicles

These include trucks, freight containers and railway box cars. An empty vehicle may contain infestation from the spoils of the previous load which may invade the next cargo and spread it to the warehouse. Fumigation will eliminate possible cross-infestation and for efficiency and speed MeBr is often used. When in good condition containers can make efficient fumigation units but many in use are less gas-tight than they appear, doorways and floors are weak points difficult to seal. It will be necessary to cover trucks and sometimes containers with gas-proof sheets to ensure adequate treatment.

Many containers are fitted with special fumigant application points for use with cans of MeBr, others are dosed via a rigid pipe which is passed through the seal of the door. Dosing from the cans and cylinders are done mostly by allowing liquid MeBr to vaporise on contact with ambient conditions.

Vaporised gas is used sometimes and in the UK a mixture containing 80% MeBr in liquid carbon dioxide is also used in freight containers to achieve better distribution. In the USA a number of containers are fumigated together under one gas-proof sheet with their doors open under the sheet, and fans are used to assist in the mixing of the gas. Venting is accomplished by removing the sheet or just by opening the doors of the containers of box-cars and may take about 1–2 hours. Detection instruments are used to check gas concentrations before the containers are declared safe for entry.

- Ships and barges

Empty holds in cargo and in store-rooms etc. in passenger ships may often have residual infestations of stored product pests and present problems not experienced in any other kind of fumigation operation. The design of a ship, configuration of holds and crews' quarters and the age of the ship are the major factors which are taken into consideration to conduct an effective fumigation. Insecticidal spray, mist or smoke cannot be relied upon for adequate control (Monro 1969), whereas MeBr penetrates well into the cracks, crevices and hidden areas of the ships and facilitates control. Infestations of quarantine insect pests or of rats come under the international plant health and human health regulations. Fumigation on board ships is governed by the 'Recommendations for the Safe Use of Pesticides in Ships' (Anon 1993c). Doses of MeBr vary from $16\,\mathrm{g\,m^{-3}}$ to $32\,\mathrm{g\,m^{-3}}$ in temperatures ranging from above or below $10\,^\circ\mathrm{C}$. For *Trogoderma granarium* (khapra beetle) which is highly tolerant to MeBr, a dose of six times higher has been recommended (Slabodnik 1962). Fumigation with MeBr are carried out as in mills and warehouses. Vaporised gas is introduced and fans are used in the holds for proper distribution.

- Aircraft

    When infestations of mice or cockroaches are reported, aircraft are grounded immediately. In such circumstances, a quick but reliable control measure is called for and MeBr currently fulfils this requirement.

    As liquid MeBr reacts with aluminium, pure vaporised gas is used at the rate of $8-20\,\mathrm{g\,m^{-3}}$ for 4 hours depending on the kind of pests. A mixture of MeBr in liquid carbon dioxide is found to be very useful (Wainman *et al* 1983) for aircraft fumigation because of its instantaneous vaporisation, without employing any heating equipment. Aircraft need thorough checking for residual MeBr before being put into the service.

6.3.1 CONCLUSIONS

Infestation is a problem in food production and in many other commodities which are stored and transported. Apart from the enormous food loss and destruction of commodities by pests, the consumer is now neither ready to accept infested and contaminated food, nor does he like to live with infestation of any kind. Depending on the circumstances the pest problems are selectively controlled by the use of many available insecticides or one of the few fumigants along with a system of pest management.

Methyl bromide is a major tool in the pest control process and in its 60 years of use, there have been no reports of development of pest resistance against it. The importance of MeBr is due to its effectiveness against a broad range of insect pests, rodents, mites and nematodes; its ability to penetrate inaccessible areas and into bulk food products. As the pests are susceptible to MeBr at relatively low concentrations, the gas usually does not pose residue problems in food commodities. Methyl bromide offers complete control of all life stages of insects within a short (up to 48 hours) exposure period which is advantageous for the movement of commodities in national and international trade, for the treatment of production areas and for dwellings.

For quarantine applications in many situations MeBr treatments are the only available approved procedures accepted by the regulatory authorities around the world. Concerted research efforts will be necessary to bring about reduced use of MeBr and will require many procedural changes in storage, transportation and food handling practices. Also, the implementation of anything different from the current practices will need time and co-operation of the relevant agencies.

REFERENCES

Anon, 1976, Plant protection and quarantine treatment manual. Section N. Treatment facilities—Part 2. Atmospheric fumigation chambers—construction and performance standards—fittings for pressure—leakage test and fumigant concentration sampling. MAFF, p. 4.

Anon, 1985, Australian Quarantine and Inspection Service Plant Quarantine Treatment Schedule, AQIS, p 21.

Anon, 1988, Australian Quarantine and Inspection Service Plant Quarantine Manual, Sec. E. AQIS, p 1.

Anon, 1992, Animal and Plant Health Inspection Service Plant Protection and Quarantine Treatment Manual, USDA, p 524.

Anon, 1993a, *Bulletin of European and Mediterranean Plant Protection Organisation*, **23**, 207.

Anon, 1993b, *Bulletin of European and Mediterranean Plant Protection Organisation*, **23**, 203.

Anon, 1993c, International Maritime Organization. Recommendations on the safe use of pesticides in ships, 1993 edition, p 23.

Anon, 1994a, EH 40. Occupational Exposure limits, Health and Safety Executive HMSO Publications Centre, London. p 45.

Anon, 1994b, Animal and Plant Health Inspection Service Plant Protection and Quarantine Treatment Manual, USDA, T101.

Anon, 1994c, Animal and Plant Health Inspection Service Plant Protection and Quarantine Treatment Manual, USDA, T108.

Anon, 1994d, Animal and Plant Health Inspection Service Plant Protection and Quarantine Treatment Manual, USDA, T200.

Anon, 1995, UNEP 1994. Report of the Methyl Bromide Technical Options Committee, UNEP, Ozone Secretariat, Nairobi. ISBN No. 92-807-1448-1.

Anthon EW, Moffit HR, Couey HM and Smith LO, 1975, *J. Econ. Entomol.*, **68**, 524.

Armitage HM, 1956, *J. Econ. Entomol.*, **49**, 490.

Armstrong JW and Couey HM, 1984, *J. Econ. Entomol.*, **77**, 1229.

Armstrong JW, Schneider EL, Garcia DL and Couey HM, 1984, *J. Econ. Entomol.*, **77**, 680.

Armstrong JW and Garcia DL, 1985, *J. Econ. Entomol.*, **78**, 1308.

Armstrong JW, Harvey JM, Garcia DL, Menezes TD and Brown SA, 1988, *J. Econ. Entomol.*, **81**, 1120.

Banks HJ, 1984, *Proc. Aust. Dev. Asst. Course on Presentation of Stored Cereals*, **2**, 533.

Banks HJ, 1986, in: Champ BR and Highley E (eds), *Proc. on Pesticides and Humid Tropical Grain Storage Systems*, Manila, Philippines, May, p 291.

Banks HJ, 1994, in: Highley E, Wright EJ, Banks HJ and Champ BR (eds), *Proc. 6th Int. Wkg Conf. Stored Prod. Prot.*, April 1994, Canberra, Australia, CAB International, p 2.

Banks HJ, Desmarchelier JM and Elek JA, 1976, *Pestic. Sci.*, **7**, 595.

Banks HJ and Annis PC, 1977, Canberra, Commonwealth Scientific and Industrial Research Organisation, Division of Entomology, Technical Paper No. 13, p 23.

BPCA, 1992, British Pest Control Assn. Submission to the Dept. of Trade and Industry, May 1992.

Bell CH, 1978, *Pestic. Sci.*, **9**, 529.

Benezet HJ, 1989, *Proc. 43rd Tobacco Chemist's Research Conference*, **15**, 1.

Benschoter CA, 1979, *Proc. Fla. State Hort. Soc.*, **92**, 166.
Blackith RE and Lubatti OF, 1965, *J. Sci. Food Agric.*, **16**, 455.
Bond EJ, 1984, Manual of fumigation for insect control. FAO Plant Production and Protection Paper No. 54, FAO, Rome, p 432.
Brecht JK, Huber DJ, Sherman M and Lee J, 1986, *J. Plant Growth Regulations*, **5**, 29.
Brown WB and Heuser SG, 1953a, *J. Sci. Food Agric.*, **4**, 48.
Brown WB and Heuser SG, 1953b, *J. Sci. Food Agric.*, **4**, 378.
Brown WB and Heuser SG, 1956, *J. Sci. Food Agric.*, **7**, 595.
Bruch CW, 1961, *Ann. Rev. Microbiol.*, **15**, 245.
Burden JH and McMullen MJ, 1951, *Aust. J. Sci.*, **14**, 57.
Calderon M and Carmi Y, 1973, *J. Stored Prod. Res.*, **8**, 315.
Carmi Y, 1985, *OEPP/EPPO Bulletin*, **15**, 15.
Chapman PJ, 1940, *J. Econ. Entomol.*, **48**, 483.
Couey HM, 1983, *Hortiscience*, **18**, 45.
Cowley JM, Baker RT, Engleberger KG and Langi TG, 1991, *E. Econ. Entomol.*, **78**, 879.
Dumas T, 1982, *J. Assoc. Off. Anal. Chem.*, **65**, 913.
Dumas T and Bond EJ, 1985, *J. Agric. Food Chem.*, **33**, 276.
Fairall RF and Scudamore KA, 1980, *Analyst, London*, **105**, 251.
FAO/WHO, 1967, FAO-PL: CP115; Geneva, WHO/Food Add. 67. 32.
FAO/WHO, 1980, FAO Plant Production and Protection Paper 20 sup.
Galetti GL and Berger SH, 1987, *Simiente*, **57**, 201.
Gaur SN, Raha SK, Masood M and Chattopadhyay D, 1984, *Plant Prot. Bull. India*, **36**, 133.
Gaunce AP, Madsen HF, McMullen RD and Hall JW, 1980, *Canadian Entomologist*, **112**, 1033.
Goodey T, 1945, *J. Helminthol.*, **21**, 45–59.
Greve PA and Hodgendoorn EA, 1979, *Meded. Fac. Landbouwwet. Rijksuniv. Gent.*, **44**, 877.
Hanson PR, Wainman HE and Chakrabarti B, 1987, *Seed Science and Technology*, **15**, 155.
Health and Welfare, 1975, Canada, Information Letter No. 454, 29 December
Helyer NL and Ledieu MS, 1989, *Annals of App. Biol.*, **114**, 9.
Heuser SG and Scudamore KA, 1968, *Analyst, London*, **93**, 252.
Heuser SG and Scudamore KA, 1969, *J. Sci. Food Agric.*, **20**, 566.
Heuser SG and Scudamore KA, 1970, *Pestic. Sci.*, **1**. 244.
Hill EG, 1978, London, H.M.S.O., Ministry of Agriculture, Fisheries and Food. p 137.
Hinsch RT and Harris CM, 1992, *Int J. Refrigeration*, **15**, 59.
IMO, 1993, International Maritime Organization. Recommendations on the safe use of pesticides in ships, 1993 edition, p 1–30.
Japan Ministry of Agriculture, Forestry and Fisheries, 1987, Collection of Ministry Notifications of Plant Production Laws and Regulations. Plant Protection Division, Japan Plant Quarantine Association. Notification No. 187.
Johnson AC, Bowen CH and Phillips GL, 1947, USDA Publication, E-711, p 44.
Junaid AHM and Nasir MM, 1956, *Agric. Pak.*, **6**, 80.
Khanna SC and Yadar TD, 1987, *Seed-Research*, **15**, 183.
King JR and Benschoter CA, 1991, *J. Agric. Food Chem.*, **39**, 1307.
Lepigre AL, 1949, Alger, Insectarium, Jardin d'essai.
Liese W and Ruetze M, 1985, *EPPO Bulletin*, **15**, 29.

Lipton WJ, Tebbets JS, Spitler GH and Hartsell PL, 1982, USDA Marketing Res. Report. No. 1125.

Lubatti OF and Smith B, 1948, *J. Soc. Chem., Ind., London*, **67**, 297.

MAFF, 1974, Ministry of Agriculture, Fisheries and Food. Fumigation with methyl bromide under gas proof sheets 41.

MAFF, 1982, Ministry of Agriculture, Fisheries and Food Surveillance Paper No. 9. HMSO. p 14.

Matthews RH, Fifield CC, Hartsing TF, Storey CL and Dennis NM, 1970a, *Cereal Chem.*, **47**, 579.

Matthews RH, Fifield CC and Hartsing TF, 1970b, *Cereal Chem.*, **47**, 587.

Meheriuk M, Gaunee AP and Dyck VA, 1990, *Hort. Science*, **25**, 538.

Moffit HR, 1971, *J. Econ. Entomol.*, **64**, 1258.

Moffit HR, Burditt AK Jr and Sell CR, 1983, *Proc. Oregon State Hort. Soc.*, **74**, 38.

Monro HAU, 1956, *Down to Earth*, **11**, 19.

Monro HAU, 1958, Thesis, McGill Univ.

Monro HAU, 1969, Canada Department of Agriculture, Publication, No. 855 Rev.

Monro HAU and King JE, 1954, *J. Sci. Food Agric.*, **5**, 619.

Morgan CVG, Gaunce AP and Jong C, 1974, *Can. Entomol.*, **106**, 917.

Nair SM, 1986, in: Barry S and Haughton DR, (eds), Proc. 6th Int. Biodeterioration Symposium, Washington DC. August, 1984, 337.

New Zealand Ministry of Agriculture and Fisheries, 1992a, Border protection specifications: clearance of fresh produce, NASS: 152.02. Sec. 2,0; T2–11.

New Zealand Ministry of Agriculture and Fisheries, 1992b, NASS: 152:02. Border Protection specifications. Clearances of fresh produce, section 2.0, T8.

Phillips WR and Monro HAU, 1939, *J. Econ. Entomol.*, **32**, 344.

Pimental D, 1991, in: J Richard Gorham (ed), Ecology and Management of Food-industry Pests; AOAC, 5–10.

Pinniger D, 1994, Archetype Publications. Insect Pests in Museums, 3rd edition, p 8–22.

Rasmussen S, 1967, *Hylotropes bajulus. Mater. Org.*, **2**, 65.

Reeves RG, McDaniel CA and Ford JH, 1985, *J. Agric. Food Chem.*, **33**, 780.

Rhodes AA, 1986, Interagency Agreement No. DE-A 104-83 AL 24327.

Richardson HH, 1955, *J. Econ. Entomol.*, **48**, 483.

Richardson LT and Monro HAU, 1962, *Appl. Microbiol.*, **10**, 448.

Scott HG, 1991, in: J Richard Gorham (ed), Ecology and Management of Food-industry Pests; AOAC, p 463.

Slabodnik M, 1962, *Pest Control*, **30**, 30.

Smith CP, 1988, Proc. 8th Brit. Pest Control Conf. Stratford-upon-Avon, UK., 2nd Session, Paper 3, p 15.

Soma Y, Sunagawa K, Kurokawa K, Nakamura M, Misumi T and Kawakani F, 1991, *Japanese Plant Protection Service Res. Bulletin*, **27**, 83.

Spalding DH, King JR, Benschoter CA, von Windeguth DL, Reeder WF and Burditt AK, 1978, *Proc. Fla. State Hort. Soc.*, **91**, 156.

Spitler GH and Couey HM, 1983, *J. Econ. Entomol.*, **76**, 547.

Stechmann DH, Englberger K and Langi TF, 1988, *Anzeiger für schadlingskunde, Pflanzenschutz, Umweltschutz*, **61**, 125.

Steinweden JB, Mackie DB, Carter WB and Smith SS, 1942, *Bull. Dep. Agric. Calif.*, **31**, 31.

Stijve T, 1977, *Dtsch. Lebensm. Rundsch.*, **73**, 321.

Storey CL, 1967, USDA Marketing Res. Report No. 794, p 16.

Storey CL, 1971a, USDA Marketing Res. Report No. 915, p 17.

Storey CL, 1971b, USDA Marketing Res. Report No. 929, p 8.

Strong RG and Lindgren DL, 1961, *J. Econ. Entomol.*, **54**, 764.

Sun YP, 1946, St. Paul, Minnesota Agriculture Experiment Station Technical Bulletin No. 177.

Tebbets JS, Vail PV, Hartsell PL and Nelson HD, 1986, *J. Econ. Entomol.*, **79**, 1039.

UNEP, 1992, Methyl Bromide Technical Options Workshop, Washington DC. UNEP Ozone Secretariat, Nairobi, Kenya..

USEPA, 1989, Letter dated July 7.

Waddell BC, 1993, *The Orchardist of New Zealand*, **66**, 25.

Waddell BC, Clare GK, Maindonald JH and Petry RJ, 1993, Cook Islands Report-3 Hort. Research Client Report No. 93/270.

Wainman HE and Chakrabarti B, 1973, *Int. Pest Control*, **15**, Nov/Dec., 6–10.

Wainman HE, Chakrabarti B and Warre PR, 1983, *Int. Pest Control*, **25**, 174.

Wirtz RA, 1991, FDA Technical Bulletin 4. Arlington V A., AOAC, p 469.

Wontner-Smith T and Chakrabarti B, 1994, *Int. Pest Control*, **36**, 15.

Yokoyama VY, Miller GT and Hartsell PL, 1987a, *J. Econ. Entomol.*, **80**, 1226.

Yokoyama VY, Miller GT and Hartsell PL, 1987b, *J. Econ. Entomol.*, **80**, 840.

Yokoyama VY, Miller GT and Hartsell PL, 1990a, *J Econ. Entomol.*, **83**, 466.

Yokoyama VY, Miller GT and Hartsell PL, 1990b, *J Econ. Entomol.*, **83**, 2335.

Yokoyama VY, Miller GT and Hartsell PL, 1994, *J Econ. Entomol.*, **87**, 730.

7

Alternatives—Chemicals

P. C. ANNIS AND C. J. WATERFORD
CSIRO Division of Entomology, Canberra, Australia

C.H. Bell, N. Price and B. Chakrabarti: The Methyl Bromide Issue

1.1 INTRODUCTION

Methyl bromide is a fast acting fumigant with a long history of use (Bond 1984). Since its introduction in the early thirties (Mackie 1938) it has become an accepted part of normal treatment procedures for pest control. Methyl bromide has a broad range of efficacy and affords rapid treatments in durable and perishable commodities; soil sterilisation; and structures and transport; with particular use in quarantine. Restrictions on its use because of ozone depletion, has set an interesting scientific, technical and political challenge. A rare degree of co-operation between all three sectors is required, if viable solutions are to be in place for many uses, by the time methyl bromide is withdrawn.

   No single 'plug-in' replacement for methyl bromide appear to exist because of its broad spectrum of activity amongst the likely chemical alternatives. Adoption of most replacements will require significant changes in pest management practice. Furthermore, many of the options are not yet at the stage where they could be used extensively without considerable further laboratory and field evaluation. In addition new chemicals will face the hurdle of registration and older chemicals that have fallen into disuse will have to be re-registered. Those chemicals that have a limited scope, either by use or country, will need further work to extend their use into other areas. Whatever the mix that finally replaces the many uses of methyl bromide, that mix will be far more complex to administer and control if the expected efficacy is to be achieved and maintained. The choice of alternative may change from year to year depending on the target pest and purpose for control. In this environment the need for integrated pest management (IPM) tools and decision support systems (DSS) will be one way of ensuring that the body of knowledge and expertise necessary is not lost and is passed on to farmers, producers, marketers and consumers.

It can be anticipated that many alternatives will be more expensive and less convenient than methyl bromide has been in the past. However, any cost considerations must take account of other positive factors including environmental and health benefits that will arise from phasing out methyl bromide (Lubulwa *et al* 1995).

A number of potential alternative chemicals are identified and these are briefly reported in this chapter. They include fumigant and non-fumigant materials. However, many of them are under study for their potential to cause negative health effects as well as possible detrimental effects on the environment. It is likely that regulatory restrictions on alternatives to methyl bromide will be as severe as for other agrochemicals. This means that the conditions placed on the use of chemical treatments for agricultural pests (including alternatives to methyl bromide) will become much harder to meet than was the case in the past.

This chapter presents a review of these chemical alternatives to methyl bromide and gives some indication of their current availability and applicability as replacements for methyl bromide in any of its applications. Potential replacement chemicals and treatment applications fall into one or more of the following broad categories of availability and readiness for use.

- Currently in use, widely accepted and efficacious. Such alternatives may be considered as 'off the shelf' methods.
- Under active development and possibly in limited commercial use. These would need extensive evaluation for each new application or jurisdiction.
- Under active research with a very high potential for use. These alternatives require registration in addition to evaluation for each new application.
- Have a high potential but in the very early stages of research. These may provide long-term answers but would not be expected to be available during the initial phase-out of methyl bromide.
- Widely used in some applications and not in others, extensive evaluation will be required to extend the use to other commodities and situations.
- Older materials, apparently efficacious but now fallen into disuse, they would have to face registration hurdles as any new materials under development.
- Older materials whose use has been discontinued for a range of well documented reasons including residues, environmental issues and human toxicity or carcinogenicity.

In this text we discuss uses of methyl bromide in four broad treatment categories. These form the basis for further discussion on individual chemicals and chemical groups.

1. **Soil fumigation, and sterilisation** accounts for approximately 75% of current annual use (UNEP 1994) of methyl bromide. The aim is to control insect, nematode, and fungal pests, weeds in the production of strawberries, tobacco, cucurbits, tomatoes and peppers; hence this sector requires the broadcast range of efficacy in any replacement strategy. Chemical alternatives exist, mainly as fumigants, or non-fumigants such as herbicides, fungicides, nematicides and insecticides. Individually they do not have the broad range of efficacy required. Mixtures and combinations, with non-chemical alternatives to achieve a similar broad spectrum, are used commercially and are regarded as effective in many production systems (Thomson 1989). However, more work is required for full understanding of these combinations.

2. Disinfestation of **durable commodities** accounts for 13% of annual use (UNEP 1994). Alternative chemicals, both fumigants and non-fumigants, are well developed, effective and in place in this sector. Methyl bromide plays a small but significant role because of its rapid action and reliability. In a few important situations it remains the treatment of choice or is mandatory particularly where quarantine treatment is required at point of import. Development of acceptable alternative quarantine treatments is the major challenge in this sector.

3. Disinfestation of **perishable commodities** uses 8.6% (UNEP 1994). Chemical alternatives in this sector are generally of limited commercial use because they are either difficult to apply, have a narrow spectrum of activity, damage the commodity, or lack registration approval in some countries. Some suggested alternatives are still in the experimental stage or lack registration.

4. Disinfestation treatment of **structures and transport** accounts for most of the remainder (UNEP 1994). Though alternative chemicals exist that can be used in this sector, mainly other fumigants, along with insecticides, residual dusts and a range of application methods, none can replace methyl bromide in all cases.

Alternative chemicals fall into two major categories. Those that are used as fumigants and those that are used as contact treatments. Fumigants are generally active in the gaseous phase whereas contact treatments, including insecticides and fungicides, are generally active in the solid (and very rarely in the liquid) phase.

1.1.1 FUMIGATION TREATMENTS

Fumigation is the act of releasing and dispersing a pesticidal chemical so that it reaches a pest completely, or partially, while in the gaseous state. Effective fumigation depends on achieving a combination of concentrations, and times (Ct) at differing temperatures designed to kill pests without damage to the commodity, artefact or structure that they infest.

Methyl bromide is effective in all the four treatment categories listed above. However the range of alternative fumigants does not include any one that has the same broad range of efficacy.

Phosphine, except for soil fumigation and perishables, is currently the most widely used alternative. It is easy to use in many situations, and in some, may be preferable to methyl bromide, with lower residue and taint potential for consumables. However, its different chemical, physical and toxicological properties make it technically inappropriate as a substitute in other situations. It has very little activity against fungi and requires long exposures to control the life stages of some insect pests, particularly at low temperatures. Electrical equipment containing copper may be corroded and precautions must be taken to avoid flammability and explosion hazards.

Other chemicals such as ethylene dibromide (EDB) and ethylene oxide, though highly effective, have been largely withdrawn because of health and safety issues. However EDB should not be discounted. Maybe researchers should revisit it and reassess the perceived hazards associated with it in the light of the demise of methyl bromide. Hydrogen cyanide and carbon disulphide, which were previously in wide use, will require registration or re-registration if they are to be used as alternatives for methyl bromide.

Newer fumigants such as sulphuryl fluoride and carbonyl sulphide either do not have the broad range of application or still face significant research and financial hurdles before they can be registered to cover their wide range of possible applications. It is interesting to note that in their 1938 paper, Fisk and Shepard (1938), dealing with efficacy of the 'new' fumigant methyl bromide, compared it with hydrogen cyanide, chloropicrin, ethylene oxide and carbon disulphide. These were then the chemicals for which methyl bromide was to be the alternative.

1.1.2 CONTACT TREATMENTS

Contact treatments include insecticides, nematicides, insect growth regulators (IGRs), fungicides, certain botanicals preperations with useful active properties, and inert dusts. Application is by:

- admixture with commodities,
- spraying solutions onto commodities,

- fogging into the airspace where protection is required,
- structural treatments where surfaces are coated, and
- chemical dips.

Unlike fumigants, contact treatments can provide persistent protection against reinfestation as long as the active ingredient retains potency (Desmarchelier 1978). This is particularly true of insecticides applied either directly to grain for protection against insect pests, or to storage buildings and transport vehicles to reduce the likelihood of re-infestation of commodities being stored or transported. Contact treatments are not normally registered for use on processed commodities. One disadvantage of contact treatments is the possibility that particular chemicals may become ineffective with frequent use because of build-up of resistance by the target organism (Champ 1984). Though resistance to both phosphine and methyl bromide (Champ and Dyte 1976) is known, the risk does not appear to be as great.

Recent trends to move away from insecticides on food and food precursors because of consumer sensitivity to residues, makes selection of alternatives to insecticides more difficult. However much of this commercial or consumer resistance is a general perception that all residues are bad and not necessarily based on hard data. The difficulties that the loss of methyl bromide may present could force a reassessment of the relative hazards of residues when protection of foodstocks becomes more pressing. A recent discussion paper from GASGA (1995) summarises the current issues that pertain to pesticide residues in grain.

All treatment categories have alternatives to methyl bromide that are more or less effective and fall into the category of contact treatment. Durables have the best developed range of alternatives registered and in general use (Snelson 1987). Although most insecticides applied to raw grains degrade or are removed before they reach the consumer (Desmarchelier 1978), residue levels in processed and unprocessed food are a sensitive issue (Kerin 1994). Tolerance levels can vary from country to country for particular materials (GASGA 1995). At the international level, an *ad hoc* committee examines scientific data submitted in accordance with a protocol of requirements established by the Joint WHO/FAO/Meeting of Experts on Pesticide Residues (JMPR) in order to determine safe and acceptable levels of residues of chemicals in raw agricultural commodities and foods (Snelson 1987).

Maximum residue levels (MRLs) are based on experiments designed to determine the nature and level of residues resulting from the application of the chemical in accordance with good agricultural practice (GASGA 1995). The safety and acceptability of these residues is determined by comparison with extensive toxicological studies carried out on laboratory animals. Studies for carcinogenicity, mutagenicity, teratogenicity, and effects on reproduction are also required. Recommendations for MRLs are made by the Codex Alimentarius as a result of these studies, but many countries also make their own legislation, which vary from these limits (Snelson 1987).

Inert dusts may act as alternatives in some applications (McLaughlin 1994). However, they require dry conditions for efficacy (r.h. < 80%) (Aldryhim 1993), may cause excessive wear in machinery, give rise to dust problems in the work space and can alter handling characteristics of stored commodities (Jackson and Webley 1994).

Generally, fumigants and contact chemicals act differently on target pests and play different roles in stored product protection. Such differences, when used on stored grain, are summarised in Table 7.1. Despite these differences, and where permitted by regulatory authorities and market preference, both techniques can provide a pest-free end product.

Table 7.1. Basic differences between fumigants and contact insecticides

Fumigants	Contact insecticides
Lasting protection only possible in insect proof enclosures	Lasting protection possible
Grain can be treated *in situ*	Normally grain must be moved to apply insecticide
Can be used to treat most commodities	In most countries, only permitted on commodities before processing
Disinfestation possible within 1–15 days, dependent on temperature	Complete disinfestation requires a longer period dependent on emergence of adult species that develop within the grain before they are exposed to insecticide deposits
Requires skilled, certified personnel for application	Semi-skilled operators can apply contact insecticides
Effective, generally, against all insect species but efficacy may be dependent on temperature	Compounds may be selectively effective against different insect species, also dependent on moisture content and grain temperature
No incidence of substantial (> 5×) methyl bromide tolerance known, but development of resistance to phosphine a current concern	Most insect pests have developed resistance to particular insecticides or groups of insecticides, with continued use
Good penetration of grain bulks	Virtually no penetration of grain bulks

1.1.3   CONSTRAINTS

Registration of any new chemical is time consuming and expensive. Full registration may take up to ten years. The costs related to registration are extremely high (Reichmuth 1990) as comprehensive evaluations are demanded by consumers and regulators. However, in the USA the EPA has indicated that it will speed up the registration process for alternatives.

There are also other· barriers to the introduction of substitutes or alternatives. These include technical and practical matters such as the need for training, requirement for suitable application equipment, cultural barriers caused by having to do things differently and logistical problems including the timing of alternative treatments in the established flow of produce from production centres to the market. Furthermore, most alternatives and substitutes do not have the same broad spectrum of activity as methyl bromide. Some are regarded as inferior in effectiveness requiring greater exposure time to be fully effective or may need to be used in combination with other chemicals, or with other measures in an integrated system to provide an adequate level of pest control.

Overall, the disadvantages of any particular alternative must be assessed against recognised operational disadvantages of methyl bromide, in addition to its effect on the ozone layer. These include: the need for strict safety precautions to ensure worker safety; safety in the local environment; provision of gastight enclosures for fumigation treatments; lack of access to the commodity under treatment; problems of excessive bromide ion residues and taint on some commodities; and problems of phytotoxicity with some seeds.

Irrespective of any possible phase-out of methyl bromide, alternatives will have to be assessed case-by-case, taking into account environmental issues, substrate treated, target pests, and the market and use for which the commodity is destined. Economics, logistics and engineering of installations, *inter alia*, will all need to be considered when introducing alternatives.

Other promising non-chemical alternatives, notably temperature manipulation and controlled atmospheres are available and will need consideration, as well as the chemical alternatives discussed in this chapter.

2.1   SOIL FUMIGATION

Application of methyl bromide to fumigate soil has been used successfully, under a wide variety of conditions, to enhance crop productivity in locations where a range of soil-borne pests, including plant pathogenic fungi, bacteria, viruses, invertebrate organisms and weeds, would otherwise limit crop production to uneconomic levels.

Use of methyl bromide for soil fumigation has been successfully replaced in some areas. In the Netherlands, a former major user, methyl bromide use was phased out between 1980 and 1992. The phase-out shows use of methyl bromide for soil fumigation is not a pre-requisite for financial viability in agriculture. However, none of the specific alternative methods based on chemicals discussed here, used alone, have the same broad spectrum of activity or efficacy of methyl bromide. However, mixtures of soil fumigants may provide broad activity approaching that of methyl bromide. The development of comparable agriculture systems without the use of methyl bromide will, in many cases, require integration of multiple alternative technologies and extensive research to achieve a similar spectrum of efficacy and reliability.

There are a reasonable number of chemicals that have actual or potential roles as partial replacements for methyl bromide in soil fumigation (Table 7.2). Most alternatives to methyl bromide will need to be part of an IPM system. IPM utilises pest monitoring techniques, establishment of pest injury levels, and a mix of strategies and tactics to prevent or manage pest problems in an environmentally sound and cost-effective manner (van Alebeek 1989).

2.1.1 CONSTRAINTS

While this list of potential replacement chemicals includes many compounds and mixtures, it can be seen that most are not immediately available for wide-spread use. Constraints to their immediate use are many but include: current registration status, toxicological data related to accidental exposure of humans, effects on target and non-target organisms, environmental and economic factors. Many of these may be overcome by research. However, in most cases, resource barriers, such as time, money and skills, are likely to slow down this process. Most alternatives are toxic in their own right (UNEP 1994) with a range of possible effects that need to be considered including acute toxicity, skin and eye irritation, reproductive effects, genotoxicity and oncogenecity, and any new, increased or modified use of these chemicals is likely to be scrutinised closely. Soil treatments present a difficult ecological challenge because they represent a potential starting point for contamination of the food chain and important elements of environmental systems such as aquifers. This care is required to ensure that such treatments, or breakdown products resulting from them: lose their toxicity quickly; will not accumulate with repeated application; and will not migrate rapidly to places where they may present an unintended hazard.

Table 7.2. Alternative soil treatments as replacements for methyl bromide

Common name	Status	Reference
1,3-dichloropropene	5	Johnson and Feldmesser (1987)
1,3-dichloropropene + Chloropicrin	5	Rodríguez-Kábana et al (1977), Thomson (1989)
1,3-dichloropropene + MITC	5	Rodríguez-Kábana et al (1977)
Anhydrous ammonia	5	Rodríguez-Kábana et al (1981, 1982)
Bromonitromethane	4	Rodríguez-Kábana et al (1991)
Carbon disulphide	6	Rodríguez-Kábana et al (1977), Thomson (1989)
Chloropicrin	3	Wilhelm et al (1974)
Dichlor-isopropyl ether	5	Thomson (1989)
Ethylene dibromide	7	Johnson and Feldmesser (1987)
Formaldehyde	4	Rodríguez-Kábana et al (1977)
Furfuraldehyde	4	Rodríguez-Kábana and Walters (1992)
Methyl isothiocyanate (MITC), MITC generators	2	Rodríguez-Kábana et al (1977)
Dazomet and Metam-sodium	5	Alsphach (1989)
Potassium azide	8	Kelley and Rodríguez-Kábana (1975), van Wambecke et al (1983, 1984)
Propargyl bromide	7	Rhode et al (1980)
Sodium azide	8	Kelley and Rodríguez-Kábana (1979a, b)
Sodium tetrathiocarbonate	4	Young (1990)
Sulphur dioxide	4	Rodríguez-Kábana et al (1977)

Status description of methods that are:

1 widely accepted, efficacious and currently available,
2 under active development and possibly in limited use,
3 undergoing active research with a very high potential for use,
4 of potential but in very early stages of research,
5 widely used for some applications and not at all for other,
6 efficacious older materials now fallen into disuse,
7 older material, use discontinued for well documented reasons,
8 has had limited use.

3.1 DURABLE COMMODITIES

Durable commodities are those that, when sufficiently dry, can be stored safely without deterioration for long periods. They include a variety of dry foodstuffs, principally cereal grains, grain products, oilseeds and legumes, beverage commodities, dried fruit and nuts, logs, timber and timber-containing products, and wood artefacts. Although such commodities are usually relatively dry, they are all subject to infestation by a range of pests which can breed and survive on them during storage. The necessity for disinfestation includes the risk of physical loss, quality loss and spread of exotic, and damaging pests from country to country.

Infestation control in durable commodities is most effectively undertaken by fumigation. Methyl bromide and phosphine are the two fumigants currently used for this purpose. Methyl bromide provides rapid and complete control of pests, within 24 hours at temperatures of 5 °C and above. It has been the fumigant of choice for disinfestation and quarantine treatments, particularly against khapra beetle and wood-destroying insects. In over 60 years of use, there have been no reports of development of technical resistance to methyl bromide (Champ 1985). However, because of the high dosage rates required for efficacy, combined with bromide residue limits small changes in resistance could alter this picture dramatically.

It is estimated that approximately 13% of the annual (1992) world non-feedstock use of methyl bromide is for disinfestation of durable commodities (UNEP 1994). It provides the principal means of pest control of some economically important industries. These industries include the dried fruit and the nut industry, export and import trade in unsawn timber and storage of bagged food grain stocks in some countries. Additionally substantial quantities of grain are fumigated, in some countries, with methyl bromide on import or export to meet phytosanitary requirements (FAO/NRI 1994).

Most uses of methyl bromide on durable commodities have existing or potential alternatives. However, there are no direct in-kind replacements for methyl bromide and all alternatives require some changes in practice. Although there are several potential alternatives, only phosphine is extensively in use, principally for cereals and legumes. There are technical barriers to extend its use further. Insect resistance is an emerging problem with this fumigant, particularly where it is used under poor containment conditions or with short exposure periods. Further investigation may allow application of phosphine in other areas where methyl bromide is currently used. However, complete substitution by phosphine is unlikely. For example, where methyl bromide is used to control some insects strains resistant to phosphine particularly when exposure time is a constraint. In addition, as indicated in Table 7.3 phosphine requires longer exposure periods and higher commodity temperatures, for effective action. It is generally accepted that for higher temperatures the exposure periods required for control are likely to be shorter. However, unpublished research indicates that a longer period of phosphine exposure was required at 35 °C to control resistant *Rhyzopertha dominica* than at 25 °C (personal communication, Winks and Hyne).

Certain commodities in long-term storage infested with mites can be disinfested by two phosphine fumigations with an interval dependent on ambient temperature, to allow eggs surviving the first fumigation to hatch. This interval varies from 2 weeks at 20 °C to 6 weeks at 10 °C (Bowley and Bell 1981).

Table 7.3. Minimum exposure periods (days) for phosphine required to control all life stages of some stored product pests*

Species	Common names	Temperature	
		10–20 °C	20–30 °C[†]
Acanthoscelides obtectus	Dried bean beetle	8	5
Caryedon serratus	Groundnut borer	10	8
Corcyra cephalonica	Rice moth		
Cryptolestes ferrugineus	Rust-red grain beetle		
Cryptolestes pusillus	Flat grain beetle	5	4
Ephestia cautella	Tropical warehouse moth	10	5
Ephestia elutella	Warehouse moth		
Ephestia kuehniella	Mediterranean flour moth		
Lasioderma serricorne	Cigarette beetle	5	5
Oryzaephilus mercator	Merchant grain beetle		
Oryzaephilus surinamensis	Saw-toothed grain beetle	3	3
Plodia interpunctella	Indian-meal moth		
Ptinus tectus	Australian spider beetle		
Rhyzopertha dominica	Lesser grain borer		
Sitophilus granarius	Grain/granary weevil	16	8
Sitophilus oryzae	Rice weevil		
Sitophilus zeamais	Maize weevil		
Sitotroga cerealella	Angoumois grain moth		
Tribolium castaneum	Rust-red flour beetle		
Tribolium confusum	Confused flour beetle		
Trogoderma granarium	Khapra beetle		

*Based on a phosphine concentration of $1.0 \, \mathrm{g \, m^{-3}}$. This dosage is as recommended for good conditions and the dosage applied needs to be increased considerably in leaky situations (EPPO 1984)
[†]All species listed succumb to a 4-day exposure at this dosage level at 30 °C or above.

There are other fumigants that may have potential as alternatives for methyl bromide. Hydrogen cyanide, once widely used to treat durable commodities, was superceded by methyl bromide and phosphine. As a result, interest in this fumigant waned resulting in the maximum residue limit (MRL) established in Codex Alimentarius to lapse due to lack of support (FAO/WHO 1995). Ethyl formate is used in some countries to disinfest packed dried fruit (Tarr *et al* 1994). Similarly, carbon disulphide was once widely used but its use has been discontinued in most countries. Ethylene oxide, as a fumigant for food, has been withdrawn in most countries because of the production of carcinogenic residues (Wesley *et al* 1965, Heuser and Scudamore 1969).

Two potential new fumigants for durables are carbonyl sulphide (Desmarchelier 1994) and methyl isothiocyanate (Ducom 1994). Presently however, insufficient data are available and neither compound has yet been registered as a fumigant.

Contact insecticides are used extensively in certain situations to protect raw durable commodities (Champ and Highley 1985). These chemicals include conventional pesticides, inert dusts, insect growth regulators and plant extracts or their analogues. One main constraint associated with contact insecticides is the possible presence of chemical residues in the treated commodities. This may prevent their use in processed products. Although many of these chemicals are registered for use on durable commodities, resistance is a major problem leading to loss of effectiveness. As is the case for all chemical treatments cost of registration constrains the development of new products.

One option, in the short term, is reduction of emissions of methyl bromide to the atmosphere from treatments. Introduction of better sealing techniques to improve fumigant retention should permit the lowering of dosage rates and subsequent emissions of gas to the atmosphere. Use of methyl bromide and protectant insecticides on durables can be reduced substantially by the introduction of Integrated Commodity Management (ICM) systems (Evans and van S. Graver 1987). This will require improvements in design of storage structures or retrofitting the required improvements to existing storages. Proper training will be required to successfully introduce an ICM system, and will require consistent maintenance in order to succeed. Difficulties in the transfer of technologies to reduce or replace the use of methyl bromide will arise wherever changes are made to existing systems. However, use of decision support systems and associated training support systems (Longstaff 1994) may provide a means to facilitate this process. Introduction of technology to use such new treatments may be time consuming and costly. Other long-term issues include the maintenance and servicing of complex equipment required for the alternative treatment techniques.

Disinfestation of artefacts is already being undertaken by a number of alternatives, each with particular advantages and disadvantages (Wudtke and Reichmuth 1994). Choices for control of pests of artefacts has recently been reviewed (Pinniger 1991). However, here the choice is critical, since once the artefact is destroyed its heritage value is lost forever.

3.1.1 CONSTRAINTS

Durable commodities include the staple food grains of many communities, and as such, represent an easy route for pesticide residue consumption. This means that extra care is needed in introducing new alternatives. The seriousness of this limitation is usually reduced when the longer storage time associated with these commodities makes possible treatments where residues of more labile chemicals will have reduced to an acceptable level by the time of consumption. Constraints due to residues need to be based on a realistic and safe compromise between consumer worries in wealthier countries and

Table 7.4. Chemical alternatives to control pests infesting durables

Common name	Status	Reference
Azadirachtin	2	Yamasaki and Klocke (1987)
Biflourides	5	UNEP (1994)
Bioresmethrin	1	
Bromophos		Snelson (1987)
Carbaryl	1	
Carbonyl sulphide	4	Banks and Desmarchelier (1993)
Chlorpyrifos-methyl	1	
d-Phenothrin		Snelson (1987)
Deltamethrin		
Diazinon	7	
Dichlorvos	1	
Diflubenzuron	5	Ahmed (1992)
Ethyl formate	5	Tarr et al (1994)
Ethylene oxide	7	Wesley et al (1965), Heuser and Scudemore (1969)
Etrimfos		Snelson (1987)
Fenitrothion	1	Snelson (1987)
Fenoxycarb		Samson et al (1990)
Fenvalerate		Snelson (1987)
Hydrogen cyanide	6	Anon (1989)
Inert dusts	5	McLaughlin (1994)
Iodofenphos		
Lindane	7	Snelson (1987)
Malathion	1	
Mancozeb + monocrotophos		Prasad (1992)
Methacrifos	5	Snelson (1987)
Methoprene	5	Snelson (1987), Ryan (1995)
Methyl isothiocyanate	4	Ducom (1994)
Ozone	4	Yoshida (1975), Erdman (1979) and Mills (1992)
Permethrin		Snelson (1987)
Phosphine	1	Anon (1982)
Phoxim and Phoxim-Methyl		
Piperonyl Butoxide	1	
Pirimiphos-Methyl		Snelson (1987)
Pyrethrum	1	
Sulphuryl fluoride	5	Dow (1963)
Tetrachlorvinphos		Snelson (1987)

Status description

1 widely accepted efficacious methods that are currently available,
2 under active development of possible limited use,
3 under active research with very high potential for use,
4 of potential but in very early stages of research,
5 widely used for some applications,
6 efficacious older materials but fallen into disuse,
7 older material; use discontinued for well documented reasons.

Table 7.5. Durables with alternatives for pest control

Commodity group	Treatment	Status
artefacts	carbonyl sulphide	potential
artefacts	phosphine	routine but sensitivity for pigments and metals
artefacts	hydrogen cyanide	colour sensitivity for pigments
artefacts	ethyl formate	potential
beverage crops	hydrogen cyanide	potential
beverage crops	phosphine	routine
dried fish	pyrethrins + piperonyl butoxide	recommended
dried fish	pirimiphos methyl	recommended
dried fruit	phosphine	routine
dried fruit	hydrogen cyanide	no longer in routine use
dried fruit	ethyl formate	routine some places
dried fruit	IGRs	not approved
grain bagged	phosphine	routine
grain bulk	phosphine	routine
grains	inert dusts	routine for some purposes
grains	dichlorvos	routine for some purposes
herbs and spices	ethylene oxide	used but unlikely to continue
herbs and spices	phosphine	routine
nuts	phosphine	routine except walnuts
seeds	phosphine	routine but doesn't kill nematodes
seeds	mancozeb + monocrotophos	potential
seeds	methyl allyl chloride	routine
tobacco	methoprene	increasing
tobacco	phosphine	routine
wood	sulphuryl fluoride	rarely
wood products	phosphine	routine some places
wood products	sulphuryl fluoride	routine some places
wood products	borate	routine some places
wood products	biflourides	routine some places

the need to preserve stocks of food for poorer countries (Desmarchelier 1994).

4.1 PERISHABLE COMMODITIES

Perishable commodities are those with very high water activity and characterised by short storage or shelf life. Examples include fresh fruit and vegetables, cut flowers, ornamental plants, fresh root crops and bulbs. Out of eleven alternative treatments approved for commercial use on specific commodities, only two are chemical. However very few of the alternatives are in commercial use (UNEP 1994).

Alternatives must be fully tested, efficacious, non-phytotoxic and economical. For perishables, this narrows the possibilities for chemicals either because of phytotoxic responses or pesticide residues.

Compared to methyl bromide, potential alternatives require more data (than is the case for durables) demonstrating efficacy and phytotoxicity responses. Some methods will be more commodity specific than methyl bromide therefore requiring more research at the plant species and cultivar levels (Weller *et al* 1995). The application methodologies for alternative treatments are generally more complex than fumigation with methyl bromide which increases their costs of implementation.

Currently, no alternative chemical treatments exist for apple, pear, and stonefruit exports that are host to codling moth; for berry fruits; for grapes from Chile exported to the United States; and for many rootcrops that are exported to developed countries from Article 5 (1) countries (UNEP 1994). Urgent international co-operation is needed if alternative treatments based on globally-accepted phytosanitary standards are to be developed before the loss of methyl bromide to those countries.

Only a small percentage of the global production of methyl bromide, $< 10\%$, is used for treating perishables, either on arrival in an importing

Table 7.6. Chemical alternatives for methyl bromide currently used on some perishable commodities

Common name	Target organism	Status	Reference
acetaldehyde	insecticide	3	Aharoni *et al* (1979)
carbaryl dip	insecticide	5	Anon (1992)
dichlorvos aerosol	insecticide	5	Carpenter and Stocker (1992)
dimethoate	insecticide	5	Heather *et al* (1987)
ethyl formate	insecticide	3	Stewart and Mon (1984)
ethylene dibromide	insecticide	7	Bond (1984)
hydrogen cyanide	insecticide	5	Bond (1984)
malathion dip	insecticide	5	UNEP (1994)
malathion-carbaryl dip	insecticide	5	Anon (1992)
methomyl	fungicide	5	UNEP (1994)
methyl formate	insecticide	3	Stewart and Mon (1984)
permethrin aerosol	insecticide	5	UNEP (1994)
pirimiphos-methyl	insecticide	5	UNEP (1994)
pyrethroids	insecticide	1	Carpenter and Stocker (1992)
sulphur dioxide	fungicide	5	Vail *et al* (1991), Vota (1957)

Status description

1   widely accepted methods that are efficacious and currently available,
2   under active development and possibly limited use,
3   active research with a very high potential for use,
4   of potential but in very early stages of research,
5   widely used for some applications and not at all for other,
6   efficacious older materials that have fallen into disuse,
7   older material use discontinued well documented reasons.

country if undesirable live pests are intercepted, or occasionally prior to export if the importing country deems the pest to be a serious threat to its agricultural security. Quarantine treatment will remain one of the major difficulties for withdrawal of this chemical because in many cases national legislation specifically requires use of methyl bromide.

Many countries take a precautionary approach to meet export requirements. For example, countries sometimes fumigate perishable commodities prior to export to ensure their immediate release onto retail markets, and to avoid more expensive fumigation costs on arrival. On average, almost half the tonnage of methyl bromide used on perishable commodities was for disinfestation of exported fruit (UNEP 1994). A minor quantity of methyl bromide was used to prevent the spread of pests within the same country (UNEP 1994).

Until recently there has been little perceived need to investigate alternatives to methyl bromide because it is:

- toxic and efficacious to a broad spectrum of pests,
- easily applied,

Table 7.7. Application of alternatives on perishable commodities

Commodity	Treatment	Status
bulbs	dichlorvos aerosol	potential
tulips	pirimiphos methyl	potential
lilies	methomyl	potential
bulbs	hydrogen cyanide	limited
cut flowers	carbaryl dip	limited
cut flowers	hydrogen cyanide	limited
cut flowers	chemical dips	limited
cut flowers	malathion-carbaryl dip	limited
cut flowers	dichlorvos aerosol	limited
cut flowers	permethrin aerosol	limited
cut flowers	malathion	limited
tropical fruits	insecticide	potential
berryfruit	hydrogen cyanide	potential
citrus	fungicidal film	some
grapes	sulphur dioxide	potential but sulphite residue risk
berryfruit	sulphur dioxide	potential
citrus	hydrogen cyanide	routine
tropical fruits	wax	routine
rambutan	sulphur dioxide and heat	limited
tomatoes	dimethoate	limited
asparagus	hydrogen cyanide	limited
root crops	insecticide dipping	potential
leafy vegetables	methyl formate	potential but flammability risk
leafy vegetables	ethyl formate	potential but flammability risk

- cost-effective, and
- accepted by most countries.

There has been, therefore, insufficient time and resources dedicated to generating scientific data to support the ability of potential alternatives to control pests of perishables without damage to the commodity. Most alternative treatments are more specific to commodities and pests, require more complex equipment or procedures to apply, and are often more expensive than methyl bromide. However methyl bromide does cause some injury to products (e.g. cut flowers Weller *et al* 1995), and less harmful, though still effective, alternatives would be desirable.

Treatment of perishables presents a special problem in terms of human consumption. By definition, perishables are not generally stored for long periods. Therefore, the time during which residues can decay is limited and much more important than in the case of durables. Furthermore, many perishables are purchased on the basis of superficial appearances (for example fruit and flowers) and the effect of the treatment on these is paramount.

5.1   STRUCTURAL FUMIGATION

Structural fumigation is a disinfestation technique in which a fumigant is applied to an entire structure or a significant portion of it. The structures may contain raw agricultural commodities, raw products in process or finished food products awaiting delivery to distribution points. Fumigation is utilised whenever the infestation is so widespread that localised treatments may result in reinfestation or when the infestation is within the walls or other inaccessible areas. The primary intention of a structural fumigation is to disinfest the structure and installed equipment, rather than the commodities contained within it.

Methyl bromide is currently used as a structural fumigant in three types of facilities:

- food production and storage (mills, food processing, distribution warehouses),
- non-food facilities (dwellings, museums), and
- transport vehicles (trucks, ships, aircraft, railcars).

Target pests include stored product pests; cockroaches, silverfish, psocids, flies, spiders; wood destroying insects; and rodents.

Pest management in these facilities is best achieved through the use of integrated pest management (IPM) procedures. A reduction in the use of fumigation in IPM programs can be accomplished, depending on the degree to which other procedures are implemented. Even the best IPM programs may occasionally require a full site treatment; currently fumigation with methyl bromide. There are occasions when total elimination of pests is essential, particularly where these are timber boring insects within structural timber, or in structures important because of their heritage value.

With regard to treatment of structures for pests other than wood destroying insects, in most situations there is currently no alternative treatment to methyl bromide for eradication. In some cases the most efficacious fumigant is phosphine but currently its corrosive properties (Bond *et al* 1984), resistance and time required for fumigation limit its use. However, strategies that incorporate lower concentrations of phosphine, in combination with carbon dioxide and heat are being tested as an alternative (Mueller 1994). These techniques do not appear to have the corrosive potential of the higher concentrations of standard dosages. Hydrogen cyanide is another efficacious fumigant, however it is only used on a limited basis in a few countries due to its acute toxicity and reaction with some pigments (Unger and Unger 1986).

Although methyl bromide is often used in some countries to treat wood destroying insects, other methods have been available for many years, including sulphuryl fluoride, non-fumigant pesticides and non-chemical methods (UNEP 1994). In non-food situations sulphuryl fluoride is a substitute full site fumigant for these pests and its use is increasing (UNEP 1994). However, the control of some insect life stages requires a significantly higher fumigant concentration (7-30x) (Su and Scheffrahn 1990).

Ships, aircraft and other transport vehicles pose particularly difficult pest management problems because they often contain sensitive equipment and innumerable harborages. Frequently, it is economically impossible to keep them out of operation for extended periods of time. Presently, there are no acceptable chemical alternatives to methyl bromide for rodent and insect elimination aboard aircraft. In many cases the release of methyl bromide during fumigation could be significantly reduced by improved sealing and scrubbing or reclaiming the gas at the end of the fumigation.

Research is needed in many areas before alternative strategies can be adopted. As new technologies are developed there will be an increased need for technology transfer to developing countries. Programs designed to transfer knowledge and train operators must be developed to bring about a successful transition to reductions in methyl bromide use. Particularly crucial is training in pest identification, biology and habits, monitoring and use of new technologies. The success of this training is dependent on economic, social and cultural conditions in developing countries.

Various constraints exist which affect the adoption of substitutes and alternatives. Presently there are situations where no feasible chemical alternatives exist for food processing structures or structures containing sensitive metals, particularly aircraft, when pest eradication is the goal. A critical element in determining applicability and effectiveness of alternatives is the establishment of economic thresholds for the various pests. Differing regulatory requirements and use constraints throughout the world will determine the suitability of alternative pesticide products and fumigants.

The economical constraints of substitute and alternative strategies must also be considered. For example, the use of sulphuryl fluoride is not allowed where food is exposed though recent research suggests that plastic films may protect food from residues. Phosphine also has some limitations on its use. High demurrage rates for ships, non-operational time for aircraft and extended closure time for mills can significantly increase pest management costs and the cost of goods sold.

6.1 PROSPECTIVE FUMIGANT ALTERNATIVES

6.1.1 CARBON DISULPHIDE

Grains. Formerly carbon disulphide was widely used to fumigate bulk and bagged grain. Typically it was applied as a liquid mixed with carbon tetrachloride or alone. In most countries its use has been discontinued and registration has lapsed. However, it is still used in parts of Australia, where it is applied to farm-stored grain. Application to large bulks is restricted by the potential fire hazard of the material and safe methods for large-scale use need to be developed. Work is currently in progress to examine whether carbon disulphide can be used in a similar fashion to phosphine with longer exposure periods and lower concentrations that are below the explosion limit. If this approach is feasible, then it could be a candidate as an alternate fumigant to phosphine in large-scale storage.

Soils. Carbon disulphide can be injected into soil with no covering required as a soil fumigant. In addition to its insecticidal properties it has limited fungicidal and nematicidal activity (Rodríguez-Kabána *et al* 1977, Thomson 1989). However, because of phytotoxicity, carbon disulphide, may require long waiting periods (> 2 weeks) before soil can be planted.

6.1.2 CARBONYL SULPHIDE

This gas has insecticidal properties (Desmarchelier 1994), but is not yet registered as a fumigant. Its use as a potential replacement for methyl bromide has recently been patented (Banks and Desmarchelier 1993). Efficacy has been demonstrated in laboratory studies with activity against most stored grain pests at about $200\text{--}600\,\mathrm{g\,h\,m^{-3}}$ at $25\,°\mathrm{C}$ (Desmarchelier

1994). Potential problems with residual odours caused by the hydrolysis of carbonyl sulphide to H_2S or by its reaction to form other sulphur compounds, along with toxicity, corrosion and flammability must be addressed. Carbonyl sulphide may be an alternative to methyl bromide for the fumigation of artefacts, but no practical data are available on its use for this application at this time. It has good penetration properties (Desmarchelier 1994). Development of carbonyl sulphide as a fumigant for durables, including timber, is being actively pursued (van S. Graver 1994).

6.1.3 ETHYL FORMATE

Ethyl formate was formerly used as a fumigant for grain (Back and Cotton 1937) but was superceded by methyl bromide. However, its use is now restricted to dried fruit in Australia (Tarr *et al*. 1994) and processed cereal products.

The action of ethyl formate against pests of durable foodstuffs is quite rapid, with optimum exposures of a few hours, but because this fumigant is highly sorbed problems of distribution usually lead to the need for exposures of several days. Typical dosages on dried vine fruits are 3 to 6 ml per 15 kg, with action complete within 24 hours. Ethyl formate may be an alternative to methyl bromide for disinfestation of artefacts, but no practical data are available on its use in this application.

Although ethyl formate is efficacious against stored product pests and can penetrate packaging material, it is flammable and explosive. It is not registered for use in USA and Europe, and requires a 72-hour minimum exposure for maximum efficacy. It is also corrosive to unpainted metals; especially iron and steel.

6.1.4 ETHYLENE OXIDE

Ethylene oxide was used extensively for many years to reduce microbial contamination spoilage in food commodities such as spices, cocoa beans and some processed foods. It has also been used for insect control on grain. In terms of toxicity to the target organism it is the fumigant of choice where sterilisation, combined with pest control, is required. It was withdrawn from use in 1980, within the European Community, but it is still used in many other parts of the world.

Because of its flammability, ethylene oxide is generally supplied in mixtures with inert diluents such as CO_2 or HCFCs. Ethylene oxide reacts with chemical constituents of some food commodities producing potentially carcinogenic compounds. Detection of the reaction product, ethylene chlorohydrin, was reported by Wesley *et al* 1965. Similarly, ethylene bromohydrin was found in flour and wheat previously treated with methyl bromide followed by ethylene oxide (Heuser and Scudamore 1969). Ethylene oxide should not be used on foods that include salt because of formation of potentially carcinogenic compounds.

Ethylene oxide may potentially replace methyl bromide in some non-food uses, where health and environmental regulations permit. Applications include treatment of certain artefacts, such as manuscripts and other archive and museum materials.

6.1.5 HYDROGEN CYANIDE

Hydrogen cyanide has been used as a fumigant for almost a century and still has some uses in a few countries (Bond 1984). It is the only fumigant gas that is lighter than air. It acts rapidly in suitable locations, particularly against rodents, and is easy to apply leaving very little residue if used correctly. Hydrogen cyanide is now little used for a variety of reasons including the fact that it is perceived to be more toxic to humans than other fumigants because skin absorption alone can cause death at the concentrations normally used. In practice, inadequate distribution can occur and localised concentrations can reach explosive levels. Its water solubility can interfere with penetration and cause exposed water to become toxic. It can react with some foods to produce toxic compounds. Severe restrictions are imposed on its transport. In addition liquid hydrogen cyanide is unstable and cannot be stored for long periods. However, it can be developed *in situ* (Anon 1989).

Availability and registration, or re-registration, difficulties may prevent immediate substitution of hydrogen cyanide for methyl bromide should this be required.

Durable commodities. Hydrogen cyanide was formerly widely used as a fumigant for durable commodities until it was largely superseded by methyl bromide and phosphine. Both are much more convenient and, in many cases, more effective to use. Modern instructions for use of HCN are given in AFHB/ACIAR (1989). The Codex Alimentarius approved residue limits for hydrogen cyanide residues in grain and flour have recently lapsed, due to lack of support (FAO/WHO 1995).

Dried fruit and nuts. Hydrogen cyanide is not currently in use on dried fruit and nuts and is no longer registered in most countries. Dried vine fruit tend to absorb hydrogen cyanide with formation of quite stable cyanhydrins because of their high sugar content.

Beverage crops. The potential to use hydrogen cyanide on beverage crops as an alternative to methyl bromide exists, but is not currently used for this purpose.

Fresh fruit. It was commonly used on fresh commodities for control of pests such as thrips, white flies, scale and aphids (Bond 1984) and still has use in quarantine treatments in Japan.

Artefacts. Hydrogen cyanide has also been used for pest control in artefacts, with a recommended dosage of $20 \, g \, m^{-3}$ for 72 hours exposure. However this use is very limited because of its high solubility in water, low fungicidal effect and slow desorption, as well as possible reaction with the treated material, particularly with pigments (Unger and Unger 1986).

6.1.6 METHYL ISOTHIOCYANATE

Soils. Methyl isothiocyanate (MITC) and compounds such as dazomet and metam-sodium, which generate MITC, are highly effective for control of some soil pests. These compounds are highly dependent on soil preparation and moisture for activation and uniform distribution. MITC was introduced in 1959 by Schering AG as a soil nematicide under the trade name Trapex. In addition to nematodes, it effectively controls arthropods, some weeds and soilborne pathogens, principally fungi and a limited number of plant parasitic nematode species (Rodríguez-Kábana *et al* 1977). In soil fumigation it is mostly used in combinations with 1,3-dichloropropene (1,3-D), which enhances its nematicidal activity. In practice it is injected into the soil as a liquid. Product stability and corrosion have limited the use and distribution of this compound.

The toxicological profile of MITC (toxic, skin and eye irritant, sensitiser) may constrain its use as an alternative. More importantly MITC has physical characteristics which indicate its potential to contaminate ground water, however, monitoring has not revealed groundwater contamination.

There are a number of compounds or their formulations which, when incorporated into moist soil, decompose to produce methyl isothiocyanate. The activity of MITC generators is similar to MITC (Rodríguez-Kábana *et al* 1977). These materials, commercially available for 40 years, can give inconsistent results because uniform distribution of the formulated product and release of MITC in the soil are highly dependent upon application method, adequate soil moisture, temperature, pH, and the effectiveness of the soil surface sealing method (e.g. waterseal, covering with plastic or compaction).

Metam-sodium is formulated as a liquid and may be applied to the soil either by injection, drip or sprinkler irrigation, or sprayed onto the soil surface prior to tilling. It is mildly toxic, an eye irritant, a teratogen and genotoxin. These properties may constrain its use as an alternative to methyl bromide.

Dazomet is supplied as a granular formulation which is generally incorporated into the soil by roto-tilling (Anon 1989). Although Dazomet is a MITC generator, the major breakdown products are MITC and formaldehyde, plus some carbon disulphide. Persistence in soil of MITC, or its breakdown products, is influenced by temperature and moisture, which if sub-optimal such as cool and wet, often require longer waiting periods before planting to prevent crop phytotoxicity.

Durable commodities.   The potential of MITC as a grain fumigant and protectant is currently being studied (Ducom 1994). Preliminary studies of its biological efficacy indicate that MITC is very active against all stages of *Sitophilus granarius* at a very low *ct*-product of $8\,\mathrm{g\,h\,m^{-3}}$. For optimal results, this compound has to be very well mixed with the grain.

Timber and bark.   Recent work suggests that basamid as a MITC generator may have value as a potential wood fumigant (Forsyth and Morrell 1995). Mixtures of MITC with carbon disulphide are more effective against wood colonising fungi than each chemical alone (Canessa and Morrell 1995).

6.1.7 OZONE

Ozone shows some potential as a fumigant. Apart from its sterilising action against bacteria and viruses, there is only limited information about its toxicity to insects and to stored product pests in particular. Activity has been found against *Sitophilus oryzae* and *Oryzaephilus surinamensis* (Yoshida 1975), *Tribolium* spp. (Erdman 1979) and *Ephestia elutella* (Mills 1992). However, it is known to break down rapidly and damage organic materials such as rubber. If it were to be selected as an alternative, it would require the normal regulatory approval process to be acceptable for use.

6.1.8 PHOSPHINE

Phosphine and methyl bromide are currently the only fumigants that are registered worldwide and effective against a wide range of pest species. Phosphine is one of the most toxic fumigants known and is used at low concentrations. Its action against pests tends to be much slower than that of methyl bromide with long exposures required, particularly at low temperatures.

   Penetration into commodities is rapid and removal by aeration after treatment is equally quick. Sorption by most commodities is low and residues from normal fumigation practice are well below $0.01\,\mathrm{g\,t^{-1}}$, the current Codex Alimentarius limit for processed cereals.

   Phosphine forms an explosive mixture with air when the concentration exceeds 1.8% by volume. However, this level would not be reached in normal fumigation practice. This figure is lower at reduced pressure and care needs to be taken in designing recirculation and vacuum systems using phosphine to ensure the limit is not exceeded (Green *et al* 1984).

Phosphine reacts with copper, silver and gold, especially in humid atmospheres, which in some situations may preclude its use because of detrimental effects on electrical equipment and machinery (Bond *et al* 1984).

Other important factors restricting use of phosphine include:

- commodity temperature which should be greater than 10 °C, preferably 15 °C;
- prolonged exposure period; for effective action against all development stages of pests, typically this should extend for 5 to 15 days, depending on the temperature; and
- development of resistance; prevention requires effective containment for the required exposure period.

Very dry commodities ($< 10\%$ moisture content in the case of cereal grains) may be difficult to fumigate because of restricted evolution of phosphine where solid metal phosphide formulations are used. This problem is overcome when the gas is obtained from 'bottle' formulations (Winks 1990) and generators (Waterford *et al* 1994).

The toxicity of phosphine is well researched and dosage schedules are available for many stored product insects and mites (EPPO 1984).

Resistance is an immediate, practical problem. High levels are known to occur in several species of stored-product beetle pest (Tyler *et al* 1983, Taylor 1989, Price and Mills 1988). High levels of resistance have been measured in laboratory tests on field strains collected in several countries, particularly from parts of Africa, the Indian subcontinent, and recently Brazil and China. The selection pressure is presumed to result from repeated use of the fumigant under conditions of poor gastightness. There have also been control failures attributable to this resistance. However, these were usually treatments of short duration. Resistance to phosphine can be managed, provided that an effective gas concentration can be maintained over the longer exposure periods required by more tolerant strains. In leaky situations such as open topped silos, insect control may be carried out by a continuous input of fumigant using a phosphine–carbon dioxide mixture delivered from a pressurised cylinder (Winks 1990). However, it is preferable to improve the degree of gastightness of the enclosure, as described for enclosures around stacks of bagged grain in AFHB/ACIAR (1989). By this means gas may be retained for a sufficient peroid. Multiple dosing may also assist. Whatever the application method, control of resistant populations requires a higher dosage and longer periods of exposure (Price and Mills

1988). Alternating phosphine use with other fumigants is one potential way of slowing development of resistance (Rousch 1989).

The duration of exposure period has a much more important role than concentration levels in the toxicity of phosphine. Thus use of *ct*-products as a measure of dosage for phosphine is not valid unless the exposure period over which it applies is stated. All stages in the life cycle of stored-product insects have a broadly similar tolerance to methyl bromide (a factor of $3\times$ or so). However, there is a high degree of variation in tolerance to phosphine, with eggs and pupae being much more tolerant than larvae and adults (Mills *et al* 1990). Mites are particularly difficult to control with phosphine since their egg stage is highly tolerant. In addition, for resistant strains, it seems that sufficient time to overcome the proposed active exclusion mechanism (Price and Mills 1988) must pass before the insect will succumb.

There is a danger of fire, or even explosion, when phosphine-generating formulations are misused or resulting residues improperly handled. Aluminium phosphide formulations, the most common form of phosphine-generating product, may contain 3% or more of undecomposed phosphide, presenting a disposal problem and further potential hazard.

The application of phosphine to stored grain and other durable commodities is described in numerous publications (e.g. Bond 1984, Banks 1986, AFHB/ACIAR 1989).

Various proprietary formulations of phosphine are available worldwide. Most contain aluminium phosphide or, less commonly, magnesium phosphide, formulated with ammonium carbamate or urea to reduce the risk of flammability. Phosphine is normally generated *in situ* by reaction of atmospheric moisture with the metallic phosphide (Bond 1984). Phosphine in pressurised gas cylinders as a non-flammable 2% mixture in carbon dioxide, as currently utilised commercially in the SIROFLO® process (Winks 1990), is available in limited quantities. In addition, phosphine generators using the controlled addition of phosphide formulations to water are being developed (Waterford *et al* 1994).

Artefacts. Phosphine is used to fumigate wooden objects, paper and other materials of vegetable origin. With some materials, for example, furs, phosphine may be preferred over methyl bromide, because of the reduced risk of taint. Because the gas may adversely effect metals and pigments in paintings it is rarely used for treating objects of this type.

Beverages. Phosphine is widely used to control insects in beverage commodities (Clifford and Wilson 1985, Wood and Lass 1985). It is claimed, in certain circumstances, that phosphine may taint cocoa beans. However, there is no firm evidence of such occurrences when the fumigation is carried out in a well controlled manner according to established procedures (personal communications, Jan van S. Graver).

Dried fruit and nuts. Phosphine is used to control stored product pests in natural raisins (Nelson 1970, Hartsell *et al* 1991). In most countries phosphine can be used on dried fruits and nuts, with the exception of walnuts (UNEP 1994). It is already in use for controlling pests of dried vine fruit in storage. Most pests of dried fruit and nuts are highly susceptible to phosphine and shorter exposure times can be used than with stored grain.

Durables. Phosphine is widely used for treating infestations in bulk and bagged grain and grain products. However, methyl bromide is still used, particularly at point of import or export (e.g. into Japan) or on stacks of bagged grain (e.g. parts of Africa, Singapore). In the first case, the speed of action and its recognised efficacy against pests of quarantine significance makes methyl bromide the preferred fumigant compared to phosphine. In the second case, methyl bromide, when properly used, has been developed into an efficient, effective and reliable fumigation system, with no apparent need to change until now.

In-transit fumigation on board ship can replace disinfestation with methyl bromide at point of export. Shipboard in-transit fumigation with phosphine is now a well developed technology (Leesch *et al* 1978, Redlinger *et al* 1979, Zettler *et al* 1982). It requires ships of appropriate design and stringent safety precautions (Snelson and Winks 1981, IMO 1993). Several grain-exporting countries, including Canada and Australia, require grain to be free of infestation at point of export and thus do not use the system. Technically, however, it presents a method where the longer exposure periods required for phosphine fumigations do not hinder trade through ports. Thus the technology is a feasible alternative to methyl bromide treatment prior to export.

A number of countries do not have established populations of khapra beetle (*Trogoderma granarium*). Currently the only permitted treatment to quarantine standards in grain (and most other durable commodities) is by methyl bromide. Phosphine is not registered as a quarantine treatment for this pest and further studies may be required to provide data for approval of phosphine treatments by quarantine authorities. However, it is highly effective against all stages of this noxious pest (Bell *et al* 1984, 1985), including the normally tolerant diapause larva, provided a sufficiently long exposure period can be used and the commodity temperature must be greater than 15 °C.

Recent developments in phosphine fumigation technology, including use of surface application in sealed systems and supply of non-flammable phosphine formulations in cylinders (Winks 1990, Chakrabarti *et al* 1987, Chakrabarti 1994) have increased the competitiveness and effectiveness of phosphine with respect to methyl bromide. Recent advances in phosphine treatment of grain against infestation are described in Navarro and Donahaye (1993) and Highley *et al* (1994).

Dried fish. This fumigant appears to provide effective control of the common insect pests infesting this commodity (Friendship 1990).

Herbs and spices. Phosphine is often used to disinfest herbs and spices, particularly where there may be possibility of excessive bromide residues which would preclude the use of methyl bromide.

Nematodes. There is little evidence of phosphine efficacy against seed-infesting nematodes. However, an experimental application of phosphine was effective in controlling nematodes in water suspension (Rout 1966).

Structural treatment. Phosphine can be used in some structural treatments. Its penetration and efficacy against some important pests is well researched and understood. It was used effectively against wood-boring insects in Norwegian churches (UNEP 1994). However, it can cause corrosion problems on copper, or gold and silver alloys (Bond *et al* 1984) and components that contain these metals must be protected from exposure. Damage to electrical and other equipment occurs more frequently in hot humid weather. The extent of this problem is not fully understood and needs to be investigated further under varying conditions (Mueller 1994). Use of phosphine is not recommended below 10 °C because it provides limited control at such temperatures, hence heating during treatment may be beneficial. It is not approved for dwellings and other structural fumigation uses in the USA. Depending on temperature and humidity, fumigations with phosphine may take five or more days, in contrast with one day for methyl bromide. In dealing with leaky structures, as with methyl bromide, it has been recommended that the phosphine dosage rate should be increased to compensate for leakage (Mills *et al* 1990). Phosphine may be added during the treatment. Phosphine ($0.09-0.14 \, \mathrm{g \, m}^{-3}$) combined with heat at $32-37 \, °C$ and CO_2 (4–6%) has been reported to provide good penetration and a rapid treatment time, similar to that for methyl bromide (Mueller 1994). Additional data are needed on efficacy, and the advantages and limitations of these techniques, particularly damage to sensitive metals in equipment.

Timber and bark. Fumigation of logs with phosphine effectively controls bark beetles, wood-wasps, longhorn beetles and platypodids at a dose of $1.2 \, \mathrm{g \, m}^{-3}$ with a 72 hour exposure period at 15 °C or more. Presently, however, commercial acceptability of this treatment is restricted by the extent of the exposure period.

Tobacco. Phosphine is the fumigant of choice in areas where the ambient temperature exceeds 15 °C. It is applied at atmospheric pressure, has excellent penetration and is effective against all developmental stages of all pests of tobacco. Application rates vary between 1 and $4\,g\,m^{-3}$ with an exposure period varying from 5–15 days according to the temperature of the tobacco (Geneve 1972, Geneve *et al* 1986).

Transport. Disinfestation of empty ships and barges with phosphine to control infestations of insects is feasible. Phosphine is rapidly lethal to rodents but slow in action against insect pests. The delays and consequent demurrage costs caused by its slow action against insects may limit its usefulness in this application.

6.1.9 SULPHUR DIOXIDE

Fresh fruit. This is used mainly for fungus control in cool stored grapes, and gives effective disease control in lychees (Coates *et al* 1994). Recent research has shown a potential to control mealybug and lepidopteran insects (Vail *et al* 1991). The surface fungi on fresh longans and lychees can be controlled effectively by sulphur dioxide (Tongdee 1994).

6.1.10 SULPHURYL FLUORIDE (VIKANE)

This chemical is an effective fumigant used mainly for termite control. Its efficacy is well researched and understood. It provides good penetration, requires a short exposure period of approximately 24 hours and aerates in 6–8 hours. It is very toxic to all post-embryonic stages of insects (Kenaga 1957, Bond and Monro 1961). However, the eggs of many species are very tolerant. In laboratory and field tests sulphuryl fluoride, unlike methyl bromide, produced no objectionable colour, odour or corrosive reactions to photographic supplies, metals, paper, leather, rubbers, plastics, clothes or many other articles fumigated (Gray 1960). Sulphuryl fluoride is widely used to control wood-destroying insects, but at present has limited application for artefacts. It is not registered for use on food and grain commodities and is unlikely to be because no food residue tolerances have been established (UNEP 1994). A recent study showed that nylon film enclosures are effective in protecting foodstuffs from exposure to both sulphuryl fluoride and methyl bromide fumigation (Schreffrahn *et al* 1995). This might extend the use of this fumigant to structures containing food.

Durable foodstuffs. Vikane is not registered for use on foodstuffs. Guidelines for use, issued by the sole American manufacturer, specifically state (Dow 1963) that: 'under no conditions should sulphuryl fluoride be used on raw agricultural food commodities, or on foods, feed or medicinal products destined for human or animal consumption or on living plants'.

Structures. Sulphuryl fluoride is available for use in several countries including the Caribbean, Germany, Japan, Sweden, and the United States. In California its use has led to virtually a 100% reduction in methyl bromide use in the disinfestation of dwellings.

The requirement for up to a 7–30× increase in dosage to control the egg stage of many pests (Su and Scheffrahn 1990) makes the cost of the higher dosage a consideration in evaluating this treatment. Good sealing techniques, which enhance fumigant containment, may make lower dosages more efficacious by extending the exposure period.

Artefacts. Because it is non-reactive, Sulphuryl fluoride could be used to disinfest libraries and museums.

Timber. Use of this chemical for plant quarantine purposes requires investigation regarding suitability for timber treatment. Its low efficacy against eggs of insect pests is the principal hindrance.

6.1.11 ETHYL OR METHYL FORMATE AND ACETALDEHYDE

There is interest in use of ethyl or methyl formate and acetaldehyde (Aharoni *et al* 1979, Stewart and Mon 1984), but none are currently registered as fumigation treatments. Ethyl and methyl formate are inflammable and explosive when mixed with air at concentrations required to kill pests and may require application in an inert diluent such as CO_2. Ethyl formate is less pesticidal than methyl formate. Acetaldehyde is more effective as a fungicide than a pesticide, and its safety to humans has been questioned (Woutersen *et al* 1984).

6.1.12 FUMIGATION AND CHEMICAL SOAKING

Promising results were obtained in the control of the nematode *Aphelenchoides besseyi* (Fortuner and Orton Williams 1975) by soaking rice seed in an aqueous solution of systemic organophosphorous compounds. The compounds are not phytotoxic. Prasad (1975) eliminated *A. besseyi* in rice seeds by soaking them in a solution of mancozeb and monocrotophos followed by fumigation with phosphine at a dose of $9.3 \, \text{g} \, \text{m}^{-3}$.

7.1 CONTACT TREATMENTS

7.1.1 AEROSOL FORMULATIONS

Formulations in liquid carbon dioxide of natural plant products (pyrethrins) or insecticides (pyrethroids, dichlorvos fenitrothion) alone, or in combination are used to disinfest certain cut flower species prior to export from countries such as New Zealand (Carpenter and Stocker 1992), Australia, and Malaysia. However, aerosols alone cannot provide complete disinfestation due to their inability to penetrate the dense inflorescences of some plant species. Thus they have to be used in combination with other treatments as a component of an IPM system extending from the grower to consumer.

7.1.2 CHEMICAL DIPS

Horticultural products can be dipped in a very dilute pesticide solution after harvest to kill targeted pests present in or on the commodity. For example, Australian tomatoes exported to New Zealand are dipped in insecticide to control Queensland fruit fly (*Bactrocera tryoni*) (Heather *et al* 1987); and some cut flowers are immersed in insecticide to control pests on the surface (Hansen *et al* 1992, Hata *et al* 1992). The residues remaining after the treatment should not exceed established MRLs. However, chemical dip treatments are usually acceptable on non-food products such as ornamental plants, bulbs, nursery plants and cut flowers.

Some countries discourage use of chemical dips because of consumer concern for chemical residues, and because disposal of the spent pesticide solution after treatment is environmentally unacceptable.

8.1 CONTACT INSECTICIDES

8.1.1 ORGANOPHOSPHORUS COMPOUNDS

This is an important group of chemical protectants in current use. The stability of deposits on commodities varies widely with substrate and ambient conditions, while their rate of degradation increases with temperature and water activity (moisture content). Furthermore, toxicity to insects increases with temperature. Consequently, persistance of biological efficacy depends upon the insecticide used. For example, dichlorvos typically becomes ineffective within several days, while malathion takes several weeks, and pyrimiphos methyl many months.

The principal chemicals used world-wide include: chlorpyrifos methyl, dichlorvos, fenitrothion, malathion, and pyrimiphos methyl. Registration details vary from country to country. Most are poorly effective against bostrichids (*Rhyzopertha dominica* and *Prostephanus truncatus*).

Dichlorvos is unique amongst the commonly used grain protectants because of its rapid action against pests and lability on grain. In the absence of resistance, and where approved, it can be sprayed onto bulk grain within a few days of export to disinfest a cargo. Subject to an adequate withholding period for residues to decay to acceptable levels, such a treatment can provide a direct alternative to disinfestation with methyl bromide on stored products. While dichlorvos is currently approved under the Codex Alimentarius for application to raw cereal grains with a maximum residue level of $2\,g\,t^{-1}$, its registration is subject to debate in some countries and its long-term future use is uncertain.

8.1.2 SYNTHETIC PYRETHROIDS

Synthetic pyrethroids are a group of insecticides with chemical constitution based on that of the active ingredients of natural pyrethrum. Deposits are quite stable on grain and their insecticidal activities may persist up to two years (Snelson 1987). Their action is much less sensitive to temperature than organophosphorus insecticides. Pyrethroids are active against bostrichid beetles at a much lower dosage than for most other insect pests of durables. Most are of low acute toxicity to human beings. A disadvantage of these pesticides is their relatively high cost. In many situations pyrethroids are added in combination with a synergist, piperonyl butoxide, to increase effectiveness and reduce cost.

8.1.3 INSECT GROWTH REGULATORS (IGRs)

The term insect growth regulator (IGR) is used to describe compounds which interfere with the life-cycle of pests. They are not normally directly toxic to adult pests. IGRs are considered to be more pest-specific than conventional contact insecticides. However, one potential disadvantage of IGRs is their long persistance on foodstuffs. This may limit their use in some potential applications. The earliest IGRs developed were juvenile hormone analogues, and include methoprene and hydroprene.

Some IGRs act against insects by ingestion and/or contact (e.g. methoprene), whilst others act only by ingestion (e.g. diflubenzuron). They tend to have low toxicity to vertebrates (Menn *et al* 1989) and thus to have a substantial margin of safety in use. The major disadvantages are their high cost and inability to control adult stages.

Methoprene is registered for use to protect a variety of stored agricultural commodities. It is effective against many stored product pests, including *L. serricorne*, *E. cautella*, *P. interpunctella*, *T. granarium*, *R. dominica* and *O. surinamensis*, but not against *Sitophilus* spp. (Snelson 1987, Mkhize 1986).

Diflubenzuron and fenoxycarb have also been evaluated as a grain protectant but are not yet registered (Samson *et al* 1990).

8.1.4 BOTANICALS

These compounds are derived from plants and include pyrethrum and azidaractin. Botanicals, as natural products, are not readily patented and there is little incentive for companies, and other organisations, to incur the cost involved to gain registration for use. The debate concerning patenting issues, for which the neem extract azidaractin is an example, is currently raging (Dickson and Jayaraman 1995) and points out the difficulty in developing promising botanicals. A comprehensive bibliographic database listing on alternative control methods to conventional pesticides has been compiled by Rees *et al* 1993.

8.1.5 CONTACT PESTICIDE ALTERNATIVES FOR SPECIFIC TREATMENTS

Beverage crops. Contact insecticides are not applied directly to these products where residues are a limiting factor. However, they may be used as a component of an IPM program to protect bagged products from insect infestation, using surface application to the bags and the store fabric.

Dried fish. Pyrethrins synergised with piperonyl butoxide are recommended as an aqueous dip for protecting dried fish from insect infestation (Proctor 1972). More recently pirimiphos methyl was recommended for the same purpose (Golob 1987), with a maximum residue limit of $10 \, \text{mg} \, \text{kg}^{-1}$ recommended by the FAO/WHO Committee on Pesticide Residues (FAO 1986). In Indonesia pirimiphos methyl is successful in controlling blowfly strike during processing and against dermestid infestation during storage (Esser *et al* 1990). Deltamethrin, a pyrethroid insecticide, applied as a 0.003% dip, similarly controlled pests during processing and storage (Rattagool *et al* 1990). In this application Madden *et al* (1994) demonstrated the effectiveness of cycloprothrin, a substituted pyrethroid of low mammalian toxicity, pirimiphos methyl, deltamethrin and Startox, a mixture of allethrin and dichlorvos. They also reported that extracts of white pepper, garlic and star fruit (*Averrhoa bilimbi*) gave control equivalent to that obtained from pyrethroids.

Grains. Grain protectants, typically organophosphate and pyrethroid insecticides and IGRs, do not readily penetrate bagged or bulk grain. This restricts their utility substantially as normally they must be applied to the grain during handling, e.g. prior to bagging or on to grain on conveyers or elevators. They are also used as sprays on storage structures and the surfaces of bagged or bulk grain as part of a sanitation program (Webley 1986).

Use of grain protectants varies widely with country, market preference and local regulations. Where permitted, and where pest resistance is not a problem, they can provide a useful means of avoiding the circumstances where fumigation, including methyl bromide, may be otherwise used. At present, the only botanical in widespread use in developed countries for protection of durables (grain) is pyrethrum extract. Others, such as azadirachtin, are under active investigation. A wide variety of botanicals are traditionally used by subsistance farmers in developing countries.

Herbs and spices.   High value products such as herbs and spices are often directly used for human consumption without further processing. These are, therefore, not normally treated with contact insecticides because of potential residue problems.

Structures.   Contact insecticides, including dichlorvos, may be used as part of pest management strategy in museums and repositories.

Timber.   There is an approved treatment for logs to be kept immersed in water for more than 30 days in order to control pests by suffocation. The surface of the logs above water is sprayed with an insecticide mixture. In Japan, approximately 14% of the logs imported in 1992 were treated in this manner. In the USA and Japan, dip-diffusion treatment in a solution of borate is registered.

Tobacco.   Methoprene, an insect growth regulator, is finding increasing use in stored tobacco against the principal pest, *Lasioderma serricorne*, removing the need for methyl bromide and other treatments (Ryan 1995).

8.1.6 CONTACT FUNGICIDES ('WOOD PRESERVATIVES')

These are effective against surface-living fungi but do not penetrate sufficiently deep into the wood to kill all spores. Research is necessary to investigate their combined effectiveness with heat and/or fumigation.

8.1.7 BIFLUORIDES

Bifluorides are commercially used in Europe and are a component of some preservatives used in the USA. In practice, timber is immersed in a 10% solution of the chemical for 5 to 10 minutes and treatment is relatively inexpensive. No monitoring equipment is required. However temperatures must be above freezing. The treatment is not registered in the USA, but is accepted in many European countries.

Overall, compared to methyl bromide, bifluorides provide a very effective treatment but they are not approved in all countries.

9.1 INERT DUSTS

Inert dusts (e.g. diatomaceous earth formulations) can provide a direct alternative to chemical protectants where their adverse effects on grain characteristics and handling are acceptable. In particular, they provide good protection against insect infestation in dry grain stored long term for animal feed. They form a useful part of IPM strategies for grain protection in sprays applied to the storage fabric to minimise residual infestation and migration of pests into bulks of stored grain.

Various inert dusts are registered in some countries for treatment of grain and grain legumes against insect pests. They are particularly effective in dry conditions as a means of controlling pests resident in storage structures. They lose effectiveness at high relative humidity, greater than about 75% r.h. (Le Patourel 1986). Inert dusts may be useful in an integrated system that controls pests to a level where methyl bromide use is not required. Their use on durables, particularly grain, has recently been reviewed (Banks and Fields 1994).

Inert dusts, such as those based on diatomaceous earth e.g. Dryacide (Desmarchelier and Dines 1987), can provide effective pest control in dry grain. However, though direct admixture can give long-term protection over several years from infestation, avoiding the need for fumigation, dusts have adverse effects on the handling qualities of grain and can cause excessive wear in handling machinery. These factors tend to constrain their use in many large-scale storage facilities. They have some particular use as admixtures in seed storage and in small-scale farm stores for animal feed.

Inert dusts, such as Dryacide, find particular application as prophylactic sprays to control insect pests in grain storage structures as part of an integrated control program. They are widely used for this purpose in Australia (Bridgeman 1994).

There are four basic types of inert dust:

- Clays, sands and earths. These materials are traditionally used as a protective layer on top of stored seed.
- Diatomaceous earth consists of a fossilised remains of diatoms, containing mainly silica with small amounts of other minerals. Proprietary insecticidal formulations are available. The dusts are effective against a wide range of pests when admixed to grain, even at rates of $1 \, kg \, t^{-1}$ or less.
- Silica aerogels which are very light, non-hydroscopic powders effective at slightly lower doses than diatomaceous earth formulations.
- Non-silica dusts, such as phosphate and lime. Phosphate has been used in traditional stores in Egypt (Fam et al 1974).

Inert dusts such as ash and lime have a long history of use for grain protection (Ebeling 1971, Golob and Webley 1980, Ross 1979, Quarles 1992a, b). Dryacide, an activated diatomaceous earth, is in widespread use in Australia in the grain handling industry using slurry application for structural spray as a prophylactic treatment against storage pests. There is limited use, in Australia, directly admixed for preserving stock feed and storing seed for planting.

Under favourable conditions, inert dusts can be quite rapid in their lethal action, with complete mortality of adult insect pests achieved within seven days at low dosage rates. Available data on responses of immature stages of grain pests is limited, although the success of inert dusts in suppressing population growth suggests that they are likely to have a strong effect on free living immature stages. It is not necessary for insects to be completely covered with the inert dust for it to be active (Maceljski and Korunic 1971).

The main advantages of inert dusts are that they do not require capital equipment, are relatively non-toxic, provide continued protection, and do not affect baking quality (Desmarchelier and Dines 1987, Aldryhim 1990). Some inert dusts are accepted as 'organic'. Diatomaceous earths are widely used as food and processing additives and there are no obvious environmental hazards. The main disadvantages are reduced flowability of grain, visible residues that can affect grading, and decreased bulk density of grain. They can also give rise to dust problems in the workspace and there are concerns about worker exposure to uncontrolled dust levels. To alleviate the workplace dust problem in Australia, inert dusts are applied as an aqueous slurry for surface treatments, although this has been reported to reduce their effectiveness (Maceljski and Korunic 1971).

10.1 QUARANTINE

Undertaking quarantine treatments without methyl bromide is a particularly challenging issue in all applications except soil treatments.

There are currently very few disinfestation treatments for durable commodities that may be used for quarantine purposes and to disinfest at the point of importation which are as effective as methyl bromide within the same timescale. With the alternatives now available, commodities may fail to get customs clearance resulting in reshipment or destruction, or they may need much more time to disinfest to the level required by the regulatory agency. While there are a variety of potential substitutes, research is required to establish them as satisfactory treatments that meet standards required by quarantine authorities.

Most currently available alternatives do not have the speed of action currently needed for materials in the transportation pipe-line. All that are widely available and fully registered, require the commodity to be held in treatment for at least five days (and often significantly longer) to ensure quarantine levels of disinfestation. This is generally considered too long in the current system. That there are no practical alternatives is recognised by the current Montreal Protocol exemptions on quarantine and pre-shipment use of methyl bromide. However, there is no guarantee that these exemptions will remain (Banks 1994). There is, therefore, urgent need for either a change in the logistic requirements for this type of treatment or development of a new treatment method with the short time requirement similar to that of methyl bromide.

New approaches to quarantine issues are being developed worldwide. In Australia the focus is shifting from end point inspection to a self regulating 'systems approach' termed Certification Assurance (Heinrich and Dean 1994). By this means the exporting industry implements a quality assurance system with controls that ensure that the importing country's regulations are satisfied. Audits by quarantine authorities check that the system is in place and effective. An integrated approach to produce a 'clean pipeline' of commodity from grower to exporter has been sucessfully developed for the Western Australian grain industry (Dean 1994). Farm hygiene and storage improvements, management of resistance to both insecticides and fumigants and a well controlled central storage system, produce a product ready for export, free of residues and insects.

11.1 REGISTRATION

The issue of registration is so extensive that we can barely address the matter here. However, it is perhaps the most important part of seeking and providing speedy availability of new, or rediscovered, alternatives when an industry discovers that methyl bromide is no longer permitted or too expensive to control its particular pest problem. Desmarchelier (1994) gives a concise view of the problems facing the registration of new protectants for grain and discusses possible solutions.

When restrictions of supply or increased price for a diminishing supply begin to be felt, users and manufacturers will look for opportunities to register alternatives. The list of possibilities is extensive, the number of applications huge and size of market for produce or application ranges from minute to significant parts of a country's gross domestic product.

Normally, any new product has to undergo critical evaluation to establish its safety and efficacy before any form of registration can be granted. A full data package, to secure registration, can run to many millions of dollars and companies may not be willing to attempt registration if the return does not justify the expense. This is already a problem associated with many of the natural products and will certainly be a problem for many of the 'small sector' applications where methyl bromide has been in use for many years with no incentive to seek an alternative. The way this dilemma is handled at a national level will vary from country to country. However, the underlying principles will doubtless be the same whichever country requires to register an alternative.

As an example, in Australia the National Registration Authority (NRA) is the agency charged with responsibility for overseeing registration of agricultural and veterinary chemicals. It provides a range of other registration options which include:

- *Extension of use*—designed for companies that wish to extend the use of a product to another state, or include other target pests on the label. These applications are made by the holder of the original product registration and may take from 3–15 months to grant, depending on the number of other agencies involved.
- *Off-label permit*—which allows the legal use of an unregistered product in a manner that is inconsistent with the label. This is not intended to circumvent normal registration and covers minor use or emergency situations. Application for off-label permits can be made by individuals, end-user groups, advisers, consultants, suppliers and manufacturers.
- If the data required is insufficient, funds to generate data may be made available through the *Minor Use Supplementary Grants Program* a scheme in its formative stages at NRA.
- Simplified procedures for biopesticides and microbial products reflect their nature and data requirements for registration are available from the NRA.

Whichever system is used there will be need for some leeway and cooperation during the period of transition or many industries will experience hardship as they adjust to life without methyl bromide. More importantly, they may be forced to choose inappropriate alternatives.

12.1 CONCLUSIONS

The difficulty, in selecting an alternative for any of the uses of which methyl bromide has been put, is the lack of a single general alternative. Chemical alternatives are available for many of the individual targets but there is no single broad spectrum replacement. The selection from the suite of alternatives is further complicated when several different target organism groups must be controlled, preferably in a single treatment.

The limited range of alternatives ready for use or 'plug in alternatives' is clearly apparent. Many chemicals have been shown to be effective in one area of use against one specific target. However, research required to extend use to other commodities, organisms or sectors is lacking.

Methyl bromide provides a rapid treatment method unlike many chemical alternatives that have been identified so far. This means that time constraints for commodities in trade and transport will be an important consideration when selecting alternative treatment strategies. The convenience of the '24 hour fumigation' will become a matter of history, and lateral thinking will be required to prevent unacceptably long delays in movement of product to market. Products in long-term storage will not be subject to the need for speed of action and other slower acting alternatives may be used in this situation.

The need to register, or re-register those that have fallen into disuse, is a serious issue for researchers and regulatory authorities. If registration of substitute chemical alternatives, or the systems which use them, does not occur immediately they become available, then there is a real risk of inappropriate or ineffective strategies being put in place which could have serious and unexpected implications. To counter this, all agencies responsible need to cooperate on this issue actively to select and fast-track the most promising alternatives.

Costs of registration, or re-registration, are a major impediment to the development of safe and effective alternatives. It will be 'political or commercial will' which determines which particular chemical alternatives are developed. The scope exists for governments, through international agencies and regulators, to provide incentives to ease the cost burden where this is needed.

Attitudes are a major impediment to the acceptance of alternative strategies. Innovation to allow existing systems to be altered so that alternatives can be introduced effectively must be encouraged. The advantages of alternatives must be widely promoted if acceptance of new control measures is to be achieved within any time frame established for withdrawal of methyl bromide.

ACKNOWLEDGEMENTS

We are grateful to Dr Jonathan Banks for his assistance in the preparation of this chapter. The work presented here draws heavily on the 1994 Report of the UNEP United Nations Methyl Bromide Technical Options Committee, 1995 assessment, which provided the starting point for our work. Many thanks to Jan van Someren Graver and David Rees for reading and making many useful suggestions to improve the draft.

REFERENCES

AFHB/ACIAR, 1989, *Suggested recommendations for the fumigation of grain in the ASEAN region, Part 1. Principles and General Practice*, ASEAN Food Handling Bureau, Kuala Lumpur and Australian Centre for International Agriculture Research, Canberra, 131 p.

Aharoni Y, Stewart JK, Hartsell PL and Young DK, 1979, Acetaldehyde—a potential fumigant for control of the green peach aphid on harvested head lettuce, *Journal of Economic Entomology*, **72**, 493–5.

Ahmed ME, 1992, Effect of Dimilin (diflubenzuron) on the fecundity, fertility and progeny of *Dysdercus ungulatus* (Hemiptera, Pyrhocoridae), *Journal of Applied Entomology*, **114**, 138–142.

Aldryhim YN, 1990, Efficacy of amorphous silica dust, Dryacide, against *Tribolium confusum* Duv. and *Sitophilus granarius* (L.) (Coleoptera: Tenebrionidae and Curculionidae), *Journal of Stored Products Research*, **26** (4), 2207–10.

Aldryhim YN, 1993, Combination of classes of wheat and environmental factors affecting the efficacy of amorphous silica dust, Dryacide®, against *Rhyzopertha dominica* (F.), *Journal of Stored Products Research*, **29** (3), 271–5

Alspach LK, 1989, Dazomet use for seedbed fumigation at the PFRA Shelterbelt Centre, Indian Head, Saskatchewan. General Technical Report RM-184, Rocky Mountain Forest Range Experiment Station, United States Department of Agriculture Forest Service, Fort Collins, Colorado, USA. 40–2.

Anonymous, 1982, Phosphine-generating products: Guidelines for the use, directions and similar requirements for registration in TCAC clearances, Document PB441, Pesticides Section, Department of Primary Industry, Canberra. (Revision in preparation 1995, National Registration Authority, Canberra Australia).

Anonymous, 1989, *Basamid-Granules*, Limburgerhof, Germany, BASF Aktiengesellschaft, 111 p.

Anonymous, 1992, USDA Animal and Plant Health Inspection Services, 1992, Plant Protection and Quarantine Treatment Manual, Interim Edition, 30 November 1992, 386 pp.

Back EA and Cotton RT, 1937, Industrial fumigation against insects, U.S. Department of Agriculture, Circular Number 369, 59 p revised 1939, 1942.

Banks HJ, 1986, The application of fumigants for the disinfestation of grain and related products, in: *Pesticides and Humid Tropical Storage*, eds BR Champ and E Highly, ACIAR Proceedings **14**. 291–8.

Banks HJ and Desmarchelier JM, 1993, Carbonyl sulphide fumigant and method of fumigation, International Application Published under the Patent Cooperation Treaty, International Publication Number, WO 93/13659, 43 p.

Banks HJ, 1994, Fumigation—an endanged technology? in: *Stored-Product Protection*, Proceedings of the 6th International Working Conference on Stored-product Protection, eds E Highley, EJ Wright, HJ Banks and BR Champ, CAB International, Oxon, 2–6.

Banks J and Fields P, 1994, Physical methods for insect control in stored-grain ecosystems, Chap. 11. in: (eds) DS Jayas, NDG White and WE Muir, *Stored-grain Ecosystems* (New York: Marcel Dekker) pp 353–409.

Bell CH, Hole BD and Wilson SM, 1985, Fumigant doses for the control of *Trogoderma granarium*. OEPP/EPPO *Bulletin*, **15**, 9–14.

Bell CH, Wilson SM, Banks HJ and Smith RH, 1984, An investigation of the tolerance of stages of Khapra beetle (*Trogoderma granarium* Everts) to phosphine, Proceedings of the 3rd International Working Conference on Stored-Product Protection, Manhattan, 23–28 October, 375–390.

Bond EJ, 1984, 'Manual of fumigation for insect control,' *FAO Plant Production and Protection Paper* No. 54, FAO, Rome, 432 p.

Bond EJ, Dumas T and Hobbs S, 1984, Corrosion of metals by the fumigant

Bond EJ and Monro HAU, 1961, The toxicity of various fumigants to the cadelle, *Tenebriodes mauritanicus*, *Journal of Economic Entomology*, **54**, 451–4.

Bowley CR and Bell CH, 1981, The toxicity of twelve fumigants to three species of mites infesting grain, *Journal of Stored Products Research*, **29**, 277–82.

Bridgeman RW, 1994, Structural treatment with amorphous silica slurry: An integral component of GRAINCO's IPM strategy, in: *Stored-Product Protection*, Proceedings of the 6th International Working Conference on Stored-product Protection, eds E Highley, EJ Wright, HJ Banks and BR Champ, CAB International, Oxon, 628–30.

Canessa EF and Morrell JJ, 1995, Effect of mixtures of carbon disulphide and methyl isothiocyanate on the survival of wood-colonizing fungi, *Wood and Fibre Science*, **27** (3), 207–24.

Carpenter A and Stocker A, 1992, *Envirosols as postharvest fumigants for asparagus and cut flowers*, Proceedings, New Zealand Plant Protection Conference, 21–6.

Chakrabarti B, 1994, Methods of distributing phosphine in bulk grain, London, Home Grown Cereals Authority Research Review, No. 27, 44 p.

Chakrabarti B, Wontner-Smith T and Hurt AD, 1987, Investigations on the use of 2% phosphine in carbon dioxide as a potential fumigant mixture, Slough Laboratory Report **19**, 5.

Champ BR, 1985, Occurrence of resistance to pesticides in grain storage pests, in *Pesticides and Humid Tropical Storage*, eds BR Champ and E Highley, ACIAR Proceedings **14**, 228–55

Champ BR and Dyte CE, 1976, Report of the global survey of pesticide susceptibility of stored grain pests, *FAO Plant Production and Protection Series* No. 5.

Champ BR and Highley E, 1985, *Pesticides and humid tropical grain storage systems*, ACIAR Proceedings **14**, ACIAR, Canberra, 364 p.

Clifford MN and Wilson KC (eds) 1985, *Coffee Botany, Biochemistry and Production of Beans and Beverage*, Beckenham, Kent, Croom Helm, 216 p.

Coates LM, Johnson GI, Sardsud V and Cooke AW, 1994, *Postharvest diseases of lychee in Australia, and their control*, ACIAR Proceedings, **58**, 68–9.

Dean KR, 1994, An integrated approach to stored-grain protection in Western Australia, in: *Stored-Product Protection*, Proceedings of the 6th International Working Conference on Stored-product Protection, eds E Highley, EJ Wright, HJ Banks and BR Champ, CAB International, Oxon, 1179–81.

Desmarchelier JM, 1978, Mathematical examination of availability to insects of aged insecticide deposits on wheat, *Journal of Stored Product Research*, **14**, 213–22.

Desmarchelier, JM, 1994a, Carbonyl sulphide as a fumigant for control of insects and mites in: *Stored-Product Protection*, Proceedings of the 6th International Working Conference on Stored-product Protection, E Highley, EJ Wright, HJ Banks and BR Champ, CAB International, Oxon, 78–82.

Desmarchelier JM, 1994b, Keynote Address, Grain Protectants: trends and developments, in: *Stored-Product Protection*, Proceedings of the 6th International Working Conference on Stored-product Protection, eds E Highley, EJ Wright, HJ Banks and BR Champ, CAB International, Oxon, 722–8.

Desmarchelier JM and Dines JC, 1987, Dryacide treatment of stored wheat: its efficacy against insects, and after processing, *Australian Journal of Experimental Agriculture*, **27**, 309–12.

Dickson D and Jayaraman KS, 1995, Aid groups challenge to neem patents, *Nature*, **377**, 95.

Dow Chemical Company, 1963, *Vikane; technical data bulletin*, Midland, Mich., Dow Chemical Company.

Ducom V, 1994, Methyl isothiocyanate as a grain fumigant, in: *Stored-Product Protection*, Proceedings of the 6th International Working Conference on Stored-product Protection, eds E Highley, EJ Wright, HJ Banks and BR Champ, CAB International, Oxon, 91–7.

Ebeling W, 1971, Sorptive dust for pest control, *Annual Review of Entomology*, **16**, 123–58.

Erdman HE, 1979, Ecological aspects of control of a stored product insect by ozonation, in: Proceedings of the second International Working Conference on Stored-Product Entomology, 10–16 September, Ibadan, Nigeria, 75–90.

Esser JR, Wiryante J, Sunarya and Tausin S, 1990, Prevention of insect infestation and losses of salted-dried fish in Indonesia by treatment with an insecticide approved for use of fish, FAO Fisheries Report No. 401, Supplement. Rome, FAO, 168–79.

European Plant Protection Organisation, 1984, Phosphine fumigation of stored products, *EPPO Bulletin*, Vol 4, No. 4, pp 598–9.

Evans DE and van S Graver J, 1987, Integrated pest and commodity management: Putting the first things first, in: *Grain Postharvest Systems*, ed. BM de Mesa, ASEAN Crops Postharvest Program, Bangkok, 356 p.

Fam EZ, El-Nahal AKN and Fahmay H, 1974, Influence of grain moisture on the efficacy of silica aerogel and katelsous used as grain protectants, *Bulletin of the Entomological Society of Egypt, Economic Series*, **8**, 105–14.

FAO, 1986, Pesticide residues in food—1985, *Evaluations, Part 1. Residues*, FAO Plant Production and Protection, Paper 72/1, 372 p.

FAO/NRI, 1994, *Facilitating regional trade of agricultural commodities in eastern, central and southern Africa*, Dar Es Salaam University Press.

FAO/WHO, 1995, Residues of pesticides in foods and animal feeds, Codex Alimentarius Commission Joint FAO/WHO food standards programme code committee on pesticide residues 27th session, the Hague, Netherlands, 24th April, 1995, Part A, p 28.

Fisk FW and Shepard HH, 1938, Laboratory Studies of Methyl Bromide as an Insect Fumigant, *Journal of Economic Entomology*, **31** (1), 79–84.

Forsyth PG and Morrell JJ, 1995, Decomposition of Basamid in Douglas-fir heartwood—Laboratory studies of a potential wood fumigant, *Wood and Fibre Science*, **27** (2), 183–97.

Fortuner R and Orton Williams KJ, 1975, Revue de la litérature sur *Aphelenchoides besseyi* nématode causant la maladie 'white-tip' du riz, *Helminthological Abstracts*, **44**, 1–140.

Friendship R, 1990, The fumigation of dried fish, *Tropical Science*, **30**, 185–93.

GASGA, 1995, *Problems of pesticides residues in grain*, GASGA executive series No. 3.

Geneve R, 1972, Les insectes parasites des tabacs entreposés et manufacturés en France, Note Interne Seita, 18 p.

Geneve R, Ducom P, Branteghem V and Delon R, 1986, Désinsectisation d'un entrepôt de tabac de 220 000 m³ par le phosphure d'hydrogéne, *Annales du Tabac*, **20**, 93–104.

Golob P, Cox JR and Kilminster K, 1987, Evaluation of insecticide dips as protectants of stored dried fish from dermistid beetle infestation, *Journal of Stored Products Research*, **23**, 47–56.

Golob P and Webley DJ, 1980, The use of plant and minerals protectants of stored products, London, *Tropical Products Institute*, G138, 32.

Gray HE, 1960, Vikane, a new fumigant for control of dry wood termites, *Pest Control*, **10**, 43–6.

Green AR, Sheldon R and Banks HJ, 1984, The flammability limit of phosphine-air mixtures at atmospheric pressure, in: eds BE Ripp *et al. Controlled Atmosphere and Fumigation in Grain Storages* (Amsterdam: Elsevier) 433–49.

Hansen JD, Hara AH and Tenbrink VT, 1992, Insecticidal dips for disinfesting tropical cut-flowers and foliage, *Tropical Pest Management*, **38**, 245–9.

Hartsell PL, Tebbets JC and Vail PV, 1991, Citrus, nut and avocado, *Insecticide and Acaricide Tests*, **16**, U.S.D.A., A.R.S., 42 p.

Hata TY, Hara AH, Jang EB, Inaino LS, Hu BKS and Tenbrink VL, 1992, Pest management before harvest and insecticidal dip after harvest as a systems approach to quarantine security for red ginger, *Journal of Economic Entomology*, **85**, 2310–6.

Heather NW, Hargreaves PA, Corcoran RJ and Melksham KJ, 1987, Dimethoate and fenthion as packing line treatments for tomatoes against *Dacus tryoni* (Froggatt), *Australian Journal of Experimental Agriculture*, **27**, 465–9.

Heinrich D and Dean J, 1994, The changing role of AQIS in the regulation of grain exports from Australia, in: *Stored-Product Protection*, Proceedings of the 6th International Working Conference on Stored-product Protection, eds E Highley, EJ Wright, HJ Banks and BR Champ, CAB International, Oxon, 1183–5.

Heuser SG and Scudamore KA, 1969, Formation of ethylene bromohydrin in flour and wheat during treatment with ethylene oxide, *Chemistry and Industry* (London), **31**, 1054–5.

Highley E, Wright EJ, Banks HJ and Champ BR (eds), 1994, *Stored-Product Protection*, Proceedings of the 6th International Working Conference on Stored-product Protection, eds E Highley, EJ Wright, HJ Banks and BR Champ, CAB International, Oxon, 1274 pp.

International Marine Organisation, 1993, Recommendations on the safe use of pesticides in ships, International Marine Organisation, 23 p.

Jackson K and Webley D, 1994, Effects of Dryacide® on the physical properties of grains, pulses and oilseeds, in: *Stored-Product Protection*, Proceedings of the 6th International Working Conference on Stored-product Protection, eds E Highley, EJ Wright, HJ Banks and BR Champ, CAB International, Oxon. 635–7.

Johnson AW and Feldmesser J, 1987, Nematicides — A historical review, in: *Vistas on Nematology*, eds JA Veech and DW Dickson, Hyatsville, Maryland, Society of Nematologists, 448–54.

Kelley WD and Rodríguez-Kábana R, 1975, Effects of potassium azide on soil microbial populations and soil enzymatic activities, *Canadian Journal of Microbiology*, **21**, 565–70.

Kelley WD and Rodríguez-Kábana R, 1979a, Nematicidal activity of sodium azide, *Nematropica*, **8**, 49–51.

Kelley WD and Rodríguez-Kábana R, 1979b, Effects of sodium azide and methyl bromide on soil bacterial populations, enzymatic studies, and other biological variables, *Pesticide Science*, **10**, 207–15.

Kenaga EE, 1957, Some properties of sulphuryl fluoride as an insecticidal fumigant, *Journal of Economic Entomology*, **50**, 1–6.

Kerin J, 1994, Opening Address, in *Stored Product Protection*, Proceedings of the 6th International Working Conference on Stored-product Protection, eds E Highley, EJ Wright, HJ Banks and BR Champ, CAB International, Oxon. xix–xx.

Leesch JG, Redlinger LM, Gillenwater HB, Davis R and Zehner JM, 1978, An in-transit ship-board fumigation of corn, *Journal of Economic Entomology*, **71**,

928–35.

Le Patourel GNJ, 1986, The effect of grain moisture content on the toxicity of a sorptive silica dust to four species of grain beetle, *Journal of Stored Products Research*, **22**, 63–9.

Longstaff BC, 1994, Decision support systems for pest management, in: *Stored-Product Protection*, Proceedings of the 6th International Working Conference on Stored-product Protection, eds E Highley, EJ Wright, HJ Banks and BR Champ, CAB International, Oxon, 940–5.

Lubulwa G, Desmarchelier J and Davis J, 1995, Incorporating atmospheric environmental degradation in research evaluation of options for the replacement of methyl bromide: A project development assessment of ACIAR PN 9046—The replacement for methyl bromide in timber for quarantine fumigation. A paper for presentation at the 39th Annual Conference of the Australian Agricultural Economics Society, University of Western Australia, Feb 14–16, 1995.

Maceljski M and Korunic Z, 1971, The results of investigation of the use of inert dusts in water suspensions against stored-product insects, *Zastita Bilja*, **23**, 376–87.

Mackie DB, 1938, Methyl Bromide—Its Expectancy as a Fumigant. *Journal of Economic Entomology*, **31** (1), 70–79.

Madden JL, Anggawati AM and Indriati N, 1994, *Impact of insects on the quality and quantity of fish and fish products in Indonesia*, Fish Drying in Indonesia, ACIAR Proceedings No. 59, 97–106.

McLaughlin A, 1994, Laboratory trials on desiccant dust insecticides, in: *Stored-Product Protection*, Proceedings of the 6th International Working Conference on Stored-product Protection, eds, E Highely , EJ Wright, HJ Banks and BR Champ, CAB International, Oxon. 638–45.

Menn JJ, Raima AK and Edwards JP, 1989, Juvenoids and neuropeptides as insect control agent: retrospect and prospects, in: *Progress and prospects in insect control*, BCPC Monograph No. 43, 89–106.

Mills KA, 1992, To assess the toxicity of ozone in low oxygen atmosphere to the warehouse moth *Ephestia ellutella*, CSL contract report, 23.

Mills KA, Clifton AL, Chakrabarti B and Savvidou N, 1990, The impact of phosphine resistance on the control of insects in stored grain by phosphine fumigation, Proceedings of the BCPC Crop Protection Conference, Brighton, 1181–87.

Mkhize JN, 1986, Activity of insect growth regulators with juvenile hormone-like affects against the rice weevil, *Sitophilus oryzae* (L.) (Coleoptera: Curculionidae), *Tropical Pest Management*, **32**, 324–6.

Mueller DK, 1994, A new method of using low levels of phosphine in combination with heat and carbon dioxide, *Fumigants and Pheromones*, **33**, 1–4.

Navarro S and Donahaye E (eds), 1993, CAF Proceedings of an International Conference on Controlled Atmosphere and Fumigation in Grain Storages, (Jerusalem: Caspit Press) 560 p.

Nelson HD, 1970, Fumigation of natural raisins with phosphine. U.S.D.A. Marketing Research Report 886, 8 p.

Pinniger DB, 1991, New developments in the detection and control of insects which damage museum collections, *Biodeterioration Abstracts*, **5**, 125–30.

Prasad S, 1992, Elimination of white-tip nematodes, *A. Besseyi*, from rice seed, *Fundamental and Applied Nematology*, **15**, 305–8.

Price LA and Mills KA, 1988, The toxicity of phosphine to the immature stages of resistant and susceptible strains of some common stored product beetles, and implications for their control, *Journal of Stored Products Research*, **24**, 51–9.

Proctor DL, 1972, The protection of smoke-dried freshwater fish from damage during storage in Zambia, *Journal of Stored Products Research*, **8**, 139–49.

Quarles W, 1992a, Diatomaceous earth for pest control, *IPM Practitioner*, **14**, 1–11.

Quarles W, 1992b, Silica gel for pest control, *IPM Practitioner*, **15**, 1–11.

Rattagool P, Methatip P, Esser JR, Hanson SW and Knowles MJ, 1990, Evaluation of insecticides to protect salt-dried marine fish from insect infestation during processing and storage in Thialand, FAO Fisheries Report No. 401, Supplement Rome, FAO, 189–204.

Redlinger LM, Zettler JL, Leesch JG, Gillenwater HB, Davis R and Zehner JM, 1979, In-transit ship-board fumigation of wheat, *Journal of Economic Entomology*, **72**, 642–7.

Rees DP, Dales MJ and Golob P, 1993, *Alternative methods for the control of stored-product insect pests: A bibliographic database*, Natural Resources Institute.

Reichmuth C, 1990, New techniques in fumigation research today, in: Proceedings of the Fifth International Working Conference on Stored-product Protection, eds F Fleurat-Lassard and P Ducom. Bordeaux, France September 9–14, 1990, 709–24.

Rhode WA, Johnson AW, White LV, McAllister, DL and Glaze NC, 1980, Dispersion, dissipation, and efficacy of methyl bromide-chloropicrin gas vs gel formulations on nematodes and weeds in Tifton sandy loam, *Journal of Nematology*, **12**, 39–44.

Ross ET, 1979, An alternative to chemical pesticides? *Mazingira*, **11**, 49–52.

Rousch RT, 1989, Designing resistance management programs: how can you choose? *Pesticide Science*, **26**, 423–41.

Rout G, 1966, Observations on hydrogen phosphide as a nematicide, *Current Science*, **35**, 577.

Rodríguez-Kábana R, 1991, Control biológico de nematodes parásitos de plantas, *Nematropica*, **21**, 111–22.

Rodríguez-Kábana R, Backman PA and Curl EA, 1977a, Control of seed and soilborne plant diseases, in: *Antifungal Compounds*, eds MR Siegel and HD Sisler (New York: Marcel Dekker) Vol. 1, 117–61.

Rodríguez-Kábana R, King PS and Pope MH, 1981, Combinations of anhydrous ammonia and ethylene dibromide for control of nematodes parasitic of soybeans, *Nematropica*, **11**, 27–41.

Rodríguez-Kábana R, Shelby RA, King PS and Pope MH, 1982, Combinations of anhydrous ammonia and 1,3-dichloropropene for control of root-knot nematodes in soybean, *Nematropica*, **12**, 61–9.

Rodríguez-Kábana R and Walters G, 1992, Method for treatment of nematodes in soil using furfural, U.S.A. Patent No. 5084477.

Ryan L (ed), 1995, *Post-harvest Tobacco Infestation Control*, (London: Chapman & Hall) VII, 155 p.

Samson PR, Parker RJ and Hall EA, 1990, Efficacy of the insect growth regulators methoprene, fenoxycarb and diflubenzuron against *Rhyzopertha dominica* (F.) (Coleoptera: Bostrichidae) on maize and paddy rice, *Journal of Stored Products Research*, **26**, 215–21.

Schreffrahn RH, Bodalbhai L and Su NY, 1995, Nylon film enclosures for the protection of foods from exposure to sulphuryl fluoride and methyl bromide during structural fumigation, *Journal of Agricultural and Food Chemistry*, **42** (10), 2317–21.

Snelson JT, 1987, *Grain protectants*, Canberra, ACIAR, 448 p. (ACIAR Monograph No. 3)

Snelson JT and Winks RG, 1981, In transit fumigation of large grain bulks in ships, in: *GASGA Seminar. The appropriate use of pesticides for the control of stored product pests in developing countries.* Tropical Development and Research Institute, Slough, UK, 119–130.

Stewart JK and Mon TR, 1984, Commercial-scale vacuum fumigation with ethyl formate for postharvest control of the green peach aphid (Homoptera: Aphididae) on film-wrapped lettuce, *Journal of Economic Entomology*, **77**, 569–73.

Su NY and Schreffran RH, 1990, Efficacy of sulphuryl fluoride against four beetle pests of museum (Coleoptera: Dermestidae, Anobiidae), *Journal of Economic Entomology*, **83** (3), 879–82.

Tarr C, Hilton SJ, van S Graver J and Clingeleffer PR, 1994, Carbon dioxide fumigation of processed dried vine fruit (sultanas) in sealed stacks, in: *Stored-Product Protection*, Proceedings of the 6th International Working Conference on Stored-product Protection, eds E Highley, EJ Wright, HJ Banks and BR Champ, CAB International, Oxon, 204–9.

Taylor RWD, 1989, Phosphine—A major grain fumigant at risk, *International Pest Control*, **31** (1), 10–14.

Thomson WT, 1989, *Agricultural Chemicals, Fresno*, Thomson Publications: California, p. 288.

Tongdee SC, 1994, Sulphur dioxide fumigation in postharvest handling of fresh longan and lychee for export, in: *Postharvest handling of tropical fruits*, ACIAR Proceedings, **50**.

Tyler PS, Taylor, RWD and Rees DP, 1983, Insect resistance to phosphine fumigation in food warehouses in Bangladesh, International Pest Control, **25**, 45–6.

UNEP 1994, 1994 Report of the Methyl Bromide Technical Options Committee, Montreal Protocol on Substances that Deplete the Ozone Layer. United Nations Environment Program.

Unger A and Unger W, 1986, *Holztechnologie*, **27** (5), 232–6.

Vail PV, Tebbets SJ and Smilanick J, 1991, Sulphur dioxide control of omnivorous leafroller in the laboratory, *Insecticide and Acaricide Tests*, **17**, 371.

van Alebeek FAN, 1989, Integrated Pest Management: a catalog of training and extension materials for projects in tropical and subtropical regions, Department of Entomology, Wageningen Agricultural University.

van Wambeke E, De Coninck S, Descheemaeker F and Vanachter A, 1984, Sodium azide for the control of soil borne tomato pathogens, *Mededelingen van de Rijksfaculteit Landbouwwetenschappen te Gent*, **49**, 373–81.

van Wambeke E and van den Abeele D, 1983, The potential use of azides in horticulture, *Acta Horticulturae*, **152**, 147–54.

van S Graver J, 1994, Replacements and alternatives for methyl bromide 1993–94: An interim report of work undertaken in Australia by the Stored Product Research Laboratory, Annual International Research Conference on Methyl Bromide Alternatives and Emissions Reductions, 1994.

Vota M, 1957, Preliminary study on storage of Emperor grapes in controlled atmospheres with and without sulphur dioxide fumigation, *Proceedings. American Society of Horticultural Science*, **69**, 250–3.

Waterford CJ, Whittle CP and Winks RG, 1994, New aluminium phosphine formulations for the controlled generation of phosphine, in: *Stored-Product Protection*, Proceedings of the 6th International Working Conference on Stored-product Protection, eds E Highley, EJ Wright, HJ Banks and BR Champ, CAB International, Oxon, 226–35.

Webley DJ, 1986, *Use of pesticides in bag storage of grain*, ACIAR Proceedings **14**, 303–11.

Weller GL, Van S Graver JE and Damcevski KA, 1995, Replacements for methyl bromide in quarantine treatments of cut flowers and ornamentals, in Proceedings of the Australian Postharvest Horticultural Conference, Melbourne, Australia, 16–23 September 1995.

Wesley F, Rurke B and Darbishire O, 1965, The formation of persistent toxic chlorohyrins in food by fumigation with ethylene oxide and propylene oxide, *Journal of Food Science*, **30**, 1037–42.

Wilhelm S, Storkan RC and Wilhelm JM, 1974, Preplant soil fumigation with methyl bromide-chloropicrin mixtures for control of soil-borne diseases of strawberries—A summary of fifteen years of development, *Agriculture and Environment*, **1**, 227–36.

Winks RG, 1990, Recent developments in fumigation technology, with emphasis on phosphine, in: eds, BR Champ, E Highley and HJ Banks, *Fumigation and controlled atmosphere storage of grain*, ACIAR Proceedings, **25**, 144–57.

Wood GAR and Lass RA, 1985, *Cocoa* (Harlow, Essex: Longmans) 4th edn, 620 p.

Woutersen RA, Appelman LM, Feren VJ and Vander Heijden CA, 1984, Inhalation toxicity of acetaldehyde in rats, II, Carcinogenicity study interim results after 15 months, *Toxicology*, **31**, 123–33.

Wudtke A, and Reichmuth C, 1994, Control of the common clothes moth Tineola bisselliella (Hummel) (Lepidotera: Tineidae) and other museum pests with nitrogen, in: *Stored-Product Protection*, Proceedings of the 6th International Working Conference on Stored-product Protection, eds E Highley, EJ Wright, HJ Banks and BR Champ, CAB International, Oxon, 251–4.

Yamaski RB and Klocke JA, 1987, Structure-bioactivity relationships of Azadirachtin, a potential insect control agent, *Journal of Agricultural Food Chemistry*, **35** (4), 467–71.

Yoshida T, 1975, Lethal effect of ozone gas on the adults of *Sitophilus oryzae* and *Oryzephilus surinamensis*, Scientific Report, Faculty of Agriculture, Okayama University, **45**, 9–15.

Young DC, 1990, GY-81, *A new concept in soil fumigation*, Proceedings Brighton Crop Protection Conference—Pests and Diseases, 19–22 November, Brighton, 79–85.

Zettler JL, Gillenwater HB, Redlinger LM, Leesch JG, Davis R, McDonald LL and Zehner JM, 1982, In-transit shipboard fumigation of corn on tanker vessel, *Journal of Economic Entomology*, **75**, 804–8.

8

Alternatives—Physical Methods and Emission Reduction

C. H. BELL
Central Science Laboratory, MAFF, Slough, UK

C.H. Bell, N. Price and B. Chakrabarti: The Methyl Bromide Issue
© 1996 John Wiley & Sons Ltd

1.1 HISTORY

In searching for alternatives to a globally used pesticide, the natural approach is to look towards the latest technological advances and investigate how new systems or control practices can be modified to cover for the target compound. There is also some merit, however, in looking back and reviewing the situation which existed before the compound was introduced. Methyl bromide started life as a fumigant for the disinfestation of various commodities, perishable or durable, in the 1930's (Le Goupil 1932, de Francolini 1935a, b) and came into widespread use to control insects on commodities and in structures during the 1940's and 50's (Thompson 1966), and shortly afterwards to control a range of pests and diseases in soil (Munnecke and Lindgren 1954, Hague and Sood 1963, Hague *et al* 1964). Its fungicidal properties were known from the 1930's (Richardson and Johnson 1935).

1.1.1 PERISHABLES

Immediately before the advent of methyl bromide, perishable commodity treatments relied heavily on the use of heat or cold, few chemical treatments having been developed and brought into general use. Low temperature requirements were first established for fruit fly control very early this century (Lounsbury 1907, Hooper 1907), and detailed studies followed at temperatures between −1 and 4 °C to assess effects on pests and fruit, establishing quarantine procedures by the end of the 1920's (Baker *et al* 1944).

As early as the 1890's, hot water sprays and immersion treatments were in use to clean up and disinfest fruits, vegetables, roots, bulbs and stems (Sharp 1994) but these were not recognised as suitable control methods for quarantine purposes. The use of heat on perishable goods expressly for insect control before introduction of fumigants such as ethylene dibromide and methyl bromide was by the hot vapour technique. In Florida, where the first heat-based quarantine procedure was used in 1929, heated air was mixed with steam and a fine spray of hot water for application to citrus fruit at 43.3 °C to control fruit flies (Hawkins 1930). Over the next 10 years, the use of vapour heat treatments spread to other parts of the world and other products (Weddell 1931, Latta 1932, Sein 1935, Koidsumi and Shibata 1936).

1.1.2 DURABLES

Before the arrival of methyl bromide the disinfestation of durable commodities such as grain, tobacco, dried fruit and nuts, cocoa and pulses mostly relied upon fumigants such as carbon disulphide, hydrogen cyanide, and ethylene oxide (Bond 1984), or on the use of high or low temperatures. As for perishables, cold storage treatments had been investigated since before the turn of the century with the advent of mechanical refrigeration. Before the introduction of fumigants such as methyl bromide, cold exposures had been investigated for clothing and furnishing pests (Howard 1896), mill pests (Washburn 1904), pests of cowpeas and other legumes (Duvel 1905, Larson and Simmons 1924), tobacco (Poock 1910), dried fruit (De Ong 1921) and many other products. By the 30's, however, cold storage was used commercially chiefly for shipping of fresh fruits rather than for durables.

High temperature control procedures for durables by the 1930's comprised hot air treatments with or without steam. Washburn (1905) and Goodwin (1914) gave accounts of the effectiveness of dry heat against mill pests, and Parker (1915) described the working parameters for a heated conveying belt for disinfesting dried fruit. Skaife (1918) outlined the use of heat against pea and bean weevils, and Moreira (1923) described use of a steam blast in the treatment of tobacco against cigarette beetle eggs. Other accounts exist describing the testing of heat systems against pests infesting stored foods, cereals and pulses (Bridwell 1918, Faes 1930, Grossman 1931). However, when effective fumigants such as methyl bromide appeared on the scene, the development of extreme temperature techniques for control purposes went into recession.

1.1.3 STRUCTURES

As for durables, treatment of structures relied upon other fumigants such as carbon disulphide, hydrogen cyanide, ethylene oxide and even sulphur dioxide. There was some use of heat in flour mills (Goodwin 1912, Dean 1913) and Treherne (1918) describes the use of steam to disinfest a ship of rice moth. The arrival of methyl bromide was an improvement over all previous control methods, the gas being effective, good at penetrating inaccessible areas, non-flammable, non-corrosive and easy to apply. As a result, other treatment methods diminished steadily in use, the last to disappear being the nearly exclusive use of hydrogen cyanide in flour mills, which only declined rapidly in the early 1960's (Thompson 1966).

1.1.4 SOIL

Since the establishment of plant pathology as a modern scientific discipline well over 100 years ago, soil has been treated by physical or chemical means for the control of fungal plant pathogens, nematodes, insects and weeds (Katan 1985). The first treatments were carried out using the fumigant carbon disulphide, followed by heat. Flooding was also first investigated at this time (Follet-Smith 1933, Stover 1979). Heating techniques, chiefly steam, continued to be widely used up to the advent of the 'newer' generation of fumigants, D-D (a mixture principally of 1,3-dichloropropene and 1,2-dichloropropane), ethylene dibromide and lastly methyl bromide, when for a while at least, use declined. It was the appearance of these compounds that led to the first large scale applications of nematocidal chemicals to soil after the Second World War (van Berkum and Hoestra 1979).

Where chemical control was not practised, numerous cultural processes were carried out such as crop rotation, fallowing, deep tillage and addition of organic amendments. It has to be said that yield expectations were lower and the economic climate was very different from the present time. With soil, as with all the other major use areas, we can see that many of the existing alternatives had been tried before the arrival of methyl bromide, and that some were discontinued when a more effective means of treatment was established. Those techniques that remained in use tended to undergo further development and become restricted to more specific applications.

2.1 THE USES OF METHYL BROMIDE

Methyl bromide is a broad spectrum fumigant used against a wide variety of pest and disease organisms, including pathogens (fungi, bacteria and soil-borne viruses), insects, mites, nematodes, molluscs and rodents. These pests may be in soil, in durable or perishable commodities or in buildings, storage structures and transport.

Methyl bromide has features which make it a versatile and convenient compound for use in many different applications. As a gas it is quite penetrative, reaching pests in soil, commodities and structures which would be difficult to access by most other control measures. It is non-flammable and relatively easy to apply. It is active against most pests even at low temperatures and gas concentration levels, though there is a broad range of susceptibility among the different groups of organisms requiring control. Exposure times typically range from 2 hours up to 2 days with concentrations ranging from 10 to 150 g m^{-3} for commodities and 2–4 days with application rates of 20–100 g m^{-2} for soils. Exposure periods and concentrations are set within these ranges depending on the material being treated, the target pest, and the local quarantine, contractual or regulatory requirements. Some indication of the dosages required for control of some major groups of organisms is given in Table 8.1, and the topic is discussed in further detail in Chapter 4.

Many of the diverse uses of methyl bromide each require only small quantities of the gas annually. Despite their low consumption, many of these applications are currently of considerable importance for the particular industries concerned, as well as for national economies and may involve quarantine legislation (UNEP 1995). The sheer number of individual uses of methyl bromide makes the overall consideration of replacement procedures a complex and very broad topic for discussion and research.

2.1.1 SOIL

Methyl bromide is a pre-plant soil fumigant used to enhance crop productivity in intensive horticultural practice where a broad complex of soil-borne

Table 8.1. Dosages of methyl bromide, expressed as a mean concentration achieved × time of exposure (mg/l × h) to control various pests and pathogens at ambient temperature (about 20 °C)

Pest	Dosage
Mice	12
Cockroaches, fly plant pests	65
Stored product insects	200
Mites (all stages including hypopi)	600
Nematodes	600–1200
Weed seeds and pathogens	1000–3000

pests, including plant pathogenic fungi, bacteria, viruses, nematodes, arthropods and weeds, otherwise limit economic production of certain crops locally. A wide range of salad crops such as lettuce, celery, cucumber, peppers and tomatoes, as well as flowers and ornamentals such as carnations, chrysanthemums, poinsettia and alstroemeria, in glass houses, and strawberries in open fields, are grown on methyl bromide-treated soil in temperate climates.

The widespread use of the fumigant has been encouraged by its efficacy and the availability of simple application systems and technology. The major advantages of using methyl bromide are as follows:

(1)   a rapid and consistent action against pest species;
(2)   more effective at penetrating soil than other currently used chemicals and steam;
(3)   no known pest resistance in the field;
(4)   the spectrum of activity against soil pests is wider than any other known soil treatment except steam;
(5)   can be used in soils over a wider range of moisture contents and temperatures than most other chemical treatments;
(6)   dissipates quickly after treatment and in most situations residues can be eluted out;
(7)   unlike other fumigants, methyl bromide is an effective viricide.

Soil fumigation with methyl bromide has been successfully replaced in some areas by methods and techniques that have been available for many years but have been adapted or modified to suit local requirements. Alternatives used in such systems as well as potential alternatives, are discussed here. None of these specific alternative methods used alone, has the broad spectrum of activity, efficacy or consistency of methyl bromide in achieving control and increasing yields. The development of a comparable agricultural system without the use of methyl bromide, in many cases, will require the integration of multiple alternative technologies and extensive research to achieve a similar spectrum of efficacy and reliability (UNEP 1995).

For alternatives to methyl bromide to come into widespread use, an integrated pest management (IPM) strategy will be required. The idea behind IPM is to utilise pest monitoring techniques, establish realistic pest injury levels, and to then apply a mixture of strategies and tactics to prevent or manage pest problems in an environmentally sound and cost-effective manner. This approach will require much further research to enable pest populations to be managed successfully and for future environmental problems arising from the control of soil pests to be avoided.

2.1.1.1   Cultural Practices as Alternatives to Methyl Bromide

Although cultural practices may reduce the need for, or frequency of methyl bromide applications, the prospects for their successful use without the back-up of chemical treatment are limited. Different and sometimes conflicting measures are effective for different pest or disease organisms, cropping systems, or locations. A significant commitment to applied research and technology transfer programs will be needed to take full advantage of the potential of cultural practices in reducing problems encountered in the many different situations currently using methyl bromide.

2.1.1.1.1   *Artificial Plant Growth Substrates*

The use of materials such as polyurethane foam blocks, perlite or rock wool, or a continual supply of nutrients in an aqueous medium (hydroponic systems), allow culturing of crops without soil fumigation. Pumice materials can be used to assist heat absorption and retention in carefully managed protected cropping systems and assist in the elimination of pathogens (Bello *et al* 1991, 1993). Many of these materials may be recycled using steam, dry heat, microwave or fungicidal treatments (Szmidt *et al* 1989), or solarisation (Gamliel *et al* 1989). Alternatively they can sometimes be broken up and applied to soils to improve soil structure (Anon 1992). Use of substrates is technically and economically feasible in greenhouses and even in open fields, under suitable climatic and economic conditions. In the Netherlands there has been a sharp increase in the use of soilless culture since 1975, using a range of artificial substrates or hydroponic systems. Conversion to these techniques with the assistance of government grants has helped to bring about the complete replacement of methyl bromide in greenhouse culture in the Netherlands. The cost of conversion was about $1.3 m per hectare (Anon 1992).

Artificial substrates are often based on silicaceous volcanic rock, deposits of which are common in Europe and North America. This material is processed to produce an expanded granular mineral foam of huge surface area and a high ability to retain moisture which is accessible to growing plants (Szmidt *et al* 1989). Other substrates used in soilless culture include clay granules, peat and polyphene foam (Ruijs 1992). Media such as tuff (volcanic ash), peat or compost have been used for many years in the potting industry (Gamliel *et al* 1989). Besides the between-crop sterilization of the growing media, the drain water used during growing may also require treatment by heat, ozonation or UV light for disinfection purposes so that it can be re-used. Currently about two-thirds of vegetable production and up to 25% of flower production in the Netherlands utilises soilless media. Economical considerations are restricting further increases in these methods of production and prospects are limited for the further transfer of crops still grown on soil to artificial substrates or hydroponic systems (Ruijs 1992). Elsewhere, however, the use of such systems continues to increase. In the UK the total glasshouse area utilising soilless systems has increased to over 700 ha, mostly for tomatoes, about 27% of the total glasshouse area (Table 8.2).

2.1.1.1.2 Crop Rotations

Crop rotations of various kinds are effective in controlling many soil-borne pests on crops all over the world, and some have been practised since antiquity. The absence of a suitable host leads to a reduction of pest numbers and reduces the pathogen inoculum, but rarely eliminates pest or disease problems. The practice is not effective when organisms capable of infesting multiple hosts are present. Many rotations include a season of fallow or non-cultivation for animal grazing, and here again care is needed not to proceed with a crop susceptible to diseases affecting grass, such as wheat (Rovira and Venn 1985). The literature on cropping systems is extensive and several reviews are available (Cook and Baker 1983, Rodriguez-Kabana and Canullo 1992).

Pest suppression may be increased by including with rotations plants, such as oilseed rape, that are inhibitory to some plant pests (Schmidt 1983). In some cases better results can be obtained by monoculture after a period of rotation, at least for the first few seasons, both pests and pathogens showing successive reductions (Gair *et al* 1969, Shipton 1977, Gindrat 1979). However, in general, rotations are beneficial, and long rather than narrow rotations give the best results (Schippers *et al* 1985). Limitations are, however, obvious, namely availability of land, persistent inocula of polyphagous pathogens, local pest reservoirs, lack of a sufficient diversity of economically viable crops to rotate, inability to lose a season of crop production, inadequate equipment expertise, and other socio-economic considerations. Any proposal to utilise crop rotation as an alternative to methyl bromide must take these factors into account.

Table 8.2. Protected crops grown on soilless media in the UK

Crop	Area grown (ha)	Percentage of total area
Tomato	388	14.9
Cucumber	275	10.5
Capsicum	58	2.2
Total without soil	721	27.6
Total area for protected crops	2607	–

2.1.1.1.3 Timing of Planting

The practice of moving the time of planting to coincide with the period of lowest pest density or when adverse environmental conditions for activity of soil pests prevail, has been used to limit crop damage by, for example, root-knot nematodes in Georgia, USA (Heald 1987, Trivedi and Barker 1986). However, for many crops today, there are specific windows for economical marketing and hence production, and prospects for switching growing times may be limited.

2.1.1.1.4 Deep Ploughing

Ploughing with a special moldboard type plough can reduce pathogen inoculum through burial of pathogens and stimulation of microbial antagonists. Equally, however, it is possible that such measures may inadvertantly bring dormant pathogens to the surface by breaking through the tillage pan, depending on the cropping history of the location. An inoculum of *Fusarium solani* (Martius) Appel and Wollenweber while absent from cultivated soil was shown to persist below the zone of cultivation for over five years after the last pea crop, the compacted tillage pan encouraging poor drainage and anaerobic survival (Kraft 1985). Deep ploughing has long been found useful in reducing the inoculum of *Sclerotium rolfsii* Saccardo (causative organism of many diseases including 'southern blight') in the cropping of groundnuts in the southern USA (Punja 1985).

2.1.1.1.5 Flooding/Water Management

Flooding or controlled irrigation can be used to control some pests in suitable areas but certain fungal propagules, (eg some *Phytophthora* spp.) can thrive under the low oxygen conditions produced by high water levels (Hollis and Rodríguez-Kábana 1966, Heritage and Duniway 1985). The technique is effective against *Verticillium* wilt of cotton, *Sclerotinia* rot of celery and against many nematodes (Cook and Baker 1983, Pullman and DeVay 1981, Moore 1949, Rodriguez-Kabana and Hollis 1965, Muller and van Aartrijk 1989, Muller *et al* 1992). In general long periods (over 40 days) are necessary to achieve control of fungal pathogens, though better results may be achieved for nematodes by alternate shorter periods of flooding and drying (Stover 1979, Fishler and Winchester 1964). Flooding can, however, increase the spread of fungal pathogens and nematodes and some post flooding treatment may be necessary against surface deposits (Stover 1979).

For many crops the timing of irrigation or flooding can be critical. Potatoes, for example, are only susceptible to the powdery scab fungus *Spongospora subterranea* (Wallr.) early in development while tubers are setting and problems can be avoided by delaying irrigation for this period (De Boer *et al* 1985). The efficacy of flooding can be increased by the incorporation of organic amendments into the soil beforehand. Anaerobic microbiological metabolism gives rise to products toxic to some soil pathogens (Cook and Baker 1983).

2.1.1.1.6   Cover Crops

Cover crops are non-commercial plantings of various plant species which are grown and turned back into the soil as green or dry residues. Their presence often prevents species of noxious weeds becoming established and after turning in their decomposition can stimulate activity of micro-organisms antagonistic to soil-borne plant pests. Cover crops need to be designed into the cropping system so that they do not compete with the commercial crop. In Florida, for example, the winter vegetable crop is preceded by sorghum or other plants in the summer which successfully depress nematode and pathogen levels (Rhoades 1983, McSorley *et al* 1994). It should be recognised, however, that the use of cover crops can prove counter productive. The use of grass or legume cover crops in forest nursery plots while limiting weed colonisation actually increased the level of soil pathogens in trials in the North Western USA (Hansen *et al* 1990).

Some cover crops are planted at the same time as the main crop to supress insect pests and weeds, and are often referred to as living mulches (Thurston *et al* 1994). These suppress weeds and may improve moisture retention of the soil. They may also reduce the ability of pests to locate the crop (UNEP 1995). Examples of such cover crops are miniature brassicas and certain clovers.

2.1.1.1.7   Soil Amendments and Fertilizers

Adding materials to the soil can reduce or suppress some soil-borne pathogens and nematodes by stimulating antagonistic micro-organisms, increasing resistance of host plants, providing extra nutrients, altering the pH or by various other environmental effects (Cook and Baker 1983). Minerals have long been known to have potential for acting as agents of disease prevention and pest control (Sadasivan 1965) but to some extent results are unpredictable, depending very much on pH changes, direct toxicity to causative organisms, nutritional selectivity and enhancement of beneficial soil flora (Gindrat 1979). Calcium applied as lime or sulphate, for example, is beneficial in reducing the incidence of pod rot in peanuts (Pattee and Young 1982), but the pH change induced by lime may not be advantageous in other situations.

Organic amendments such as composts, sewage, by-products from agriculture, forest, fishing and food industries can all have the effect of controlling soil-borne pests in crops (Hoitink 1988, Quarles and Grossman 1995). For example, high nitrogen materials generate ammonia which can control nematodes (Chian 1990), while addition of chitinous materials not only generates ammonia but also stimulates the activity of chitinolytic micro-organisms effective against some pathogenic fungi (Godoy et al 1983). Bark composts are effective in suppressing a variety of soil pathogens (Hoitink et al 1985, Hoitink 1988) while oilseed cake composts can suppress nematodes (Yassin and Ismail 1994). It is often necessary for large amounts of an organic amendment to be added to the soil for an effect to be registered, and hence their application will be localised and dependent on reliable sources of raw materials for conversion into useful products that can be easily applied. The potential use of composts as suppressive media to help replace methyl bromide has recently been reviewed (Quarles and Grossman 1995).

2.1.1.1.8   Plant Breeding and Grafting

Cultivars which are resistant or tolerant to one or a few specific pathogens (and races) are already available for many crop species. The subject area has long been a major field of investigation in agriculture and horticulture and cannot be dealt with at length here. In most cases new cultivars can be developed through plant breeding techniques to address specific pest problems. Plant breeding has to be regarded as a permanent component of crop production but it has proved very difficult to develop cultivars resistant to several pathogens. Frequently the planting of cultivars resistant to some pests results in increased damage from pests to which they are susceptible. Furthermore the resistance may only be apparent under certain environmental conditions and may disappear, for example, at high soil temperatures (Trudgill 1991).

Grafting of susceptible annual or perennial crops on resistant rootstocks is possible for some crop species, and has considerable attraction as it is widely known that it is more difficult to obtain resistance to root disease than resistance to diseases of the stem or shoots (Scott and Hollins 1985). Melons or cucumbers, for example, can be grafted on wild melon or pumpkin rootstocks resistant to *Fusarium* wilt (Gomez 1993). The recent finding that certain genes may be turned on by root-knot nematodes offers new opportunities for the development of resistant plant material (Opperman et al 1994).

2.1.1.1.9   Biological Control

Implicit within any ecological system, natural or contrived by man, are balances between predators and prey, parasites and hosts and other more subtle interactions between species. There is thus the potential for the use of antagonistic, predatory or protective organisms to limit the action of a target pest(s). There are some cropping systems which have successfully used biological control agents (Parker *et al* 1985). Generally, the spectrum of activity and host specificity of biocontrol agents is very narrow, and efficacy varies under different cultural conditions.

Certain plants such as annual rye *Secale cereale* L. and some *Brassica* spp., together with a range of micro-organisms, produce metabolic bi-products, often referred to as allelopathic toxins, that inhibit the growth of weeds and pathogens (Thurston *et al* 1994). Some commercial preparations are available of soil bacteria antagonistic to fungal pathogens. Fluorescent pseudomonads have proved effective in preventing *Verticillium* wilt in potatoes and other crops (Schippers *et al* 1985, Wadi and Easton 1985). There are also many soil fungi that can act as antagonists, including benign races of some pathogenic species such as *Fusarium oxysporum* Schlechtendahl (Alabouvette *et al* 1985). Fungi of the genus *Trichoderma* applied as a seed dressing can suppress an unusually wide range of pathogens including *Sclerotium rolfsii*, *Phytophthora* spp. and *Pythium* spp. (Backman and Rodriguez-Kabana 1975, Cole and Zvenyika 1988).

Many micro-organisms develop in close association with plant roots and besides helping in the absorbing of nutrients, afford a kind of shield delaying invasion by pathogens or nematodes (Chet 1987, Calvet *et al* 1993). These micro-organisms are termed rhizobacteria or mycorrhizae depending on whether they are bacteria or fungi. The latter are split into ecto- and endomycorrhizae depending on whether or not they penetrate beyond the root surface layer. Ectomycorrhizae are important in forestry because they are commonly associated with conifers.

2.1.1.2 Physical Control Methods on Soil

Many physical methods used on soil for disease and pest control are based on raising the soil temperature to levels lethal to the causative organisms. Temperatures have been increased by a variety of methods over the years including enhancing the effect of sunlight (solarisation), the application of steam or hot water, and the use of microwaves or dielectric heating. The temperatures required for control of soil pathogens vary according to the length of exposure and the species concerned, but most plant pathogens are recognised to be mesophilic, i.e. they do not grow well at temperatures above 30 °C (DeVay 1991). Quite short exposures at higher temperatures can achieve complete kill of many species, a logarithmic relationship existing between time and temperature (Pullman *et al* 1981). With 30-min treatment times in laboratory tests, temperatures of about 45 °C proved adequate for *Verticillium dahliae* Kleb., about 50 °C was required for *Phytophthora* spp. and *Pythium* spp., while for *Fusarium* spp. temperatures below 60 °C seemed unlikely to achieve satisfactory results (Bollen 1985). Sclerotial fungi, often difficult to kill by other means, are comparatively sensitive to heat requiring similar exposures to *Verticillium* spp. (Porter and Merriman 1983). Most insect and nematode species are also readily controlled at such temperatures.

2.1.1.2.1 *Soil Solarisation*

Soil solarisation or insolation as it is sometimes termed is the use of energy from the sun by covering of moist soil with clear plastic to increase soil temperatures. Moisture is necessary to assist in the transfer of heat to the target organisms, and better results are obtained in moist soil even though higher temperatures may be reached under dry conditions (Matrod *et al* 1991). In warmer climates the technique provides opportunities for control of many soil pests (Katan 1985, Katan and DeVay 1991). Solarisation does rely on suitable environmental conditions being available for lengthy periods and cannot be used in cool cloudy weather conditions. Black polyethylene films had been used as soil mulches for many years to increase plant growth, reduce evaporation and improve soil tilth, and although a disease control effect had been noted (Hilborn *et al* 1957, Geraldson *et al* 1965), it was only comparatively recently that films began to be used specifically for disease control purposes (Katan *et al* 1976). The innovation here was to target the warmest part of the season and time of day rather than the cooler times, and clear films were used rather than the usual black sheeting. The term solarisation was introduced to describe the combination of physical, chemical and biological changes in soil (Pullman *et al* 1981). There was a widespread and rapid adoption of the technique in warmer climates in the years following (Katan *et al* 1987).

During the hottest time of the day it is not unusual for temperatures over 60 °C to be attained near the soil surface under clear polythene sheeting under dry conditions (Table 8.3). At deeper levels, temperature changes are less marked, the difference being dependent on soil type, tilth and moisture. In trials in California, Pullman *et al* (1979) recorded temperatures of over 50 °C at 5 cm depth but at 15 and 30 cm depth temperatures were 10 and 15 °C lower respectively. Under such conditions nematodes and pathogens are rapidly killed near the surface, though some weed seeds are not affected. Weed control has often been achieved by using black polythene to deprive germinated seeds of light for photosynthesis, but with solarisation treatments transparent films are used. Here control relies on the high temperatures achieved and though many species are heat sensitive, some are not (Elmore 1991).

Some trials have been performed on different types of film but clear low density polyethylene (LDPE) is the most widely used for solarisation. Films

Table 8.3. The effect of solarisation on the peak temperatures achieved near the surface of soils (A, 0–2 cm; B, 5–7 cm) covered with various sheeting materials (after Matrod *et al* 1991)

Type of sheet	Average maximum temperature (°C)			
	Moist soil		Dry soil	
	A	B	A	B
100 μ thickness transparent polyethylene	54.5	42.3	65.0	42.5
180–200 μ transparent polyethylene	54.3	41.5	65.0	42.0
400 μ transparent polyvinyl chloride	51.3	41.3	68.8	42.8
60 μ black polyvinyl chloride	48.0	35.0	56.5	37.3
Uncovered control	42.5	32.5	56.3	36.6

as thin as 1 mil (25 microns) give good results and are relatively cheap. LDPE is essentially transparent to solar radiation (280–2500 nm) but much less transparent to terrestrial radiation (5000–35 000 nm) thus reducing heat loss from the soil (DeVay 1991). Black films absorb solar radiation and are less effective in raising soil temperatures (Haroon Usmani and Ghaffar 1985, DeVay 1991, Matrod *et al* 1991). In trials on different colour LDPE films (red, green, blue and yellow as well as transparent and black), Alkayassi and Alkaraghouli (1991) obtained slightly higher temperatures under the red film than the transparent, with the lowest rises occurring under yellow and black. All films other than yellow and black achieved temperatures in excess of 50 °C at 10 cm depth. Wavelength-selective plastic mulches potentially have the dual benefits of excluding photosynthetic light waves, thus preventing weed germination, while allowing heat generating light waves to reach the soil, thus enhancing growth of the crop above. Further investigation of their use on crops such as strawberries and tomatoes is merited.

Solarisation is still mostly used for open soil treatments in warmer climates but the technique can also be used in glass or vinyl houses (Kodama and Fukui 1979, Tjamos *et al* 1989), and in such situations success may be achieved in a cooler climate. Garibaldi and Gullino (1991) list a number of procedures to be followed for use of solarisation under marginal climatic conditions. It is essential that the soil is kept moist and the solarisation period should last at least six weeks coinciding with the hottest summer weather. Two layers or a double layer of the thinnest transparent LDPE sheeting enclosing air bubbles should be employed. Opportunity should be sought for combining solarisation with biological control agents. In trials in Northern Itay Garibaldi and Tamietti (1989) obtained significant reductions of *Rhizoctonia solani* (Kuhn) on *Phaseolus vulgaris* L. in green houses using solarisation, though the addition of the pathogen suppressant *Trichoderma* spp. did not further enhance the effectiveness of the treatment. The technique is practised widely for protected crops in Japan such as strawberry

(Table 8.4), cucumber and egg plant (Horiuchi 1991), as well as in France, Belgium, Greece and Italy (Goisque *et al* 1984, Katan 1987).

2.1.1.2.2 Steam

Of all the alternative treatment methods available, steam sterilisation is the only one offering effective control within a similar timescale to that of methyl bromide. For greenhouse and potting soil mixes, autoclaves, steam boxes and cabinets are in use for soil samples that can be forklift-loaded for treatment. Control of a wide range of soil-borne pests can be achieved under appropriate conditions both for potting media and soil beds (Anon 1992). Early methods for applying steam to soil featured the use of perforated steel pipes which were buried in the soil to a depth of about 25 cm. Depending on the size of the boiler used to generate steam, an area of 20–60 square metres could be treated at one time, after which the pipes were pulled out of the soil and inserted into the adjoining plot (Nederpel 1979). The introduction of a spiked grid, with 30+ cm spikes separated by 30 cm intervals along a 4–5 m feed pipe designed to be trodden into the soil, offered some improvement but still required a great deal of heavy labour. The first significant advance was the introduction of the steam plough in the 1960's, a 3 m wide device which applied steam at 40 cm depth in the soil while being pulled through the plot by a powered winch, but problems were still encountered in achieving an even distribution of high temperatures in the treated soil, mostly because of variations in the speed at which the plough travelled through the soil. Nevertheless there have been recent developments in the production of steaming units in the USA for specific field

Table 8.4. Soil solarisation of protected strawberry cropping in the major producing regions in Japan (after Horiuchi 1991)

Region	Total area of plastic house (ha)	Solarised area (ha)	Percentage of area solarised
Nara	561	321	57
Sara	347	200	58
Fukuoka	583	187	32
Tochigi	800	97	12
Aichi	563	96	17
Okayama	140	60	43
Oita	160	54	34
Wakayama	80	50	63
Kagawa	272	42	15
Total	3506	1107*	32

*Total area for all Japanese regions 1310 ha.

applications, with satisfactory results. Grossman and Liebman (1995) reported the efficacy of a portable steaming unit that permitted the treatment of a quarter acre of chrysanthemum planting bed within a working shift. The device featured a hydraulically extensible hood of about $16\,m^2$ that permitted the covered area to be treated within 20–25 minutes.

Further development of the steaming technique in the Netherlands sought to simplify the process and sheet steaming was introduced. In this process, still widely practised, a PVC sheet, usually about 0.25 mm thick and up to 4 m wide, is layed out over the soil and anchored at the edges with chain or dunnage (Hege and Rob 1972). Steam is then generated and piped under the sheet and left to penetrate the soil. The soil type and extent of cultivation profoundly influence the efficacy of sheet steaming (Nederpel 1979). Peat soils are very difficult to sterilise by this method because of their water retaining properties. During sheet steaming operations, a positive pressure is generated which can reduce the seal on the plot. To avoid gas and heat loss, nylon nets have been employed to overlay sheeting. The presence of nets allows a pressure increase of about 10 mm water gauge from the sheet alone level of 5 mm, which greatly assists in heat being distributed to deeper soil levels. Further improvements in heat retention can be achieved by applying a layer of bubble foil over the sheet (Grossman and Liebman 1995). Unlike with soil solarisation, soil conditions at the outset of treatments can be quite dry as soil moisture contents will increase by at least 5% (10% in the top 10 cm layer) because of condensation during applications of steam. Nevertheless a careful balance needs to be struck between the boiler generation capacity and plot size to avoid over moistening from prolonged application (too small capacity) or steam breakthroughs leading to heat loss (over capacity) (Bartok 1993).

Another means of improving the penetration of heat to deeper soil levels is the installation of a permanent steam piping system. The pipes often used are plastic drain pipes 5 cm in diameter. The main feed pipe is of greater diameter (10 cm) and the interval between steam pipes has been optimised at 80 cm. (4 per 3.6 m bay in the glasshouse). Pipes are buried to about 50 cm, just below the zone of cultivation to minimise energy costs while optimising heating of the deepest cultivated layer. At the far end of the steaming pipe lines a drainage pipe transports the condensate from steaming to a land drain. The length of this condensate drain should not exceed 20 m and is a major factor controlling the design of such installations (Kaufmann and Hackbart 1967). Fixed drainage pipe installations can permit steaming times to be reduced to about 6 h rather than the 8–10 h often needed for sheet steaming (Nederpel 1979).

Because of the long times required for steaming operations and difficulties with heat distribution and transfer, systems have been developed whereby reduced pressures provided by fans have been used to assist the penetration of steam, enabling treatment times to be reduced to about half those needed for sheet steaming (Runia 1983). Underground piping systems installed for this purpose have been shown to achieve rapid temperature increases at all depths and to reduce operation costs by 30% or more (Domke 1986, van Meekeren and van Meekeren 1990). Other innovations in soil steaming technology are described by Belker (1990) and Labowsky (1990). The use of steam air mixtures, provided by blowing air into the steam supply, permits more control of the temperatures achieved, and is favoured where a soil pasteurisation procedure is desired rather than sterilisation, and in experimental work (Sylvia and Schenck 1984, Thompson 1990). The overall benefit of a strategy of complete sterilisation by negative pressure steaming, followed by selective re-introduction of beneficial micro-organisms, as against a soil pasteurisation or selective sterilisation procedure, is still a matter of debate in the industry (Grossman and Liebman 1995).

The very high temperatures that can be encountered during steaming can result in an increased mineralisation of the soil and a complete supression of beneficial micro-organisms. One notable effect is the increase in available manganese levels (Sonneveld 1979, Williams-Linera and Ewel 1984). The manganese released is fixed again on production of oxides but the process is very slow, sometimes taking over a year. Other effects are the enhanced production of ammonia from nitrates, a process which continues after the steaming operation (Sonneveld 1979), and an increase in inorganic bromide levels (Roorda-van-Eysinga, 1984).

2.1.1.2.3   Other Heat-based Methods

Several other methods have been employed to raise soil temperatures. In Japan a system of buried hot water pipes was found effective in controlling bacterial and fungal spores and weed seeds, but was expensive to instal (Hayashi and Aono 1982). Open field burning, widely practised for general agricultural purposes, has been evaluated for pathogen control but found to be ineffective (Chinanzavana et al 1986). Dry air or oven treatment of soil is used for quarantine purposes when soil is imported into the US (Liegel 1986).

The use of microwaves in raising soil temperatures, particularly for applications in the pot plant industry, has been a focal point of investigation for a number of years (Vela et al 1976, van Wambeke 1983). Microwaves are radio frequencies in the region of 300 MHz–100 GHz, representing wavelengths between 10 cm and 3 mm, longer than infrared but shorter than VHF radio transmissions (Nelson 1973, Fleurat-Lessard 1987). Most commercial microwave units operate at 2450 MHz. In tests on different soil moisture contents and granular size using a rail-mounted emitter, van Assche and Uyttebroeck (1983) identified two optima for pathogen control, one with dry soil equilibrated to 20% r.h. and one with moist soil equilibrated at 60% r.h. The effect was strongest in a sandy soil. Several studies have been performed in microwave ovens on different soil types and against different weeds nematodes and pathogens with good results after exposures of only 1–2 minutes (Benz *et al* 1984, Speir *et al* 1986, Barker and Craker 1991). Sample size influences the duration of treatment necessary to raise temperatures, a 1 kg sample requiring about 150 seconds at 650 W compared with 425 s for a 4 kg sample (Ferriss 1984).

With microwave treatments, as with steam, the problem remains that beneficial soil bacteria are also killed (Yang *et al* 1990). There are claims that the use of lower frequency radiowave emitters (10–30 MHz) selectively kill nematodes and pathogens while leaving beneficial micro-organisms intact, probably as a result of differences in their dielectric properties and a resultant differential in heating rates (Diprose and Evans 1988, Rodionova *et al* 1990).

Several other systems based on electrical heating have been tested, especially by researchers in the former Soviet Union with a view to treating soil *in situ*. Besides the high frequency radio frequency emitters already mentioned (Borodin 1989, Rodionova *et al* 1990), a disc electrode and infrared device mounted on a carriage (Baranov *et al* 1983) and a thyristor current-regulated electrode steriliser for use in greenhouses (Nikitin and Nikitin 1990) have been described. As with microwave systems, costs of such apparatus are high and hence scope for their large scale use is limited.

2.1.1.2.4 *Other Physical Methods*

Such methods include the use of ionising irradiation technologies, cool or very low temperatures, and the various mechanical treatments already discussed under section 2.1.1. on cultural alternatives. Each process may have potential for integration with management techniques designed for specific pests and production systems. Some have been researched extensively but are little used, while others will require much further research to fully explore their potential.

Ionising radiation can be provided as γ-rays, x-rays or accelerated electron beams. All require high cost installations and stringent safety precautions. Most studies have concentrated on gamma radiation as the soil sterilant, mostly for use with potting substrates and equipment (Rattink 1982, Elliott 1989, UNEP 1995) though apparatus has been designed and patented for use in tillable soil beds (Plattner 1989). The use of radiological techniques in Holland assisted the replacement of methyl bromide fumigation in greenhouses by 1991 (Anon 1992). In Denmark the use of an accelerated electron beam at 200 krad (2 kGy) was effective in eliminating a mycorrhizal infection of barley seedlings when soil samples of thickness 2 cm were irradiated (Jakobsen 1984). Dosages required for elimination of some pests and pathogens of soil are, however, relatively high. A dose of 7.5 kGy was required for complete kill of the root lesion nematode *Pratylenchus thornei* Sher and Allen exposed in a state of anhydrobiosis (Thompson 1990), while Lemanova *et al* (1983) used dosages of over 10 kGy for elimination of bacterial canker, viruses and nematodes in rooting substrates for grapevine cultivars. Hartel and Alexander (1983) stated that a dosage of 25 kGy was required for the complete sterilisation of soil samples. Use of gamma radiation on soil does not appear to affect mineralisation or nitrogen chemistry any more than other soil sterilisation techniques (Sparling and Berrow 1985, Elliott 1989).

Low temperatures limit the activity of pests and pathogens but they also limit plant growth. Nevertheless some root-knot nematodes can be successfully managed in greenhouse systems by keeping temperatures below 20 °C (Fernandez *et al* 1993, UNEP 1995). Many species are adversely affected by freezing conditions, though most micro-organisms are tolerant, at least as spores. In fact freezing techniques have been long used as means of preserving fungal cultures and subjection to sub-zero temperatures would seem to have limited potential for control. However, the technology exists in the construction industry for soil freezing and there is merit in investigating the effect of repeated freezing and thawing as a possible means of pest and disease control.

2.1.2 COMMODITIES: PERISHABLE CROPS

Perishable commodities are characterised by being unprocessed, physiologically active, respiring plant parts with high moisture contents and short shelf lives. A wide range of fresh fruit and vegetables, cut flowers, ornamental plants, fresh root crops and bulbs require treatment, principally to prevent the spread of various arthropod pests. By their very nature perishable goods require treatments to be rapid and methyl bromide fumigation has become the predominant treatment method for disinfestation purposes. Up to 8% of the methyl bromide produced worldwide is used on perishables, of which about half is for disinfestation of fruit to satisfy national and interstate quarantine requirements.

Alternatives to methyl bromide for use on perishables do exist and some have been the subject of much recent research. Besides treatment with other fumigants, there are disinfestation procedures based on cold storage, heat, controlled and modified atmospheres, irradiation and a combination of these treatments. Although these are approved for disinfestation of specific commodities, very few are in use relative to the number of different commodities treated with methyl bromide. For example, for quarantine purpose heat treatments have been approved for 14 applications, alternative chemical fumigants for 12, cold treatments for 9, pest free zones for 4 and irradiation for 2. However, approval does not mean that the technique is currently in use (UNEP 1995). Irradiation techniques for example, are little used on perishables in the West. The more widespread application of many of these techniques is restricted by their commodity and pest specificity. There are only a few examples of alternative treatments developed for commodities routinely treated with methyl bromide because until recently there was no obvious need for such development.

The use of alternatives generally requires a more complex operational procedure than treatment with methyl bromide which increases the costs of implementation. Potential alternatives which show promise need further research to determine their efficacy against pests and suitability for various commodities. Currently, there are no existing alternatives in use for apple, pear, and stone fruit exports that are host to codling moth; for berryfruit; for grapes from Chile exported to the United States; and for many root crops that are exported to developed countries from developing countries (UNEP 1995).

Most treatments are carried out on arrival in an importing country if undesirable live pests are intercepted, or occasionally prior to export if the importing country deems the risk of pest importation to be a serious threat to their agricultural economy. Some countries take a precautionary approach and apply routine treatments to meet export requirements. For example, countries sometimes fumigate perishable commodities prior to export to ensure that their release on to the retail market is not delayed and to avoid the fumigation costs which may be incurred on arrival at their destination.

2.1.2.1 Alternatives Avoiding Post Harvest Treatment

Ideally, measures which can be implemented in the growing and harvesting of crops, and in the preparation stage for marketing which avoid the need for post-harvest treatments, chemical or otherwise, to combat pest problems, are the goal in crop production. Unfortunately there are only limited prospects for reliance on such techniques and nearly all of these are restricted to specific trading between nations in specific perishable goods. All involve programmes of activity which are labour intensive. Two such measures are described here, the administration of pest-free zones and the use of various inspection techniques before and after harvest to ensure freedom from pests, a set of procedures known as the 'systems approach'.

2.1.2.1.1 Pest Free Zones

The United States, Japan and New Zeland accept certain commodities certified by government officials as originating from geographically defined regions recognised to be free of quarantine pests. Pest-free zones require justification through monitoring, reporting and continuous enforcement. A regulatory agency may require justification for many years because their acceptance has to be based on extensive knowledge and experience of both the pest and commodity biology and distribution. The first modern use of the pest-free zone strategy was employed in Texas in the early 1980's to certify citrus fruit grown in the Rio Grande Valley as free of Mexican fruit fly *Anastrepha ludens* (Loew). Following adult trapping studies using McPhail traps in the region and fruit cutting to corroborate freedom from the pest (over 60 000 fruits examined) the zone was accepted (Riherd *et al* 1994). Similar zones have since been established for Caribbean fruit fly *Anastrepha suspensa* (Loew) in Florida. Japan accepts commodities produced in Tasmania as free from Mediterranean fruit fly *Ceratitis capitata* (Wiedemann) and Queensland fruit fly *Bactrocera tryoni* (Froggatt) (Anon 1989). Other countries with designated pest-free zones with respect to various species of fruit fly are Mexico, Brazil and Ecuador.

2.1.2.1.2 The Systems Approach

Systems approaches rely on pre-harvest management of pests in the orchard or field, followed by sorting procedures to reject infested fruit. Generally, the pests are present in very low numbers and can be easily controlled. Nevertheless, pre-harvest control practices rarely exceed 90% pest mortality and are insufficient to comply with the predominant concept of a quarantine security of greater than 99.9968% mortality (Baker 1939, Couey and Chew 1986). However, the desired level of security may be achieved, for example, if pre-harvest insect reduction practices (e.g. pesticides, pest attractants, pest resistant cultivars) are combined with sorting procedures at the packing station. Currently watermelon exported from Tonga to New Zealand, feature a systems approach which includes a fumigation component using MB (Cowley *et al* 1991). Other examples which can be regarded as systems approaches involve the export of cherries from the Pacific US to Korea (Moffitt 1990) and the shipment of papaya and Sharwil avocados from Hawaii to the U.S. mainland (Armstrong 1991, Jang and Moffitt 1994). Systems approaches, demonstrating sequential pest reductions from field to packed commodity environments, require detailed monitoring and documentation in order to be acceptable to regulatory agencies.

2.1.2.2 Non-chemical Treatments After Harvest

Physical control methods for perishables today include cold storage, forced hot air, vapour heat, hot water immersion, radio frequency waves, controlled atmospheres and irradiation. Heat-based methods are unsuitable for very perishable commodities such as asparagus, cherries or leafy vegetables as shelf life is reduced. Other potential methods include the use of water or air under pressure to physically remove pests from the fruit surface.

2.1.2.2.1 Cold Storage

Cold storage is widely used for post-harvest treatments of perishable commodities. Sub-zero temperatures have a rapid effect on insects but it is not necessary for temperatures to be this low to be of use for pest control. Insects may be lethally injured by cold shock even though their body fluids do not freeze (Lee 1991). Quick freezing at temperatures below $-10\,°C$ are really only suitable for fruit pulp or slices on route for processing into juice, as extensive damage occurs to the unprocessed treated commodity (Gould 1994). Usually the exposure to cold is for a limited period, as for example the holding of fruit for 10–22 days at $-1\,°C$ to $+2\,°C$ to kill tephritid fruit flies on citrus fruit, apples, pears, grapes, stone fruit, carambola, lychees, loquats and kiwifruit (Gould 1994). Potential quarantine treatments based on cold exposure have also been studied for codling moth *Cydia pomonella* (L.) (Moffitt and Burditt 1989a, b) and oriental fruit moth *Grapholitha molesta* (Busck) (Dustan 1963, Yokoyama and Miller 1989).

Usually cold treatment is carried out commercially in-transit in containers on board ship or using land-based facilities, and precise records of the temperature and duration of exposure are required to show compliance with phytosanitary treatment specifications in order to be acceptable as a disinfestation treatment. Exposure times and temperatures are linked to the pest but need to be chosen after evaluation of effects on the fruit being treated. Many tropical and sub-tropical fruits are susceptible to cold, but chilling injury can be reduced if the commodity is conditioned at moderate temperatures prior to exposure to cold (Hatton 1990, Houck *et al* 1990). Damage can also be prevented by interruptions during the low temperature exposure. For example, cucumbers exposed at $12.5–20\,°C$ for 18–24 h every three days, or peppers exposed at $20–21\,°C$ for one day every three days, can avoid signs of damage (Paull and McDonald 1994) though treatment efficacy may be affected. Details of cold exposures required for effective control of four fruit fly species or groups are given in Table 8.5, based on information collected for the USDA-APHIS-PPQ Treatment Manual (Gould 1994).

Cold or cool storage may also be used to assist product shelf life after a heat treatment for disinfestation. Couey *et al* (1984) found a 10-day exposure at $8–9\,°C$ during transit of Hawaiian papayas to the US mainland to be effective after a hot water dip treatment for 20 minutes at $49\,°C$.

Table 8.5. USDA-APHIS-PPQ cold treatment times for different species of fruit fly (after Gould 1994)

Species/group	Cold treatment
Ceratitis capitata	10 days at 0 °C or below
	11 days at 0.55 °C or below
	12 days at 1.11 °C or below
	14 days at 1.66 °C or below
	16 days at 2.22 °C or below
Anastrepha ludens	18 days at 0.55 °C or below
	20 days at 1.11 °C or below
	22 days at 1.66 °C or below
Other species of *Anastrepha*	11 days at 0 °C or below
	13 days at 0.55 °C or below
	15 days at 1.11 °C or below
	17 days at 1.66 °C or below
Bactrocera tryoni	13 days at 0 °C or below
	14 days at 0.55 °C or below
	18 days at 1.11 °C or below
	20 days at 1.66 °C or below
	22 days at 2.22 °C or below

2.1.2.2.2 Heated Air Treatments

The use of heated air seeks to exploit the window between pest elimination and fruit damage, a window which is highly commodity specific and in some cases too narrow for effective use. Two types of heated air treatments are practised, vapour heat and forced hot air. Vapour heat was the first to be used and applied hot air saturated with water to the fruit, transferring heat by condensation (Armstrong 1994). More recently forced hot air at r.h. less than 90% (usually less than 60%) has been introduced to avoid heat transfer by water condensation, which causes damage in certain fruit.

Vapour heat treatments feature a rapid heating phase from ambient followed by a more gradual increase to the critical end point temperatures of 43–47 °C, depending on commodity and pest sensitivity. More recently, research has tended to concentrate on shorter exposures than those listed in government quarantine treatment manuals. For peppers against oriental fruit fly *Bactrocera dorsalis* (Hendel) and mangoes against melon fly *B. cucurbitae* (Coquillett) the target exposure has been identified as 3 h at 43 °C (Sugimoto *et al* 1983, Sunagawa 1987), for carambolas and mangoes against either species, 3 h at 44 °C and 30 min at 46.5 °C respectively (Kuo *et al* 1989, 1987), for zucchinis against cucumber fly *B. cucumis* (French), 30 min at 46.5 °C (Corcoran *et al* 1993) and for tropical cut flowers against surface insects, 30 min at 46.6 °C (Hansen *et al* 1992). The exposures are calculated to achieve the quarantine standard of probit 9 mortality, a standard arrived at in the 1930's (Baker 1939) whereby the expected level of survival of 32 in a million was judged to avoid the risk of pests breeding up in a new location, given a low volume of pests on the fruit. The probit 9 security level has been under increasing scrutiny over the last decade as it does not make allowance for the volume of the infested commodity passing through a trading outlet, nor the variation within the 95% confidence limits of the treatment which predict a possible 29–136 individuals surviving (Landolt *et al* 1984, Couey and Chew 1986). Recently a new quarantine standard has been adopted in New Zealand (Baker *et al* 1990). Known as a 'maximum pest limit' this process relies on an efficient sampling method and an agreed maximum number of pests allowed per day or per volume of commodity. The underlying assumption is that any emerging adult would be unable to find a mate because of the rapid separation and dispersal of the imported goods (Paull and Armstrong 1994). The actual treatment parameters used to achieve any required standard rely on data such as that developed by Jang (1986, 1991) which relate the high temperatures achieved to the rate of thermal death.

Forced hot air treatments have been the subject of active research in parallel with the development of commercial equipment incorporating software capable of giving precise temperature control (Williamson and Winkelman, 1989). The first treatment meeting a quarantine standard was developed to control fruit flies on Hawaiian papayas, and comprised a 4 phase heating schedule to bring fruit centres to 47.2 °C (Armstrong *et al* 1989, Hansen *et al* 1990). The papayas were not damaged by the treatment, which involved being held at temperatures increasing from 43 ° to 49 °C for about 10 h. Similar attempts to bring carambola fruit centres to 45.5 °C for fruit fly control resulted in a subsequent increased rate of deterioration of the fruit (Miller *et al* 1990).

A forced hot air treatment has also been proposed for mangoes to eliminate West Indian fruit fly *Anastrepha obliqua* (Macquart), exposing fruit at 50 °C until the seed surface temperature reaches 48 °C (Mangan and Ingle 1992). Hot air at 48 °C to bring fruit centres to 46.1 °C was sufficient for Caribbean fruit fly in mangoes (Sharp 1992). Sharp (1993) developed a quarantine treatment for grapefruit against Mediterranean fruit fly by holding fruit at 49 °C until the fruit centre reached 44 °C, which did not damage Marsh white grapefruit. Mangan and Ingle (1994) developed a treatment schedule for several grapefruit varieties which was effective against the more heat tolerant Mexican fruit fly. This involved exposure at temperatures increasing from 40 °C to 52 °C to bring fruit centres to 48 °C. Late season grapefruit may be more susceptible to heat treatments than earlier season fruits and applications are recommended soon after harvest (McGuire and Reeder 1992, Mangan and Ingle 1994).

Shorter exposures with forced hot air at 47 °C have been used successfully in New Zealand to control a variety of pests on the surface of persimmon fruit (Cowley *et al* 1992).

2.1.2.2.3 Hot Water Dips

Hot water is an excellent medium for heat transfer and has long been used to reduce pathogens on fruit (Armstrong 1994). From the 1940's the technique was used in conjunction with dips containing ethylene dibromide. More recently, the technique has been examined for use alone as a means of fruit fly control. The first treatment to reach a probit 9 security level for quarantine use in the USA was for bananas grown in Hawaii (Armstrong 1982), comprising a 15 minute treatment at 50 °C. Standards have since been worked up for mangoes (Sharp and Spalding 1984), comprising a 65 min dip at 46.1–46.7 °C against Caribbean fruit fly, and for guava (Gould and Sharp 1992), comprising 35 min at 46.1 °C. Attempts with other fruit have encountered problems as a very narrow window exists between pest elimination and damage with temperatures around 45 °C. Hence for papaya, two-phase and double temperature dips fell short of achieving adequate kill without fruit damage (Couey and Hayes 1986) and similar problems have been encountered for grapefruit (Sharp 1985), stone fruits (Yokoyama and Miller 1987) and carambola (Hallman 1991). Recently some promise has been shown in 22–26 min exposures at 48–50 °C as a means of treating persimmons against mealy bugs and moths in New Zealand (Lester *et al* 1995).

The adverse effects may be improved by hydrocooling fruit after dipping but investigation of the consequence of effectively shortening the treatment exposure to high temperature on subsequent insect mortality needs investigation (Sharp 1994).

Apart from tests on fruit some work has been undertaken on the use of hot water dips on plant cuttings and potential for treatments achieving quarantine security have been established for scale insects on bird of paradise plant and Cape jasmine (Hara *et al* 1993, 1994).

2.1.2.2.4 Radio Frequency Waves

Frequencies between 25 MHz and 2.45 GHz have been tested against pests of perishable commodities (Seo *et al* 1970, Hayes *et al* 1984, Del Estel *et al* 1986a, b). Problems have been encountered in achieving even temperatures within fruits sufficient to kill pests while avoiding damage. Seo *et al* (1970) reduced damage within mangoes by applying repeated 10–15 s bursts at 2.45 GHz instead of a continuous treatment, but some mango weevils *Cryptorhynchus mangiferae* (F.) were able to survive all treatments which did not damage the fruit. Hayes *et al* (1984) heated papayas by microwaves

(2.45 GHz) until centre temperatures reached 38°–45 °C prior to a hot water dip at 48.7 °C. The latter was necessary because after the microwave exposure, in other parts of the fruit temperatures varied by as much as 20 °C so that complete control of oriental fruit fly could not be obtained. There has been no commercial usage of the technique so far and further research is required to see whether uneven heating problems can be solved or whether the use of radiowaves can be usefully incorporated into a combined treatment system.

2.1.2.2.5 Irradiation

Ionising radiation (x-rays, electron beam or radioisotope) can control many pest species and has the additional advantage of being able to treat the commodity in its final packaging with no appreciable change in temperature or atmosphere. Low doses are capable of achieving quarantine security by sterilising pests but a very wide range of sensitivities to radiation is apparent among different developmental stages. Eggs of the fruit fly *Drosophila melanogaster* Meigen show 50% mortality when exposed to a dose as low as 4 Gy during cleavage, while pupae require 400 Gy for a similar level of kill (Nothel 1968). Balock *et al* (1963) found that a dose of 150 Gy reduced emergence of adult tephritid flies, but over 1000 Gy was required to prevent pupation of mature larvae. Jona and Arzone (1979) suggest a dose of 1000 Gy to prevent growth of larvae of European cherry fly *Rhagoletis cerasi* L. and stop damage to the fruit. For pests other than fruit flies a dose of 300 Gy has been quoted as adequate for quarantine purposes (Nation and Burditt 1994). Thirty-five countries have approved the use of irradiation on over 50 specific food commodities including apples, bananas, garlic, mango, onions, papaya, potatoes, stone fruit, and strawberries (UNEP 1995). United Nations agencies have recognised the similarity in response of fruit fly species on different fruit and, in order to avoid unnecessary experiments, have recommended a minimum effective dose regardless of their host commodity.

Fruit and vegetables have been classified according to their sensitivity to radiation (Table 8.6, adapted from Burditt, 1990). A considerable number have been reported to show detrimental effects on flavour, colour, vitamin levels, resistance to pathogens and ripening characteristics after exposure to doses of less than 1 kGy. In contrast, doses of about 5 kGy are required to inactivate organisms causing rot (Burditt 1994), though lower doses can reduce their effect (Morris and Jessup 1994).

Irradiation has also been investigated for treatment of cut flowers (Van de Vrie 1986, Piriyathamyong *et al* 1985). Irradiation of orchids at 500–1000 Gy, however, caused a reduction of vase life while pests were not all killed (Piriyathamyong *et al* 1985).

In spite of the positive results obtained by many researchers and approval by governments, few commercial irradiators have been developed for disinfestation purposes. Some factors delaying adoption of this technique are insufficient commercial-scale assessments of cost-benefit, suspicion over the verification of sterility of live pests, perceived public concern with the safety of isotope transport, long term storage and facility location, and a lack of general awareness of the benefits of radiation technology.

2.1.2.2.6 Controlled Atmospheres

A further potential physical disinfestation method is the use of controlled atmospheres (CA). The use of CA to extend postharvest life and quality of fruit and vegetables has steadily increased over the last 60 years (Kader 1993) and the technique is now widely used in combination with lowered temperature and raised humidity. The atmospheres used for storage (up to 10% carbon dioxide, < 5% oxygen) include those known to be effective against insect pests, but the combined use of low temperature with the CA acts against the efficacy of the treatment. For fungistatic action, addition of 5–10% carbon monoxide to the CA gives good results on commodities that cannot tolerate carbon dioxide (CO_2) levels above the 15% necessary for an appreciable effect (Kader and Ke 1994).

Table 8.6. Classification of fruits according to sensitivity to irradiation (After Burditt 1990)

Minimal detrimental effects	Inconsistent reports in literature	Significant detrimental effects	No data available
Apple	Apricot	Avocado	Kiwifruit
Cantaloupe	Banana	Cucumber	Pomegranate
Cherry	Cherimoya	Grape	
Date	Fig	Green bean	
Guava	Grapefruit	Lemon	
Honeydew	Kumquat	Lime	
Longan	Litchi	Olive	
Mango	Loquat	Pepper, bell	
Muskmelon	Orange	Pepper, chili	
Nectarine	Passion fruit	Sapodilla	
Papaya	Pear	Soursop	
Peach	Pineapple	Summer squash	
Rambutan	Plum		
Prune/dried fruit	Tangelo		
Raspberry	Tangerine		
Strawberry			
Tamarillo			
Tomato			

On a much smaller scale, atmospheres immediately surrounding the commodity can be modified to kill pests using polyfilms or coatings made from wax or cellulose-based compounds (Hallman *et al* 1994). Shrink wrapping has the effect of enabling an atmosphere to develop that is lethal to pests. The technique worked well against light brown apple moth *Epiphyas postvittana* (Walker) on Japanese persimmons (Dentener *et al* 1992), but was not suitable for papaya against oriental fruit fly (Gould and Sharp 1990).

Work on the effect of CA on different insect pests of perishable commodities was recently reviewed by Carpenter and Potter (1994). These authors also carried out the only commercial CA quarantine treatment in the export of asparagus from the US to Japan after a 4.5-day exposure to 60% CO_2 at 0–1 °C. Some insect pests can be killed by 1–2 months storage at low temperatures under low oxygen (O_2) or high CO_2 conditions (Benshoter 1987), but others, especially moth larvae in diapause, are highly tolerant requiring exposure periods of several months duration (Toba and Moffitt 1989, 1991). The time can be reduced considerably by raising the temperature of the CA treatment (Soderstrom *et al* 1990, 1991), but many products deteriorate. High CO_2 levels work better against many pests than low oxygen levels but are detrimental to the commodity (Kader and Ke 1994). This problem and the difficulty of designing large, temperature-controlled, CA disinfestation facilities with adequate gas retention, have so far limited widescale adoption of the technique.

2.1.2.2.7   Combination Methods

Disinfestation treatments can be combined to achieve the required pest mortality. For example, cherimoya fruit exported from Chile to the USA can be treated with a mixture of soapy water and a wax coating; and lychees exported from Taiwan to Japan can be treated with vapour heat followed by a cold treatment to achieve quarantine acceptance (Anon 1988). Combination treatments are rare, probably because of more extensive technical documentation required to demonstrate treatment efficacy for regulatory agencies, compared with single treatment applications. However, combined treatments or procedures will, in many cases, offer greater quarantine security than a single treatment. Additionally, combined treatments may allow a reduction in the amount of methyl bromide required for pest mortality, thus reducing the potential for commodity damage and the amount released to the atmosphere.

2.1.3   COMMODITIES: DURABLES

Durable commodities form a very wide range of products and materials loosely characterised by having less than 15% moisture content. Many durables are stored crops and plant products but also included are dried fish, animal skins, wool products, museum artifacts and timber. About 15% of the total fumigant usage of methyl bromide is employed on durables.

There are potential or existing alternatives for most uses of methyl bromide on durable commodities. However, there is no general in-kind replacement offering complete control within 24 hours, and all alternatives will require some changes in practice (UNEP 1995). Of the alternatives, only phosphine is extensively used, principally for cereals and legumes which are stored in bulk. Insect resistance to phosphine is an emerging problem, particularly in developing countries, but resistant pests can, at present, be controlled using currently available phosphine-based technology.

Some alternatives are already in industrial use for some classes of durable commodities. These include other fumigants, contact insecticides (see Chapter 7), controlled and modified atmospheres, physical methods such as heat, cold, radiation and various mechanical measures, and biological control methods. Many are limited in particular circumstances by speed of action, regulatory constraints, temperature, consumer acceptance, and lack of research data. Physical methods for insect control in stored grain have recently been reviewed (Banks and Fields 1995).

Most of the target pests of durables that are treated with methyl bromide are insects, plus a few mite species. Fungi, bacteria or nematodes are not typically target organisms, except with unsawn timber, spices and seeds for planting, respectively. Methyl bromide is sometimes specified for quarantine purposes for control of other organisms (e.g. ticks, snails) that may be transported in durable foodstuffs or timber, but which do not normally infest and damage the commodity itself.

2.1.3.1 Controlled and Modified Atmospheres

Treatments with controlled or modified atmospheres based on CO_2 and nitrogen (N_2) offer an alternative to fumigation with toxic gases for insect pest control in all durable commodities, but usually at an increased cost. They are effective against fungal pests only to the extent of restricting growth, a fungistatic rather fungitoxic action, and hence are not suitable for timber or some commodity treatments when fungal control is the primary objective.

Low O_2 atmospheres, typically created by adding N_2 to a fumigation enclosure, or by adding atmospheres generated by the combustion of propane, require that there be a maximum of 1% O_2 for effective action. For such low O_2 atmospheres to be achieved there is a need for stringent sealing of the enclosure and some facilities for topping up gas. Ideally a continuous flow of gas should be available. Propane combustion units are suited to this requirement and systems capable of providing a cooled, low ($< 1\%$) O_2 atmosphere have been developed (McGaughey and Akins 1989,

Bell *et al* 1991). Depending on the size of operation, N_2 can be transported to the site in cylinders, minitanks or in bulk cryogenic tanks. However two systems exist to provide nitrogen from compressed air on site.

The first system, which has been available since the 1970's is based on a principle for gas separation known as pressure swing adsorption or PSA. In this process N_2 is derived from compressed air passed through two beds or molecular-sieve coke. The nitrogen and oxygen are separated due to their different rates of adsorption with the nitrogen passing through the bed and into a holding tank. The two beds work alternately with one pressurised with incoming air while the other is returned to atmospheric pressure, in this application releasing the more strongly sorbed gases to waste. The oxygen content in the output gas depends upon size of plant and the air flow used. Systems exist capable of producing an O_2 content of less than 0.3% at a flow rate of over $100 \, m^3/h$.

The second system is based on filtration of air through membranes which differentiate between O_2 and N_2. The semipermeable membranes are mounted in separators containing thousands of hollow membrane fibres and capable of withstanding high pressures. Incoming air at 100–150 psi (7–10 bar) passes along the fibres and O_2 permeating through the fibre walls is removed from the vessel by reducing the pressure and venting the free space gas. Output from plants can be scaled to meet local requirements by provision of larger capacity separators and hence a similar range of treatment capacities exists to that available for PSA.

Carbon dioxide atmospheres typically are applied at about 60% CO_2 in air, as this concentration level works better than very high levels against some of the more difficult to control species that develop inside grain and seeds. The 8% oxygen present in the partially replaced atmosphere is normally sufficient to support development of most stored product pests indefinitely. CO_2 thus is regarded as having a toxic effect on insect pests (Jay 1971) and does not act just as an inert gas that reduces the oxygen level to below that supporting life. A toxic effect is also evident in anoxic atmospheres (Bell *et al* 1980). There are limited prospects for the provision of high CO_2 atmospheres on site, and bulk or cylinder-based supplies are the only viable options if there is no locally available industrial source.

Data on exposure times for control using CA are available for many species and stages of stored product pests under particular sets of conditions (Annis 1987, Bell and Armitage 1992). Most species are completely controlled by exposures ranging from 15 days with high CO_2 levels to 21 days with low O_2 at 25–29 °C (Banks *et al* 1991). As an extreme case, larvae of the khapra beetle *Trogoderma granarium* Everts in diapause easily survived exposures up to 17 days at 30 °C, with CO_2 levels at or above 60% in air (Spratt *et al* 1985). In our recent tests on diapausing larvae of the warehouse moth *Ephestia elutella* (Hubner) over 40% of larvae survived a 28-day exposure to an atmosphere containing 1% O_2 and 12% CO_2 (a typical burner gas atmosphere) at 20 °C, but atmospheres containing over 90% CO_2 achieved control in 14 days (Savvidou and Bell, in preparation). Nevertheless, even with less tolerant pests, exposure times are in general rather longer than for the fumigant phosphine and much re-working of industrial practices will be required before CA can be routinely used. A summary of the exposures needed for the control of some representative commodity, food and timber pest species at different temperatures is presented in Table 8.7.

The efficacy of CA can be improved for most developmental stages by lowering the r.h., because of a reduced ability to restrict desiccation during the exposure (Navarro and Calderon, 1974). The time of exposure required for high CO_2 atmospheres can be reduced if the treatment is combined with pressure alterations or raised temperature. Complete kill of a range of tolerant stored product pests can be obtained within 48 h by raising the temperature to 38 °C (Jay, 1986). Exposure under vacuum reduces exposure

Table 8.7. Estimated days of exposure required for control of stored product insects and mites with modified atmospheres (data sourced from Annis (1987) and Bell and Armitage (1992) with additional information as indicated)

Species and stages	60–95% CO_2			< 1% O_2		
	c. 15 °C	c. 20 °C	25–30 °C	c. 15 °C	c. 20 °C	25–30 °C
Acarus siro L., all stages	6–10a	8–14b	–	7a	–	–
Anobium punctatum Degeer, all stages*	–	–	–	–	7c	5c
C. ferrugineus, adults	7a	–	4	10a	6d	2
E. cautella, eggs and larvae	7	–	5	6	5	2
E. elutella, diapausing la.	–	14	–	–	> 28	–
H. bajalus, all stages	–	–	–	–	20c	10c
L. serricorne, all stages	–	–	6	–	9c	6c
Liposcelis bostrychophila (L.), all stages	10	8–14b	–	–	–	–
O. surinamensis, adults	5a	–	3	10a	3d	–
P. interpunctella, eggs, la.	7	< 7	–	> 14	> 4	–
R. dominica, all stages	28	–	–	> 28	–	–
S. oryzae, all stages	28	–	> 18	–	> 28	> 18
S. granarius, all stages	42–56e	–	> 9	> 49a	> 14a	–
T. castaneum, adults	6a	–	3	7a	4d	2
T. granarium, larvae in diapause	–	> 18	> 17	–	–	> 14
Tyrophagus longior (Gervais), all stages	14	–	–	14	–	–
T. putrescentiae, all stages	–	> 14b	6	–	–	–

a Bell 1993; b Newton 1993; c Valentin 1993; d Krishnamurthy *et al* 1986, e Banks and Fields 1995.

times for complete kill of the lesser grain borer *Rhyzopertha dominica* (F.) at about 20 °C from weeks to a few days (Jay, 1986, Locatelli and Daolio 1993), while exposure at the raised pressures of 10 and 15 bar achieved complete kill of all stages of the granary weevil *Sitophilus granarius* (L.) within 16 h and 8 h respectively at this temperature (Le Torc'h and Fleurat Lessard, 1991). At such high pressure O_2 also is toxic (Walter *et al* 1974), but its use may be more hazardous than CO_2 or N_2.

2.1.3.1.1 Bulk Commodities

Commodities stored in bulk, that is prior to any kind of packaging or bagging, include cereals, oilseeds and pulses. The structures employed for bulk storage must be well sealed prior to use with CA, if high rates of gas usage and expense are to be avoided. A gas tightness standard has been developed for one shot applications of CO_2 by use of a pressure test, whereby an applied pressure of 500 Pa must not decay by more than 50% within 5 min (Banks and Annis 1980). For N_2 this time needs to be extended to at least 15 min (Bailey and Banks, 1975). If a continuous flow system is available for supplying gas, atmospheres can be successfully maintained under calm conditions in much less gastight structures (Bell *et al* 1993).

The effective use of CO_2 for grain storage, was developed principally in Australia and the USA, although Australia, for preference, currently mostly uses phosphine to treat bulk grain. Until recently, the use of CO_2-based atmospheres was preferred over N_2-based ones for bulk grain for various technical reasons. Recent developments in the on-site generation of nitrogen-based atmospheres have made these atmospheres more competitive in price and convenience (Banks *et al* 1991, Bell *et al* 1993). Grain stored in bins can be held under a low O_2 burner-generated atmosphere for several weeks at a cost of less than 50p per tonne of grain. N_2-based controlled atmospheres are in commercial use in Australia in an export grain terminal in bins originally designed and equipped for methyl bromide treatments (Cassells *et al* 1994).

2.1.3.1.2 Bagged and Packaged Commodities

Bagged or packaged commodities can be treated by CA in chambers, gastight containers or in well-sealed stacks. Small packages of various dried food products are now often wrapped in an impermeable film containing an O_2 absorber (Abe and Kondoh 1991). CA is now being used for stored rice and other bagged commodities in South East Asia after initial developmental work from Australia (Annis and van S. Greve 1984, Nataredja and Hodges 1990, Sabio *et al* 1990). Success of the treatment relies on the quality of the sheeting and the completeness of the glue seal of the canopy sheet to the ground sheet. A pressure test is conducted to check for gas tightness of the enclosure by drawing a 200 Pa negative pressure and checking that 50% restoration does not occur within about 15 min (Nataredja and Hodges 1990). Dosing is accomplished by addition of CO_2 at a rate of about 2 kg per tonne from cylinders as a liquid through a copper pipe inserted below the pallets. To facilitate the atmosphere replacement within the stack a vent is cut in the roof of the canopy which is sealed when the concentration of CO_2 reaches the intended 60–70% and dosing is stopped.

Use of CA (CO_2) has also been successfully trialled for treatment of sultanas in cartons or stacks (Tarr *et al* 1994), N_2 generators are in use to treat tobacco in chambers (Ryan 1995), and CO_2 under high pressure is in limited use in Germany to treat beverages and spices (Gerard *et al* 1988, Prozell and Reichmuth 1991).

2.1.3.1.3 *Artifacts*

Nitrogen and carbon dioxide are being increasingly used in the treatment of museum artifacts (Gilberg 1990, Daniel *et al* 1993, 1994). Here the time factor for treatment is open ended and the long treatment times for CA use can easily be accommodated. Many museums have access to chambers, but treatments can also be carried out in portable plastic enclosures. The use of O_2 scavengers in small packages or in larger ones after flushing with N_2 can be effective in achieving and holding very low O_2 atmospheres (Gilberg 1990, Gilberg and Gratton 1994). The principal pests in museums include the wood boring beetles. Of these, the most tolerant species to a range of CA treatments seems to be the house longhorn beetle *Hylotrupes bajulus* (L.), which requires exposure times of 2–3 weeks at 20 °C for complete kill (Valentin 1993). A similar high tolerance is exhibited by the mite *Tyrophagus putrescentiae* (Schrank) (Newton 1993).

2.1.3.2 Inert Dusts

Inert dusts such as ash, amorphous hydrated silica, lime, various ground minerals and clays are registered in some countries for treatment of grain and grain legumes against insect pests. Sands, clays, earths, lime and phosphate dusts have been used traditionally for many years (Ebeling 1971, Golob and Webley 1980, Fam *et al* 1974). They are particularly useful in dry conditions as a means of controlling pests resident in storage structures. Silica aerogels are a refined form of diatomaceous earth and are effective against bruchids such as *Acanthoscelides obtectus* (Say) and *Zabrotes subfasciatus* (Boheman) within a few days (Giga and Chinwada 1994), and internal grain feeders such as the maize weevil *Sitophilus zeamais* (Motsch.) and larger grain borer *Prostephanus truncatus* (Horn) within two or three months (Haryadi *et al* 1994, Barbosa *et al* 1994). They lose effectiveness at high relative humidity, greater than about 75% r.h. (Le Patourel 1986). Inert dusts may be useful as part of an integrated system that controls pests to a level where methyl bromide use is not required. Their use on durables, particularly grain, has recently been reviewed (Banks and Fields 1995). Direct admixture can give long term protection in dry grain but dusts adversely affect the handling characteristics of grain and increase wear in handling machinery (Jackson and Webley 1994). The most useful applications are for seed storage prior to planting and small scale farm stores for animal feed. They are unlikely to replace any of the major methyl bromide uses on durables.

2.1.3.3 Cold Treatments

Insect development cannot proceed at low temperatures and so lowering the commodity temperature to below that supporting development is an obvious strategy for reducing infestation problems. Many stored product pests are of tropical origin and are unable to develop below 20 °C and even the more cold tolerant mite species do not increase in numbers below 5 °C. The technique is suitable for bulk storage in cooler climates where ambient air can be used to cool commodities by aeration, and for various other products in specific instances, such as small museum objects or packaged goods where a mild non-chemical disinfestation is required. For bulk grain storage, cooling is already practised where, in the absence of facilities for gas recirculation, methyl bromide has never been suitable for use. In other circumstances, cooling can sometimes present an alternative to methyl bromide use.

For elimination of insects by cold, very much lower temperatures are required than those which only prevent development. For effective action, a few days exposure (to allow for rate of cooling) at −15 °C or below are needed to ensure disinfestation, or several days to several weeks (depending on the species) at −5° to −10 °C. The rate of cooling to these temperatures must be as rapid as possible to avoid acclimation (Chauvin and Vannier 1991, Fields 1992). Temperatures need to be held as constant as possible during exposure to avoid increasing the chances of survival. Diapausing larvae of *Ephestia elutella* nearly all survived a 12-week exposure at a temperature oscillating every 12 h between 0° and −5 °C (Table 8.8), whereas very few were able to survive an unbroken 6-week exposure at −5 °C (Bell 1991). With temperatures oscillating between 0 and −10 °C, some larvae survived a 6-week exposure whereas all died after two weeks at a constant −10 °C (Bell 1992).

The biology of cold tolerance in insects is complex but two mechanisms have been demonstrated, one based on supercooling and the other on

Table 8.8. Effect of oscillating low temperatures on the survival of diapausing larvae of *Ephestia elutella* brought to low temperature by 4-day steps of 5 °C

Temperature cycle (°C, 12 h/12 h)	Exposure duration (weeks)	Survival (%)
0/−5	12	96.5
0/−10	4	76.1
	6	20.0
	8	0
−5/−10	2	46.5
	4	0
−5 continuous	4	43.2
	5	6.4
	7	0
−10 continuous	2	0

resistance to actual freezing (Zachariassen 1985). The most marked capabilities to withstand cold are usually associated with diapause. Freezing-tolerant insects often have some supercooling capability due to accumulation of glycerol but usually freezing will occur before temperatures fall to −10 °C. Protection is conferred by nucleating agents which direct crystallisation towards the haemolymph or gut and so avoid intracellular freezing (Baust and Lee 1981). Certain bacteria can interfere with this nucleation and may have potential for use to reduce the cold tolerance of some grain pests (Fields 1991). For insects relying on supercooling, freezing is avoided at much lower temperatures and higher levels of glycerol, sorbitol, trehalose and other compounds are produced as a type of antifreeze. The two mechanisms are effective in increasing survival, the freezing avoidance mechanism offering some advantage in situations of repeated freezing and thawing (Churchill and Storey 1989). The storage environment offers protection from seasonal climatic extremes but many stored product pests have retained the capacity to enter diapause, though in most the diapause intensity is low (Bell 1994). Some other pests such as the rust red flat grain beetle *Cryptolestes ferrugineus* (Stephens) are naturally cold tolerant without a diapause (Fields 1991).

2.1.3.3.1 Bulk Commodities

In temperate climates the use of cooling of grain by ambient aeration has been practised for many years. Burges and Burrell (1964) proposed that a temperature reduction to 17 °C would prevent insects that eat grain developing to significant numbers to cause the destruction of grain consignments. However, in recent years the improved standards of detection of insects in grain during trade demand a more rigorous control of temperature to reduce the risk of potential damage caused by insects, and modern aeration practices aim to achieve much lower temperatures (Armitage 1987, Gardener *et al* 1988). This is possible even in warmer climates (Arthur 1994, Bartali 1994). Arthur (1994) achieved winter daily mean temperatures below 10 °C in maize during small bin trials in the South Eastern USA by thermostatically controlled aeration, and an overall mean temperature for the period between mid October and early June of about 11.5 °C at bin centres. Very few insects developed or survived in the aerated bins in spite of several reinfestations of rust-red flour beetles *Tribolium castaneum* (Herbst), maize weevils *Sitophilus zeamais*, and the moths *Ephestia cautella* (Walker) and *Plodia interpunctella* (Hubner).

Cooling of grain by passing air over a refrigeration coil before blowing through the bulk has been proposed where ambient conditions are unsuitable for cooling and there is an urgent need to reduce temperatures (Burrell and Laundon 1967) but the practice is expensive. It is nevertheless used fairly widely in Mediterranen and sub-tropical regions (Brunner 1987), and in the USA (Maier 1994).

Below about 10 °C reproduction of most pests ceases and populations slowly decline. However, even at 4 °C adults of most species survive for many months, though their immature stages may be killed. Species of tropical origin such as the weevils *Sitophilus oryzae* (L.), *S. zeamais*, the cadelle beetle *Tenebroides mauritanicus* (L.) and the cigarette beetle *Lasioderma serricorne* (F.) tend to be cold sensitive, whereas some important pests including the granary weevil *S. granarius*, flat grain beetles *Cryptolestes* spp., bruchids, mites and some Lepidoptera are very tolerant (Armitage 1987, Lasseran and Fleurat-Lessard 1991, Fields 1992). In consequence, cooling typically is used to prevent damage and multiplication and re-invasion of pests rather than to control an existing actively developing infestation.

2.1.3.3.2 Packaged Food Materials or Other Goods

Cold storage is practised for many food commodities and other materials as part of an overall storage strategy that may also include fumigation with methyl bromide. In Israel, cooling to very low temperatures has been established as an alternative to fumigation for disinfesting dates (Donahaye *et al* 1991). The time required for treatment can be reduced by application of low pressure or low O_2, which has the effect of causing larvae to come out of the fruit (Donahaye *et al* 1992).

Domestic and commercial freezers are widely used to hold packaged foods at −15° to −20 °C, temperatures lethal to storage pests within a few hours of direct contact (Evans 1987). Commercial draft assisted freezers are in use in the tobacco industry to hold incoming leaf samples or cut tobacco in cases for 5 days at −20° to −30 °C (Ryan 1995). Library books and various museum specimens are sealed in polyethylene bags and held for 24–48 h at −20 °C or below for pest eradication (Nesheim 1984, Florian 1990). The comparatively short exposure times are permissible if the heat transfer in the sample is rapid.

2.1.3.4   Heat Treatments

Of the very few pest control options which are capable of matching the speed of treatment afforded by methyl bromide, heat treatment technologies are the most likely to come into practical use. The strategy is to heat commodities briefly to temperatures up to 70 °C and then rapidly cool them back to ambient to avoid damage to heat-sensitive constituents. The time required is strongly dependent on the temperature reached and experienced by the target pest and is only indirectly related to the temperature of the commodity. Cereals can withstand temperatures up to 70 °C for short periods without loss of germination (Sutherland *et al* 1986). Disinfestation from stored product insects (all stages) can be achieved in less than one minute above 62 °C (Banks and Fields 1995). At 45 °C the time required increases to about a day. Some lethal times at high temperatures are given in Table 8.9.

2.1.3.4.1   Bulk Commodities

Three heating systems have been developed for use on bulk grain, hot air pneumatic conveying (Sutherland *et al* 1989), the spouted bed (Claflin *et al* 1986) and the fluidised bed treatment systems. Of these, only the fluidised bed heating system for bulk grain has been developed to a commercial

Table 8.9. Response of insect pests of durable products to high temperatures (Based on Banks and Fields (1995))

Temperature range (°C)	Effect on insects
25–32	Optimum for development
30–35	Maximum temperature for reproduction of most species
35–42	Populations die out, mobile insects seek cooler zones
45–50	Death within a day
50–62	Death within an hour
Above 62	Death within a minute

prototype stage, with treatment rates up to 150 t/h (Evans *et al* 1983, Thorpe *et al* 1984, Fleurat-Lessard 1985).

Airflows through a perforated floor receiving the incoming grain are sufficiently high to circulate individual grains within the chamber. Incoming air temperatures are normally in the region of 100 °C. The facility uses fluidised beds to both heat and cool the grain and also acts as a grain drier (Thorpe 1987). Modern grain conveying systems run at up to 500 t/h throughput, and the installation of a heat treatment facility of this capacity would be highly capital intensive.

Pilot studies have also been carried out on the use of rapid heating of grain by microwaves or radio frequency radiation for the disinfestation of grain (Nelson 1972, Fleurat-Lessard 1987). Here the principle is again simply to achieve lethal temperatures quickly, and if possible differentially with pest temperatures exceeding commodity temperatures. Commodity cooling is a problem as these methods tend to heat from the centre outwards, and a fluidised bed or similar facility may be necessary to bring temperatures down quickly after treatment (Fleurat-Lessard 1987). Infrared heating systems offer another possibility but testing has not progressed beyond laboratory trials (Kirkpatrick and Cagle 1978).

2.1.3.4.2 Other Products

The opportunities for the use of heating techniques to combat infestation in products other than grain are limited. Spices and beverages are not suitable for treatment as their quality is strongly dependent on volatile constituents. There is, however, scope in the tobacco industry as leaf tobacco may be vacuum steamed and then redried in preparation for storage, and driers are also used prior to cut rag storage (Ryan 1995). These drying processes involve exposure at temperatures up to 60 °C. Microwaves at temperatures up to 55 °C have been suggested to offer an effective means of controlling tobacco pests (Hirose *et al* 1975), and have also been tested for the treatment of packaged rice (Locatelli and Traversa 1989) and edible walnuts (Wilkin and Nelson 1987), but technical problems exist in achieving an even temperature distribution within commodities and operational costs are generally uncompetitive.

Heat, either dry heat in kilns or steam heat, is also suitable for use on logs and sawn timber against fungi, a temperature of 66 °C at the wood centre needing to be held for 75 minutes (Chidester 1991). In hot climates the potential exists for use of solar heating to dry and disinfest maize cobs and other products held in simple solar cabinets, or under clear polythene sheeting (McFarlane 1989). Heat and cold are being increasingly used for various museum artifacts, but care is always necessary. Herbarium specimens, seeds and furniture can be treated at temperatures down to −20 °C or up to 60 °C, other materials such as skins, furs, books and textiles may be frozen for preservation but not heated, while some others such as painted or lacquered wood may be heated if bagged in polyethylene, but not frozen (Strang 1995).

2.1.3.5 Irradiation

Three types of ionising radiation have been evaluated against insect pests of stored foods, gamma rays (wavelength 10^{-12} to 10^{-14} m) emitted from a radioactive source such as cobalt-60, accelerated electrons emitted from a heated cathode, and x-rays, electromagnetic radiation emitted when electrons from a heated cathode are impacted under vacuum on an anode receptor (wavelength 10^{-10} to 10^{-12} m). Irradiation is a potential method of controlling pests in a wide variety of durable stored commodities. The effectiveness of treatment for insect control and effects on food quality are related to the energy delivered. Toxins are not produced in foods treated at doses up to 10 kGy (Urbain 1986) though some qualities important in food processing may be affected at doses down to 3 kGy and starch damage in wheat has been reported after exposure to only 0.5 kGy (Ng *et al* 1989). Insects may be sterilised by doses as low as 0.1 kGy but complete kill within a few days is not achieved by very much higher doses (Brown *et al* 1972, Tilton and Brower 1987). Dosages required for control of bacterial contamination of products such as spices are still higher, in the range of 5–10 kGy, but flavour in such products does not seem to be adversely affected (Urbain 1986).

Disinfestation by irradiation has a long history (since 1912) and there has been much investigation over the years. In spite of some promising results (e.g. Wohlgemuth 1973) and extensive use for pest detection, there has been no commercial development of x-ray facilities for disinfestation purposes. For the higher energy gamma radiation, Brower and Tilton (1985) and Tilton and Brower (1987) have summarised the radio-sensitivity data on forty stored-product pest species, and indicate that irradiation could be used as a quarantine measure. Their data showed that different pests, and the developmental stages of each species, differ in their sensitivity to radiation, with adults generally being the most tolerant stage. Irradiated insects may survive for several weeks at doses suitable for commodity treatment (Table 8.10) (Watters 1984, Banks and Fields 1995). Larvae of the dermestid beetles *Trogoderma inclusum* LeConte and *T. variabile* Ballion fail to develop to the adult stage when irradiated at the relatively low dose of 0.1 kGy (Brower and Tilton 1972), whereas some adult moths may remain fertile after exposure to doses up to ten times this level (Banks and Fields 1995).

Accelerated electrons are slightly less effective than gamma rays in insect

Table 8.10. Days required to achieve complete mortality of some adult stored products insects to γ-radiation (after Banks and Fields 1995)

Species	Dose		References
	0.3 kGy	0.5 kGy	
Attagenus piceus Olivier	35	35	Tilton *et al* 1966a
Cryptolestes ferrugineus	21	14	Watters and MacQueen 1967
Ephestia cautella	5	5	Burkholder *et al* 1966
Oryzaephilus mercator (Fauvel)	14	14	Brower and Tilton 1972
Oryzaephilus surinamensis	> 70	> 70	Brower and Tilton 1972
Rhyzopertha dominica	42	28	Tilton *et al* 1966a
Sitophilus oryzae	21	21	Tilton *et al* 1966b; Cornwall *et al* 1957
Sitophilus granarius	14–28	14–49+	Watters and Macqueen 1967 Brown *et al* 1972
Sitophilus zeamais	28	28	Brown *et al* 1972
Tenebrio molitor L.	21	21	Brower 1973
Tenebrio obscurus F.	7	7	Brower 1973
Tribolium castaneum	21–28	21–28	Watters and MacQueen 1967 Brower and Tilton 1973
Tribolium confusum DuVal	14–21	14–21	Watters and MacQueen 1967 Tilton *et al* 1966b
Tribolium madens (Charpent.)	35	28	Brower and Tilton 1973
Trogoderma inclusum	35	42	Brower and Tilton 1972
Trogoderma variabile	28	35	Brower and Tilton 1972

control, lacking the penetrative power of the latter (Adem *et al* 1978). However, they are inherently easier to work with as they can be switched on and off. Selection of the type of irradiation equipment to be used depends on whether the commodity is to be irradiated in packages or in bulk, the quantity of product to be treated and other factors. Gamma irradiators can treat packaged or bulk products, and accelerators can more effectively treat bulk products in thin layers (2–5 cm thickness). A plant incorporating two 1.4 MeV accelerated electron units was built on the Black Sea coast near Odessa which was capable of treating 200 tonnes of grain per hour in a shallow layer (Zaklodnoi *et al* 1982) but today the facilities are not in use. The International Consultative Group on Food Irradiation (ICGFI), under the aegis of the FAO/IAEA Joint Division, has published provisional guidelines for the irradiation of cereal grains for insect disinfestation as a recommendation to be followed when using the technology (ICGFI 1988).

In spite of the affirmation given by government research departments and regulatory authorities, there are few agreements presently that allow movement of irradiated products in international trade. This is an impediment to the more widespread use of the method and is especially critical for irradiation to be used to satisfy quarantine requirements. One problem is that there is no routine inspection system that can ascertain whether or not live pests are sterile. The food industry is also concerned about consumer acceptance of irradiated food products. There are also questions regarding the large initial capital expenditure required for plant construction and related logistics (Rhodes 1986). As a result there are no approved quarantine treatments to date, though Ahmed (1991) reports that radiation facilities are in use on a wide variety of food materials in 24 countries.

2.1.3.6 Impaction

Many situations in which agricultural products are mechanically conveyed during food processing offer the opportunity for control of insects by shock, abrasion and impaction. The principle was developed over 50 years ago for use in the flour milling industry (Cotton and Frankenfeld 1942) and machines such as the Entoleter became a routine fixture in flour mills. In the Entoleter, flour falls between two rapidly spinning discs. Centrifugal force pushes the flour to the edges of the discs where it impacts a row of steel pegs mounted on the rims, and is thrown against the outer steel casing before falling into the basal receiving hopper. The material passing through the Entoleter thus encounters two major impactions and this is responsible for the control of all free living insect stages (Bailey 1962).

Working with moving grain, Loschiavo (1978) found that dropping of adult insects into free flowing grain caused substantial mortality, while Bahr (1991) found that with a range of stored grain insects, passing through a pneumatic conveyer caused between 48 and 95% mortality of adult beetles, while four passes through a vacuum cleaning system caused between 72 and 100% kill, depending on the species, of all developmental stages. Free-living insects were easier to control than those developing inside the grain. Subjection of grain to impaction machinery could not eliminate internal grain feeders below levels causing damage to the grain (Bailey 1962, Stratil *et al* 1987).

In studies of bruchid infestation of beans, Quentin *et al* (1991) found that gentle tumbling of beans every eight hours over a two-week period, reduced population growth by 97%. The effect was explained by prevention of first instar larvae from entering the seed after egg hatch. The use of disturbance and impaction techniques merit further experimentation and development in the field of insect control.

2.1.3.7 Physical Removal (Sanitation)

Under this heading is the application of a diverse range of measures designed to remove pests or prevent their access to the product or commodity. These include cleaning and removal of harbourages for pests, including removal of food residues in which pests can multiply, redesigning machinery and refurbishing buildings. Normal good warehousing practice, e.g. stock rotation and, where applicable, insect-proof packaging, also reduce pest population pressure. The retention of polythene sheeting on a stack after a fumigation is an effective means of preventing reinfestation, as demonstrated in trials and current practice in South East Asia (Annis 1990). Other measures include sieving, screening, separation by projection and aspiration. Whereas none of these methods is capable alone of achieving sufficient control, they can be useful in combination with other measures. The topic of physical removal and exclusion was reviewed by Banks (1987) and more recently, together with all physical control methods, by Banks and Fields (1995).

2.1.4 STRUCTURES AND TRANSPORT

The use of fumigants such as methyl bromide in buildings and empty structures to control residual infestation is often referred to as space fumigation. There are three broad areas in which such treatments are carried out: firstly in food production and storage premises (mills, food processing, distribution warehouses), secondly in non-food facilities (dwellings, museums), and thirdly in transport vehicles (trucks, ships, aircraft, railcars). Target pests include the typical stored product pests (mites, beetles, moths), cockroaches, silverfish, psocids, flies, spiders; wood destroying insects and rodents (Table 8.11).

The use of fumigation in such facilities depends very much on the degree to which other control and maintenance procedures are implemented. The structure may contain raw agricultural commodities, semi-processed products or finished food products awaiting delivery to distribution points, and fumigation is necessary whenever infestation becomes so widespread that localised treatments cannot prevent reinfestation or when the infestation is within the walls or other inaccessible areas.

Alternative strategies to methyl bromide fumigation usually incorporate the use of insecticidal sprays coupled with a range of non chemical procedures. Of the physical control options, heat treatment is probably the only effective non chemical technique for pest elimination in structures, provided that equipment and construction materials can withstand the increased temperatures. Other non chemical treatments include application of liquid nitrogen, electrocution, and microwaves to localised areas within a structure, and these rely heavily on the technician's ability to locate the site(s) of infestation.

The treatment of ships, aircraft and other transport vehicles poses particularly difficult pest management problems. They often contain sensitive equipment, innumerable harbourages and it is often not economically feasible to keep them out of operation for extended periods of time. Options

Table 8.11. Pests of different premises and structures

Type of facility	Examples of pests
Food production and storage facilities	
Food processing plants	Stored product insects, rodents,
Flour and feed mills	cockroaches, psocids, mites, silverfish,
Bulk commodity storage (e.g. silos)	beetles
Warehouses	
Bakeries	
Ham smoke houses	
Cheese plants	
Refrigerated storage	
Restaurants	
Non-food facilities	
Seed warehouses	Rodents, stored product insects
Museums	Dermestid beetles, clothes moths, cigarette beetles, drugstore beetles
Poultry houses	Lesser meal worm, mites, rodents
Mushrooms houses	Mushroom flies, mites
Condemned housing or public health compliance	Rodents, cockroaches, venomous spiders, fleas
Dwellings including apartments, condominiums, trailer homes, historical buildings	Cockroaches, furniture beetles, powder post beetles, longhorn beetles
Structural elements before building or in place, e.g. beams	Powder post beetles, longhorn beetles
Museums	Wood boring beetles, house and clothes moths, dermestids
Antique vehicles	Powder post beetles, moths
Transport vehicles	
Trucks, truck trailers, vans and containers (empty)	Beetles and moths
Ships, shipholds, galley & quarters (empty)	Insect and rodents
Railcars (freight or commodity)	Insect and rodents
Buses	Insects
Aircraft	Cockroaches and other insects, rodents

for a rapid disinfestation operation are extremely limited and in many countries methyl bromide is the only fumigant allowed for various quarantine treatments on ships. There are also no acceptable alternatives to methyl bromide for rodent and insect elimination abroad aircraft (UNEP 1995).

2.1.4.1 Controlled Atmospheres

Considerable research has been conducted on disinfestation of stored commodities with controlled atmospheres (see above) and the data obtained are applicable to structures. Long exposures are required to achieve control during which CO_2 levels have to be maintained at about 40% in air or O_2 levels below 1%. There are very few buildings that can be sealed well enough to hold the required concentrations for the required amount of time, usually more than 10 days, without excessive use of gas. The structural modifications required for older food production facilities or incorporation of the technology into new construction could be expensive. Accidental breaking of the seal during long exposure periods is a greater risk to treatment efficacy than with shorter duration fumigations.

The application, distribution and maintenance of even gas distribution in large structure is a formidible problem and much specialised equipment is required. Nevertheless, Keever (1989) reports a successful 7-day bulk CO_2 treatment of a $12\,706\,m^3$ tobacco warehouse against the cigarette beetle at a mean temperature of 29 °C, although the cost was nine times that of conventional fumigation. There is active research into methods of gas generation to enable on-site provision of atmospheres for insect control. Equipment capable of generating modified atmospheres on site have already been tested for treatment of grain pests in large silos (Bell *et al* 1993, Cassells *et al* 1994). In smaller structures such as freight containers modified atmospheres have been used with success. CO_2 supplied from dry ice held in a partly sealed polystyrene container can maintain a lethal atmosphere in a well-sealed container initially purged with CO_2 from a bulk liquid source (Banks and Annis 1981).

2.1.4.2 Temperature Extremes

The intense periods of winter cold have long been used by millers in Canada and the Northern USA for a 'freeze out' of mill pests (Watters 1984) and there is seldom a need for chemical control methods in the first few months after treatment. Cold can also be used as a spot treatment by the injection of liquid nitrogen into confined spaces such as wall voids. Drywood termite eradication can, for example, be achieved by such treatment within 30 minutes. The performance of the technique against other wood pests should be comparable, but it cannot be used in inaccessible areas. Insulation in walls can affect cold distribution causing warm spots in walls. Interior surfaces can be stained and warping of wooden structural components is possible.

At the other end of the scale, heating to about 52 °C (125 °F) has been used to control insects in flour mills since the turn of the century. It is still used by some major food processors as an important part of their pest control strategy. Food plants that can be successfully heat-treated rarely require fumigation. Although some expansion of use of this technique may be expected, there are some important limitations. For example, some structures cannot tolerate the stresses caused by extreme changes in temperature and differential expansion of structural components, e.g. of concrete and steel. Insects can sometimes migrate temporarily to outer walls or into drains and so escape the effect of treatment. Heat dissipation is often slow, and may delay resumption of normal activities in the building after treatment. Some equipment must be modified or removed to avoid damage. Some greases may liquefy and need to be reapplied after treatment. Some products cannot withstand the required temperatures and may have to be removed and treated separately to prevent the reintroduction of pests. Some buildings are not constructed so that they can be uniformly heated to the required temperatures. The technique is best applied in a warm climate so energy costs can be minimised. The use of heating as a means of treating infestation in a food plant is described in detail by Sheppard (1984).

An alternative approach is to use heat in combination with other treatments such as fumigation or controlled atmospheres to reduce treatment times and improve efficacy. Under such conditions temperatures need not be increased to such an extent as when heat is used alone. In mill fumigation trials with phosphine at 65–100 ppm and 3–6% CO_2, Mueller (1994) claimed that treatment times against flour beetles in mills could be held to about a day at 38 °C.

For localised 'spot' treatments, microwave heating has been used in structures but with very limited success. Insects are destroyed by heating the moisture in wood, but the wood itself may be scorched (UNEP 1995). Further research is needed before this technique can be recommended.

2.1.4.3 Inert Dusts

Slurries of inert dusts are sometimes applied to the walls and ceilings of premises as part of an integrated control strategy (Bridgeman 1994). They are, however, unsightly and are more often applied to areas such as wall voids that are not visible. Silica gel and diatomaceous earth dusts are abrasive to insect cuticles, particularly at the joints, and act as desiccants. The formulations are non-toxic, provided they do not contain crystalline silica. They can provide long-term residual control of crawling insects in wall voids and other locations, and are most effective against insects with a thin cuticle that depend on an oily or waxy coating to preserve body moisture, such as psocids, silverfish and other active insects. Most inert dusts have little effect at high humidities (over 80%) and their effective use in buildings is restricted to drier climates. They adhere weakly to surfaces and are easily removed from the site of application. With exposed surfaces, retreatment is advised at 3 monthly intervals.

2.1.4.4 General Maintenance and Sanitation

Sanitation is a vital component in the control of structural pests, regardless of any other practices carried out. It reduces pest food and harbourages within and without a structure by regularly removing waste and debris during vacuum cleaning, sweeping and washing. Construction and maintenance also play major roles in reducing pest harbourages and denying pest access to structures. New construction should include pest preventative design as a priority. For existing facilities, problems may only be solved by changes to the structure such as repairs and closures of pest entrances and niches, including caulking and applying new surfaces, and whole sections of structures may have to be replaced at a prohibitive cost. Maintenance and construction practices can enhance other alternative treatments e.g. by improving containment, allowing heat treatment, and reducing the need for residual pesticide application.

3.1 ALTERNATIVE APPROACHES TO REDUCE METHYL BROMIDE EMISSIONS

So far this chapter has dealt with possible non-chemical alternatives for each major use area of methyl bromide. Prospects for reducing methyl bromide emissions by substitution of other control techniques or management processes appear to be limited, at least in the foreseeable future. For a significant impact on emission reduction in the short term, other lines of investigation need to be followed on gas containment, recovery and recycling of the compound (Chakrabarti and Bell 1993). This would permit fumigant usage to fall without loss of coverage of the numerous control problems which currently rely on the availability of methyl bromide.

For soil treatments, particularly in the open, most gas is lost by leakage from the edges of the sheeted area, or by permeation through the sheet itself (Chakrabarti *et al* 1995). For most commodity treatments, venting after fumigation represents the largest discharge of gas to the atmosphere. Emission of adsorbed methyl bromide is probably the next largest with leakage coming third, although this is very much dependent on the commodity, site and degree of proficiency that the treatment is carried out. Measures limiting the first and to some extent the third source can be coupled with reduced dosage. The second can only be controlled by prolonged gas recovery followed, if possible, by recycling, reclamation or destruction. Some estimates of the quantities of methyl bromide emitted from the major fumigation use categories are given in Table 8.12.

Table 8.12. Estimated usage of methyl bromide (1992 figures) and emissions to atmosphere estimated for different categories of fumigation (updated from UNEP 1995)

Type of fumigation and commodity	Amount used		Estimated emissions	
	(t)	(% of global usage)	(t)	(%)
Enclosed Space (Durables)				
Grains, etc	4601		2347–3221	51–70
Nuts	236		120	51
Dried Fruit	236		205	87
Timber	4782		4208	88
Subtotal	9855	12.96%	6880–7754	70–79
Enclosed Space (Perishables)	6537	8.59%	5556–6210	85–95
Enclosed Space (Structural)	2264	2.98%	2038–2151	90–95
Soil Fumigation				
Soil injection—shallow with tarp	31 000		10 230–24 800	33–80
Soil injection—deep with tarp	2296		689	30
Vaporised gas with tarp	22 963		11 481–19 518	40–85
Soil injection—deep without tarp	1148		918	80
Subtotal	57 407	75.47%	21 022–459 25	37–80
Total agricultural usage	76 063		35 229–617 73	46–81
Mean emission over all categories			48 680	64

3.1.1 CONTAINMENT

Better gas containment is the first step towards any method of reducing emissions from methyl bromide fumigations, but in isolation its effect may only be to delay the subsequent release of gas to the atmosphere. A well sealed enclosure ensures that a maximal amount of the gas applied is available for as long as is necessary for an effective treatment. This means that initial dosage levels can be reduced and exposure times extended. If the seal on the treatment enclosure is sufficient, then a significant proportion of the gas originally applied should be available for recovery, but in many situations recovery may not be practicable from a logistical or economic standpoint. Nevertheless all reasonable attempts to maximise the degree of seal on a fumigation enclosure should be taken in the interests of safe working practice and to keep gas recovery options open. The potential for gas recovery will now be examined for the various major use categories.

3.1.1.1 Soil Fumigation

Methyl bromide is applied to soil as a vaporised gas or from a nitrogen pressurised system (hot or cold gas dosing) under gas-proof sheets. Leakage is reduced by burying the edges of the fumigation sheets in the soil, but the reliability of this process varies widely with the conditions under which the fumigation is carried out, depending on the skills of the operator, the degree of quality control exercised and the quality and intactness of the sheeting used. Emissions to the atmosphere from soil fumigation can be reduced by using sheeting with a lower permeability to methyl bromide than poly-ethylene and by improving the techniques used for sealing the edges of the sheeting, for example by watering the soil. At present no objective test method has been developed to determine the degree of seal achieved under sheets for soil fumigation. Research is in progress to limit permeation losses by new laminated sheets incorporating materials other than polyethylene, and on the extent to which dosages can be reduced. Table 8.13 gives some early experimental results showing improved retention of gas as a result of using standard polyethylene sheets (De Heer *et al* 1981). More recent work with even less permeable sheets (Wontner-Smith and Chakrabarti 1994, Chakrabarti *et al* 1995) has demonstrated that dosages can be reduced by 50% without loss of efficacy. Indeed, with gas retention improved to such an extent, failure to apply lower doses could give rise to increased bromide residues in soil, water and crops alike, as the residue level obtained is dictated by the CTP achieved.

Another consequence of a lower dose being able to achieve the target CTP for control using the less permeable sheet is that a greater proportion of the applied methyl bromide will be bound as residue so that the amount of gas available for emission will reduce to a lower level than would be

Table 8.13. Emissions from a methyl bromide fumigation of greenhouse soil using two different covers (from de Heer *et al* 1981)

	Saranex	LDPE*
Exposure time (days)	5	5
% emitted	20	56
Airing time (days)	2	2
% emitted during airing	48	22
Total % emitted	68	78
Remaining in soil (%)	*c.* 20	*c.* 10
Transformed by reaction (%)	*c.* 10	*c.* 10

*Low density polyethylene

predicted from the amount applied. Hence if a 50% dosage reduction is employed, and a CTP, of, say, 4000 mg h/l is obtained as in a treatment with the full dose, then the bromide residue obtained may account for as much as half the dose applied instead of the usual 25%. This would mean that with the dosage reduction taken into account, a 50% dosage reduction would achieve a two-thirds reduction of emission of methyl bromide to the atmosphere.

3.1.1.2   Structural and Commodity Fumigation

The success of any fumigation depends on an adequate level of containment being achieved. Improving gas tightness and increasing the duration of the fumigation can in many cases enable lower doses to be used, thereby reducing emissions to the atmosphere. A large proportion of durable and perishable commodities are fumigated in temporary, portable enclosures such as under gas-proof sheets or tarpaulins. The remainder are treated in fixed enclosures such as freight containers and purpose-built fumigation chambers. A minimum CTP must be achieved to ensure effective fumigation. A move to more gas-tight or permanent enclosures, especially those equipped with circulation fans would make recovery of methyl bromide for subsequent recycling or destruction more feasible. In buildings and structures, improving the degree of seal enables lower initial doses to be used and the need for top-up doses prevented.

In the interests of efficacy and safety some countries have set standards of gas tightness for various structures and enclosures which are treated with methyl bromide. These are applied to ensure leakage from fumigation is reduced so that methyl bromide concentrations are maintained at a level necessary for an effective treatment while avoiding environmental or human health risks in the vicinity of the fumigation. Table 8.14 shows standards for several countries. These standards are very much less stringent than those recommended for phosphine or controlled atmosphere treatments, which

Table 8.14. Examples of pressure test standards for gas-tightness of some fumigation enclosures and treated structures for use of methyl bromide

Structure	Country	Typical pressure range (Pa)	Full or empty	Time	Reference
Flour mills, churches	Germany	10–5	empty	> 4 s	Reichmuth, 1993
Freight containers	Australia	200–100	empty	> 10 s	Banks, 1988
Fumigation chambers	USA	500–50	empty	> 30 s	USDA-APHIS, 1976

usually require a pressure half-life of at least 5 minutes (Banks 1984, Nataredja and Hodges 1990), and daily leakage rates may be high.

3.1.2 RECOVERY AND DISPOSAL OR RECYCLING

Recovery or capture methyl bromide from fumigation operations has been a topic of active investigation for a number of years. Once trapped, the methyl bromide needs to be stored for reuse, or destroyed, or the act of recovery will only result in gas being emitted to the atmosphere at some later stage, or being recycled within the fumigation facility. Three recovery techniques currently in use or under investigation are adsorption on to activated carbon or zeolite, condensation, and absorption into reactive liquids. In addition, methods of disposing of methyl bromide by combustion and scrubbing methods for combustion products have been pursued. If containment and recovery is to be the chosen method for reducing methyl bromide emissions, it will be necessary for an agreement to be reached on the maximum quantity of gas release that can be tolerated so that the required performance of recapture equipment can be specified (UNEP 1995).

3.1.2.1 Activated Carbon

Activated carbon is used worldwide to remove trace amounts of organic contaminants from numerous air circulatory systems and gas streams. It has a very high capacity for such sorption, up to 30% by weight depending on activated carbon type and the conditions prevailing. For fumigation operations with methyl bromide, a vessel containing activated carbon can be installed in the gas vent line, and after the fumigation treatment, the gas mixture containing methyl bromide and air passes through the activated carbon which adsorbs the methyl bromide. The proportion retained depends on the concentration of gas in the air stream, the flow rate, the dilution of gas during purging, the temperature, and the amount and nature of the activated carbon. At low loadings in closed recirculatory systems recovery rates of close to 100% are achievable. Attempts to extract the last fraction of gas can, however, be counter productive as methyl bromide can be desorbed if excess air with very low levels of gas is passed through the system. Hence, for most systems, it is not possible to avoid some methyl bromide being emitted to the atmosphere.

After use in one or more operations, the adsorption capacity of the activated carbon may be reached and the filter needs to be recharged or replaced. Regeneration of used filters can be achieved by passing hot gas over the activated carbon and storing the methyl bromide in a chamber, if available, prior to reuse. Alternatively the activated carbon and methyl bromide can be incinerated in a specialised facility. However concerns about emissions of toxic chemicals may prevent this from being a viable option in some areas. Chambers fitted with carbon filters are few and far between but there are five $30 \, m^3$ chambers in the Netherlands (one transportable) each with a 70 kg filter of activated carbon (UNEP, 1995). Fumigation at $30 \, g \, m^{-3}$ is carried out and a 40–50% recovery is achieved. The activated carbon lasts for 40 fumigation operations and the spent carbon containing the adsorbed methyl bromide is incinerated in a special incineration facility. There is also a $30 \, m^3$ chamber in Thailand fitted with a 72 kg bed of activated carbon capable of reducing methyl bromide concentrations in the vented gas to 5 ppm within 30 minutes. However, the fully absorbed activated carbon is disposed of in a sanitary land-fill and it is inevitable in such situations that some gas will find its way into the atmosphere.

In the first use of carbon filters on a large scale, trials were carried out in a mill in Germany in 1994 using activated carbon to recover and recycle methyl bromide. The system, which is capable of recycling gas, is based on temperature swing adsorption on activated carbon (Stankiewicz and Schreiner 1993). Built into the system is an enrichment step to obtain a localised high concentration of sorbed gas and also condensation. This reduces the size of the kit making it readily transportable, and enables high concentrations to be achieved during the desorption cycle even when the concentration of methyl bromide in the extracted air falls to a low level. Another, more simple, activated carbon system has also been developed by Rentokil UK for use with their fumigation bubble, a well-sealed plastic tent enclosure used for fumigation of small structures. A 10 kg activated carbon bed which can hold up to 1.5 kg methyl bromide (equivalent to five standard fumigations in a $30 \, m^3$ tent) is used. Regeneration of the activated carbon is achieved by blowing hot air through the beds, resulting in direct emission of methyl bromide to the atmosphere. However, the strategy was designed only to prevent emissions of methyl bromide in the treatment area that might endanger people in the immediate vicinity.

In addition to applying hot air, it is technically possible to recycle methyl bromide adsorbed on activated carbon by altering the pressure (temperature and pressure swing adsorption). Pilot scale studies have demonstrated the technical feasibility of such a process with up to 95% of the recoverable methyl bromide being available for direct reuse. Circulating air strips the desorbed methyl bromide from the activated carbon prior to reintroduction into the fumigation chamber. In this way the methyl bromide is reclaimed as a high concentration mixture in air suitable for direct reuse as a fumigant. To maintain a satisfactory fumigation concentration, some topping up of gas will be required at intervals to compensate for system losses. There is a possibility of a build-up of other gaseous impurities during prolonged use, which may be of concern both from product quality and regulatory view points, and this aspect warrants investigation for commodity treatments.

3.1.2.2 Adsorption on to Zeolite

Zeolite adsorbents are already in use to remove CFCs from vented air streams, and work is well advanced on the commercial development of this process for recovery of methyl bromide (Nagji and Veljovic 1994a). Although zeolites are more expensive than activated carbon, their adsorptive capacity is high, particularly at low gas concentrations, and they could utilise a similar technology to that being developed for carbon filters. They can be manufactured to very narrow pore size tolerances for specific applications and this offers the prospect of avoiding potential problems of contamination of the recovered methyl bromide with other volatile compounds released from commodities. In pilot scale tests, recovery in excess of 90% was achieved in a fumigation of cherries, and in this case no other volatile compounds were retained with the methyl bromide (Nagji and Veljovic 1994b). The Port of San Diego Authority have recently installed a full sized plant based on adsorption using zeolite to reduce methyl bromide emissions from a $2100 \, m^3$ quarantine chamber. After recovery, the methyl bromide can be removed from the zeolite unit by heating and condensed into bottles ready for reuse, or immediately reinjected into the chamber. Recovery is in the region of 95%. Silica-based adsorbents are, however, vulnerable to high moisture in the air stream.

3.1.2.3 Condensation

Refrigeration and condensation techniques are used to recover methyl bromide at installations where pure methyl bromide is dispensed from bulk containers into smaller ones for direct use. This option was formerly considered too complex and expensive for recovery of gas from fumigation operations, because of the low methyl bromide concentration in vented gases and its low boiling point. However, if the methyl bromide is concentrated in filters after extraction from air, then prospects for use of condensation techniques can be much improved. A plant is in operation in Los Angeles, where methyl bromide is reclaimed and recycled using condensation in partnership with an activated carbon process (UNEP 1995). The recovery/recycling plant has been fitted to two vacuum chambers. At the completion of each fumigation operation, the methyl bromide in the chamber is diluted by the addition of air from a single airwash as atmospheric pressure is restored. This diluted mixture is then drawn through vessels where liquid nitrogen cools and condenses out most of the methyl bromide. After this process the effluent gas is passed through an activated carbon bed where most of the remaining methyl bromide is adsorbed. Periodically the activated carbon bed is isolated and undergoes a pressure swing desorption. Installed in 1993, the computer-controlled plant is designed to recover 98% of the methyl bromide used in the chambers.

3.1.2.4 Absorption into Liquids

Methyl bromide readily reacts with alkalis and amines in solution. Amines typically react to produce methylated, non-volatile products which can be disposed of safely. A system based on organic amines and alkali for removing methyl bromide after fumigation of $29\,m^3$ freight containers in Russia has been described (Rozvaga and Bakhisev 1982). Another system using aqueous sodium sulphite as a neutraliser and a mixture of ethylene diamine and sodium carbonate as an adsorbent has also been tested in Russia (Mordkovich *et al* 1985), but neither technique is in general use.

In Japan research was carried out in the 1970s to develop a technique of liquid scrubbing to remove methyl bromide from fumigation operations on stacked timber (Anon 1976a). The process involved circulation of methyl bromide and air from the fumigation enclosure through a tank of aqueous monoethanolamine (50%) and back to the fumigation tent. The process achieved a 70% reduction in methyl bromide concentration levels, but required 40–60 minutes to do so. The size of the necessary equipment for full scale operation and the difficulties of handling the large volumes of contaminated liquid material have so far prevented any commercial development, but the situation could now change and further investigation is warranted.

3.1.2.5 Improved Solid Absorbents

Research in Japan has led to the development of a new adsorbent, MBAC, which is a preparation of activated carbon with surfactant amine groups, giving a greater adsorptive capacity for methyl bromide that activated carbon alone. This material can be produced as sheets and introduced into packaging to recover the slowly desorbing methyl bromide from fumigated commodities and also has potential to recover some methyl bromide from soil fumigations. The Japan Methyl Bromide Industries Association is currently conducting evaluation tests. There are other possibilities in this area for development of improved materials based on silica gel or carbon. Such research will need to encompass measures for disposal or recycling of contaminated material.

3.1.2.6 Ozone Treatment

The idea that ozone, the substance vulnerable to attack by methyl bromide in the stratosphere, could be used to remove methyl bromide after fumigation is being investigated. A system is under development in California to use ozonation to directly break down methyl bromide in the vented gas steam from a chamber fumigation operation. Following this treatment, the exhaust gas could be passed through an activated carbon bed to remove any remaining methyl bromide. This process has undergone pilot scale tests and a full scale plant with a target efficiency of 90% is planned (UNEP 1995).

3.1.2.7 Direct Combustion and Catalytic Destruction

Research into methods of eliminating methyl bromide by direct combustion and a catalytic cracking method was being carried out in Japan in the 1970s for destroying methyl bromide to prevent emissions from chamber fumigations (Anon. 1976b). Pilot plants were built to test both techniques, and each was found to be effective at reducing the concentration in vent gas streams down to ppm levels. However, neither technique was developed further because of the difficulties of handling the breakdown products (HBr and Br_2), and problems associated with costs and transportability.

4.1 CONCLUSIONS

The options for reducing emissions rather than uses of methyl bromide appear to offer considerable scope for short to medium term success in the quest to reduce the risk to the ozone layer. At present the Montreal Protocol controls and measures taken by governments relate to restrictions on the availability of the chemical, and yet the question of finding alternatives can be seen to be a long term solution only, as most substitutes will require considerable development and considerable changes in commercial practice are necessary for their widespread adoption in the many individual use areas.

Soil fumigation with methyl bromide has been successfully replaced in some areas by methods and techniques that have been available for many years but that have been adapted to suit local requirements. While there are many methods and materials that offer possibilities for replacement of methyl bromide, there is no single alternative available for immediate use by producers. Alternatives to methyl bromide will develop in direct response to financial investment in research and development. It is essential that such research on alternatives (chemical, non-chemical and integrated systems) be performed thoroughly under a wide range of practical conditions. This is essential to avoid substitution of methyl bromide by fumigation or treatment with other chemicals, or by other methods, which may be more damaging to human health and the environment than methyl bromide.

For perishable commodities, heat-based measures, possibly coupled with subsequent low temperature and atmosphere control, offer the best prospects for finding alternatives to methyl bromide. However, it is not easy to transfer new disinfestation technologies between countries because each has commodities and pests that differ in tolerance to the proposed treatment and each has adopted an individual set of phytosanitary criteria. Because each potential alternative treatment must be adapted to the new country, successful implementation requires skilled research and development, which is labour intensive, time-consuming and costly. Commercial application typically requires 3–7 years from the initial proposal to adapt an existing alternative to a further control purpose, and will take considerably longer for new alternatives.

For durable commodities, although many alternative control techniques are available and most of these are in use today, few can be regarded as true potential substitutes for fumigation with methyl bromide. All alternatives will require changes in commercial practice to accomodate longer treatment times or different presentation and organisation of the commodity. Some will require considerable investment and developmental research before they can be used commercially.

Structures and transport present their own particular problems for pest control. Methyl bromide fills a valuable niche in the provision of these needs. Replacement of methyl bromide seems best served by another fumigant, modified atmospheres (CO_2), or by the use of heat alone or in combination with other measures. In many situations, however, there is no realistic alternative to methyl bromide.

Better gas containment offers the prospect of reducing fumigant dosages applied and this possibility can be further enhanced by increasing the lengths of exposures. With a reduced availability of the fumigant, such measures could permit continued use of methyl bromide at similar to present frequencies of application and, for soil, over similar to currently treated areas.

Gas recovery and possible recycling could further conserve the amount of methyl bromide used, but at a cost. Prospects for success with recycling are best at fixed installation chamber facilities. Prospects for reducing emissions from other fumigations by absorbing methyl bromide on various filters are good but further work is needed on ways of enhancing sorptive capacity and on subsequent disposal or recycling methods for used filters.

Although many alternative control techniques are available for a wide variety of situations, only a few can be regarded as true potential substitutes for the use of methyl bromide. All alternatives will require further research and changes in commercial practice. For those of a chemical nature, implementation will have to await a commercial initiative to pursue registration for use in each country.

REFERENCES

Abe Y and Kondoh Y, 1991, in: Brody AL (ed.), *Controlled Modified Atmosphere Vacuum Packaging of Foods* (Food & Nutrition Press, Inc., Trumbull) p 149.

Adem E, Watters FL, Uribe-Rendn R and De la Piedad, 1978, *J. Stored Prod. Res.*, **14**, 135.

Ahmed M, 1991, in: Fleurat-Lessard F and Ducom P, (eds), Proc. 5th Int. Wkg Conf. stored Prod. Prot., September 1991, Bordeaux, France. Vol II, p 1105.

Alabouvette C, Conteaudier Y and Louvet J, 1985, in: Parker CA, Rovira AD, Moore KJ, Wong PTW and Kollmorgen JF, (eds), *Ecology and Management of Soilborne Plant Pathogens*. (The American Phytopathological Society: St. Paul Minnesota) p 101.

Alkayassi AW and Alkaraghouli AA, 1991, *FAO Plant Production and Protection Paper*, **109**, 297.

Annis PC, 1987, in: Donahaye E and Navarro S, (eds), Proc. 4th Int. Wkg Conf. Stored Prod. Prot. September 1986. Tel Aviv, Israel, p 128.

Annis PC, 1990, in: Champ BR, Highley E and Banks HJ, (eds), Fumigation and Controlled Atmosphere Storages of Grain. ACIAR Proceedings No. 25, p 20.

Annis PC and Greve J, van Someren 1984, in: Ripp BE, Banks HJ, Bond EJ, Calverley DJ, Jay G and Navarro S, (eds), *Controlled Atmosphere and Fumigation in Grain Storages* (Amsterdam: Elsevier) p 125.

Anonymous, 1976a, Methyl Bromide Research Society (Japan) Report No. 3, 8–16.

Anonymous, 1976b, Methyl Bromide Research Society (Japan) Report No. 3, 24–30.

Anonymous, 1988, MAFF-Japan notification number 326.

Anonymous, 1989, MAFF-Japan notification number 47.

Anonymous, 1992, Proceedings of the International Workshop on Alternatives to Methyl Bromide for Soil Fumigation. U.N. Environment Programme. 19–21 October 1992, Rotterdam; 22–23 October 1992, Rome/Latina.

Armitage DM, 1987, in: Lawson TJ (ed), *Stored Products Pest Control BPCA* Monograph No. 37, 219.

Armstrong JW, 1982, *J. Econ. Entomol.*, **75**, 787.

Armstrong JW, 1981, *J. Econ. Entomol.*, **84**, 1308.

Armstrong JW, 1994, in: Paull RE and Armstrong JW, (eds), *Insect Pests and Fresh Horticultural Products: Treatments and Responses* (Wallingford, UK: CAB International) p 103.

Armstrong JW, Hansen JD, Hu BKS and Brown SA, 1989, *J. Econ. Entomol.*, **82**, 1667.

Arthur FH, 1994, *J. Econ. Entomol.*, **87**, 1359.

Backman PA and Rodríguez-Kábana R, 1975, *Phytopathology*, **65**, 819.

Bahr I, 1991, in: Fleurat-Lessard F and Ducom P, (eds), Proc. 5th Int. Wkg Conf. Stored Prod. Prot., September 1991, Bordeaux, France. Vol II, p 1135.

Bailey SW, 1962, *J. Econ. Entomol.*, **55**, 301.

Bailey SW and Banks HJ, 1975, Proc. 1st Int. Wkg Conf. Stored-Prod. Ent., Savannah, GA, USA, p 362.

Baker AC, 1939, USDA Circular 551, 8 pp.

Baker AC, Cowley JM, Harte DS and Frampton ER, 1990, *J. Econ. Entomol.*, **83**, 13.

Baker AC, Stone WE, Plummer CC and McPhail M, 1944, USDA Miscellaneous Publication 531, p 155.

Balock JW, Burditt AK Jr and Christenson LD, 1963, *J. Econ. Entomol.*, **56**, 42.

Banks HJ, 1984, Report CSIRO Division of Entomology, Canberra, 38 pp.

Banks HJ, 1987, in: Donahaye E and Navarro S (eds), Proc. 4th Int. Wkg Conf. Stored Prod. Prot., September 1986. Tel Aviv, Israel, p 165.

Banks HJ, 1988, in: Ferrar P (ed.), *Transport of Fresh Fruit and Vegetables*. ACIAR Proceedings No 23, p 45.

Banks HJ and Annis PC, 1980, in: Shejbal J, (ed), *Controlled Atmosphere Storage of Grains* (Amsterdam: Elsevier) p 207.

Banks HJ and Annis PC, 1981, Australia CSIRO Division of Entomology Report No 26, p 13.

Banks HJ, Annis PC and Rigby GR, 1991, in: Fleurat-Lessard F and Ducom P, (eds), Proc. 5th Int. Wkg Conf. Stored Prod. Prot., September 1991, Bordeaux, France. Vol II, p 695.

Banks J and Fields P, 1995, in: Jayas DS, White NDG and Muir WE (eds), *Stored Grain Ecosystems*, (New York, Basel, Hong Kong: Dekker) p 353.

Baranov LA, Kalamkaliev MKh and Ilyukhin GP, 1983, *Mekhanizatsiyai Elektrifikatsiya Sel'skogo Khozyaistva*, **5**, 42.

Barbosa A, Golob P and Jenkins N, 1994, in: Highley E, Wright EJ, Banks HJ and Champ BR, (eds), Proc. 6th Int. Wkg Conf. Stored Prod. Prot., April 1994, Canberra, Australia, CAB International, p 623.

Barker AV and Craker LE, 1991, *Agronomy Journal*, **83**, 302.

Bartali H, 1994, in: Highley E, Wright EJ, Banks HJ and Champ BR (eds), *Proc. 6th Int. Wkg Conf. Stored Prod. Prot.*, April, Canberra, Australia, p 281.

Bartok JW, 1993, *Greenhouse Manager*, **11(10)**, 88.

Baust JG and Lee RE, 1981, *J. Insect Physiol.*, **27**, 485.

Belker N, 1990, *Deutscher-Gartenbau*, **44**, 672.

Bell CH, 1991, *Post Harvest Biol. Technol.*, **1**, 81.

Bell CH, 1992, Presentations from the Methyl Bromide Seminar, 27th November, Central Science Laboratory, Slough, p 30.

Bell CH, 1993, *Food Science and Technology Today*, **7**, 212.

Bell CH, 1994, *J. Stored Prod. Res.*, **30**, 99.

Bell CH, Spratt EC and Mitchell DJ, 1980, *Bull. Ent. Res.*, **70**, 293.

Bell CH, Llewellin BE, Sami B, Chakrabarti B and Mills KA, 1991, in: Fleurat-Lessard F and Ducom P, (eds), Proc. 5th Int. Wkg Conf. Stored Prod. Prot., September 1990, Bordeaux, France, p 1769.

Bell CH and Armitage DM, 1992, in: Sauer DB, (ed), *Storage of Cereal Grain and their Products* 4th edition. (American Association of Cereal Chemists: St. Paul Minn) p 249.

Bell CH, Chakrabarti B, Conyers ST, Wontner-Smith TJ and Llewellin BE, 1993, in: Navarro S and Donahaye E, (eds), Proc. Int. Conf. Controlled Atmosphere and Fumigation in Grain Storages, Winnipeg, June 1992, Caspit Press, Jerusalem, p 315.

Bello A, Rodríguez CM, López Cepero J, González JA, 1991, Comunicaciones de la XVII Reunión Nacional de Suelos. Islas Canarias, Universidad de la Laguna, p 153.

Bello A, González JA, Bun M, Domiuguez J, López Cepero J, Rodríguez CM and Tello J, 1993, Actas del XII Congreso Latinoamericano de la Ciencia del Suelo, p 1608.

Benshoter CA, 1987, *J. Econ. Entomol.*, **80**, 1223.

Benz W, Moosmann A, Walter H and Koch W, 1984, *Mitteilungen aus der Biologischen Bundesanstalt für Land und Forstwirtschaft Berlin Dahlem*, **223**, 128.

Bollen GJ, 1985, in: Parker CA, Rovira AD, Moore KJ and Wong PTW (eds) *Ecology and Management of Soilborne Plant Pathogens*, (The American Phytopathological Society, St Paul Minnesota) p 191.

Bond EJ, 1984, Manual of Fumigation for Insect Control. FAO Plant Production and Protection Paper No. 54, FAO, Rome, p 432.

Borodin JF, 1989, *Elektrichestvo*, **6**, 1.

Bridgeman BW, 1994, in: Highley E, Wright EJ, Bansk HJ and Champ BR, (eds), Proc. 6th Int. Wkg Conf. Stored Prod. Prot., April'1994, Canberra, Australia, p 628.

Bridwell JC, 1918, *Proc. Hawaii Ent. Soc.*, **3**, 506.

Brower JH, 1973, *J. Econ. Entomol.*, **66**, 1175.

Brower JH and Tilton EW, 1972, *J. Econ. Entomol.*, **65**, 250.

Brower JH and Tilton EW, 1973, *J. Stored Prod. Res.*, **9**, 93.

Brower JH and Tilton EW, 1985, in: Proceedings FAO/LAEA Symposium, Radiation Disinfestation of Food and Agricultural Products, Vienna, Austria, p 75.

Brown GA, Brower JH and Tilton EW, 1972, *J. Econ. Entomol.*, **65**, 203

Brunner H, 1987, in: Donahaye E and Navarro S, (eds), Proc. 4th Int. Wkg Conf. Stored Prod. Prot., September 1986, Tel Aviv, Israel p 219.

Burditt AK Jr, 1990, *FAO Plant Prot. Bull.*, **39**, 25.

Burditt AK Jr., 1994, in: Sharp JL and Hallman GJ, (eds), *Quarantine Treatments for Pests of Food Plants*, (Boulder Colorado: Westview Press).

Burges HD and Burrel NJ, 1964, *J. Sci. Food Agri.*, **15**, 32.

Burkholder WE, Tilton EW and Cogburn RR, 1966, *J. Econ. Entomol.*, **59**, 976.

Burrell NJ and Laundon HJ, 1967, *J. Stored Prod. Res.*, **3**, 125.

Calvet C, Pera J and Barea JM, 1993, *Plant and Soil*, **148**, 1.

Carpenter A and Potter M, 1994, in: Sharp JL and Hallman GJ, (eds), *Quarantine Treatments for Pests of Food Plants*, (Boulder, Colorado: Westview Press) p 171.

Cassells J, Banks HJ and Allanson R, 1994, in: Highley E, Wright EJ, Banks HJ and Champ BR, (eds), Proc. 6th Int. Wkg Conf. Stored Prod. Prot., April 1994, Canberra, Australia, p 56.

Chakrabarti B and Bell CH, 1993, *Chemistry and Industry*, 1993, **24**, p 992.

Chakrabarti B, Wontner-Smith TJ and Bell CH, 1995, Reducing methyl bromide emissions from soil fumigations in greenhouses. Paper given at the Annual International Conference on Methyl Bromide Alternatives and Emissions Reduction, San Diego, California, November 1995, 3 pp.

Chauvin G and Vannier G, 1991, in: Fleurat-Lessard F and Ducom P, (eds), Proc. 5th Int. Wkg Conf. Stored Prod. Prot., September 1991, Bordeaux, France, vol II, p 1157.

Chet I, 1987, *Innovative Approaches to Plant Disease Control* (New York: Wiley and Sons) p 372.

Chian Ru-Jo, 1990, Inorganic nitrogen compounds as amendments to soil for nematode control. Auburn University, 143 pp.

Chidester S, 1991, Proceedings of the 23rd Annual Meeting of the American Wood-Preservers Association, p 316.

Chinanzavana S, Grosshandler WL and Davis DC, 1986, *Transactions of the ASAE American Society of Agricultural Engineers*, **29**, 1797.

Churchill TA and Storey KB, 1989, *J. Insect Physiol.* **35**, 579.

Claflin JK, Evans DE, Fane AG and Hill RJ, 1986, *J. Stored Prod. Res.*, **22**, 153.

Cole JS and Zvenyika Z, 1988, *Plant Pathology*, **37**, 271.

Cook RJ and Baker KF, 1983. American Phytopathological Society, St. Paul Minnesota, 539 pp.

Corcoran RJ, Heather NW and Heard TA, 1993, *J. Econ. Entomol.*, **86**, 66.

Cornwall PB, Crook LJ and Bell JO, 1957, *Nature*, **179**, 670.

Cotton RT and Frankenfeld JC, 1942, *American Miller*, **70**, 36.

Couey HM and Chew V, 1986, *J. Econ. Entomol.*, **79**, 887.

Couey HM and Hayes CF, 1986, *J. Econ. Entomol.*, **79**, 1307.

Couey HM, Linse ES and Nakamura AN, 1984, *J. Econ. Entomol.*, **77**, 984.

Cowley JM, Baker RT, Englberger KG and Lang TG, 1991, *J. Econ. Entomol.*, **84**, 1763.

Cowley JM, Chadfield KD and Baker RT, 1992, *New Zealand J. Crop and Hortic. Sci.*, **20**, 209.

Daniel V, Maekawa S and Preusser FD, 1993, ICOM Committee for Conservation 10th Triennial Meeting, Washington DC, p 863.

Daniel V, Hanlon G and Maekawa S, 1994, IIC 15th International Congress, Preventative Conservation, Practice, Theory and Research, Ottawa.

De Boer RF, Taylor PA, Flet SP and Merriman PR, 1985, in: Parker CA, Rovira AD, Moore KJ, Wong, PTW and Kollmorgen JF (eds) *Ecology and Management of Soilborne Plant Pathogens*. (The American Phytopathological Society: St Paul Minnesota) p 197.

De Francolini J, 1935a, *Revue Path. Veg. Ent. Agric. Fr.*, **22**, 1.

De Francolini J, 1935b, *Revue Path. Veg. Ent. Agric. Fr.*, **22**, 9.

De Heer H, Hamaker P and Tuinstra LGMT, 1981, Report on an optimization study after the use of the fumigant methyl bromide in horticulture. Report 10B, Wageningen.

De Ong ER, 1921, *J. Econ. Entomol.*, **14**, 444.

De Vay JE, 1991, *FAO Plant Production and Protection Paper*, **109**, 1.

Dean GA, 1913, *J. Econ. Entomol.*, **6**, 40.

Del Estel P, Viñuela E, Camacho C and Page E, 1986a, in: Economopoulos AP, (ed), Proceedings II International Symposium on Fruit Flies/Crete, Athens: Elsevier, p 115.

Del Estel P, Viñuela E, Page E and Camacho C, 1986b, *J. Appl. Entomol.*, **102**, 45.

Dentener PR, Peetz SM and Birtles DB, 1992, *New Zealand J. Crop Hort. Sci.*, **20**, 203.

Diprose MF and Evans GH, 1988, Engineering advances for agriculture and food. Proceedings of the 1938–1988 Jubilee Conference of the Institution of Agricultural Engineeers. Co-sponsored by the Fellowship of Engineering, Robinson College, Cambridge, 12–15 September, 1988 [edited by Cox, SWR]. London, UK; Butterworths p 363.

Domke O, 1986, *Gb-+-Gw*, **84**, 520.

Donahaye E, Navarro S and Rindner M, 1991, *J. Appl. Entomol.*, **111**, 297.

Donahaye E, Navarro S, Rindner M and Dias R, 1992, *J. Econ. Entomol.*, **85**, 1990.

Dustan GG, 1963, *J. Econ. Entomol.*, **56**, 167.

Duvel JWT, 1905, *Bull. U.S. Bureau Ent.*, **54**, 49.

Ebeling W, 1971, *Ann. Rev. Ent.*, **16**, 123.

Elliott JC, 1989, *Acta Horticulturae*, **238**, 173.

Elmore CL, 1991, *FAO Plant Production and Protection Paper*, **109**, 129.

Evans DE, 1987, in: Donahaye E and Navarro S, (eds), Proc. 4th Int. Wkg Conf. Stored Prod. Prot. September 1986. Tel Aviv, Israel, p 149.

Evans DE, Thorpe GR and Dermott T, 1983, *J. Stored Prod. Res.*, **19**, 125.

Faes H, 1930, *Annu. Agric. Swisse*, **31**, 287.

Fam EZ, El-Nahal AKN and Fahmy H, 1974, *Bull. Ent. Soc. Egypt, Econ. Series*, **8**, 105.

Fernández C, Pinochet J and Felipe A, 1993, *Nematropica*, **23**, 195.

Ferriss RS, 1984, *Phytopathology*, **74**, 121.

Fields PG, 1991, in: Fleurat-Lessard F and Ducom P, (eds), Proc. 5th Int. Wkg Conf. Stored Prod. Prot. September 1991, Bordeaux, France, p 1183.

Fields PG, 1992, *J. Stored Prod. Res.*, **28**, 89.

Fishler DW and Winchester JA, 1964, *Proc. Soil Crop Sci. Soc. Fla.*, **24**, 150.

Fleurat-Lessard F, 1985, *EPPO Bulletin*, **15**, 109.

Fleurat-Lessard F, 1987, in: Lawson TJ (ed), Stored Products Pest Control BPCA Monograph No. 37, 209.

Florian ML, 1990, *Collection-Forum*, **6**, 1.

Follet-Smith RR, 1933, *Trop. Agric. (Trinidad)* **10**, 91.

Gair R, Mathias PL and Harvey PN, 1969, *Ann. Appl. Biol.*, **63**, 503.

Gamliel A, Katan J, Chen Y and Grinstein A, 1989, *Acta Horticulturae*, **255**, 181.

Gardener RD, Harein PK and Subramanyam B, 1988, *Am. Entomol.*, **34**, 22.

Garibaldi A and Gullino ML, 1991, *FAO Plant Production and Protection Paper*, **109**, 253.

Garibaldi A and Tamietti G, 1989, *Acta Horticulturae*, **255**, 125.

Geraldson CM, Overman AJ and Jones JP, 1965, *Proc. Soil and Crop Sci. Soc. of Florida*, **25**, 18.

Gerard D, Kraus J and Quirin K-W, 1988, *Pharm. Ind.*, **50**, 1298.

Giga DP and Chinwada P, 1994, in: Highley E, Wright EJ, Banks HJ and Champ BR, (eds), Proc. 6th Int. Wkg Conf. Stored Prod. Prot., April 1994, Canberra, Australia, p 631.

Gilberg M, 1990, *Bulletin of the Australian Institute for the Conservation of Cultural Materials*, **16**, 27.

Gilberg M and Gratton D, 1994, IIC 15th International Congress, Preventative Conservation, Practice, Theory and Research, Ottawa.

Gindrat D, 1979, in: Mulder D (ed), *Developments in Agricultural and Managed-Forest Ecology* (Amsterdam: Elsevier) p 253.

Godoy G, Rodríguez-Kábana R, Shelby RA and Morgan-Jones G, 1983, *Nematropica*, **13**, 63.

Goisque MJ, Louvet H, Marten C, Lagier J, Davet P and Couteaudier Y, 1984, *Plasticulture*, **64**, 32.

Golob P and Webley DJ, 1980, Tropical Products Institute, London, G138, p 32.

Gómez AM, 1993, El injerto herbaceo como métado alternativo de control de nfermedades telúricas y sus aplicaciones agronomicas. Spain, Universidad Politécnica de Valencia, Doctoral Thesis, p. 494.

Goodwin WH, 1914, *J. Econ. Entomol.*, **7**, 313.

Goodwin WH, 1912, *Bull. Ohio Agric. Exp. Sta.*, **234**, 171.

Gould WP, 1994, in: Sharp JL and Hallman GJ, (eds), *Quarantine Treatments for Pests of Food Plants.* (Westview Press) p 119.

Gould WP and Sharp JL, 1990, *J. Econ. Entomol.*, **83**, 2324.

Gould WP and Sharp JL, 1992, *J. Econ. Entomol.*, **85**, 1235.

Grossman EF, 1931, *Rev. Appl. Ent.*, A **20**, 224.

Grossman J and Liebman J, 1995, *The IPM Practitioner*, **XVII(7)**, 1.

Hague NG and Sood U, 1963, *Pl. Path.*, **12**, 88.

Hague NG, Lubatti OF and Page ABP, 1964, *Hort. Res.*, **3**, 84.

Hallman GJ, 1991, *HortScience*, **26**, 286.

Hallman GJ, Nisperos-Carriedo MO, Baldwin EA and Campbell CA, 1994, *J. Econ. Entomol.*, **87**, 752.

Hansen EM, Hara AH and Tenbrink VL, 1992, *HortScience*, **27**, 139.

Hansen EM, Myrold DD and Hamm PB, 1990, *Phytopathology*, **80**, 698.

Hara AH, Hata TY, Hu BKS and Tenbrink VL, 1993, *J. Econ. Entomol.*, **86**, 1167.

Hara AH, Hata TY, Hu BKS, Kaneko RT and Tenbrink VL, 1994, *J. Econ. Entomol.*, **87**, 1570.

Haroon Usmani SM and Ghaffar A, 1985, in Parker CA, Rovira AD, Moore KJ,

Wong PTW and Kollmorgen JF, (eds), *Ecology and Management of Soilborne Plant Pathogens*. (The American Phytopathological Society, St Paul Minnesota) p 285.

Hartel PG and Alexander M, 1983, *Soil Biology and Biochemistry*, **15**, 489.

Haryadi Y, Syarief R, Hubeis M and Herawati I, 1994, in: Highley E, Wright EJ, Banks HJ and Champ BR, (eds), Proc. 6th Int. Wkg Conf. Stored Prod. Prot., April 1994, Canberra, Australia, CAB International, p 633.

Hatton TT, 1990, in: Wang CY, (ed), *Chilling Injury of Horticultural Crops*. (Boca Raton, Florida: CRC Press) p 269.

Hawkins LA, 1930, *Plant Quarantine and Control Administration Service Regulation Announcements*, **102**, 25.

Hayashi I and Aono N, 1982, *Bulletin of the Kanagawa Horticultural Experiment Station*, **29**, 62.

Hayes CF, Chingon HTG, Nitta FA and Wang WJ, 1984, *J. Econ. Entomol.*, **77**, 683.

Heald CM, 1987, in: Veech JA and Dickson DW, (eds), *Classical Nematode Management Practices, Vistas on Nematology*, (Hyatsville MD, Society of Nematologists) p 100.

Hege H and Rob H, 1972, Kuratorium für Technik und Bauwesen in der Landwirtschaft e.V., Frankfurt am Main p 65.

Heritage AD and Duniway JM, 1985, in: Parker CA, Rovira AD, Moore KJ, Wong PTW and Kollmorgen JF, (eds), *Ecology and Management of Soilborne Plant Pathogens* (The American Phytopathological Society, St Paul Minnesota) p 199.

Hilborn MT, Hepler PR and Cooper GR, 1957, *Maine Farm Res.* V, 11.

Hirose T, Abe I, Kohno M, Suzuri T and Oshima K and Okakura T, 1975, *J. Microwave Power*, **10**, 181.

Hoitink HAJ, 1988, *Ann. Rev. Phytopathology*, **24**, 93.

Hoitink HAJ and Kuter GA, 1985, in: Parker CA, Rovira AD, Moore KJ, Wong PTW and Kollmorgen JF, (eds), *Ecology and Management of Soilborne Plant Pathogens*. (The American Phytopathological Society, St Paul Minnesota) p 237.

Hollis JP and Rodriguez-Kabana R, 1966, *Phytopathology*, **56**, 1015.

Hooper T, 1907, *J. Dept. of Agr., Western Australia*, **15**, 252.

Horiuchi S, 1991, *FAO Plant Production and Protection Paper*, **109**, 16.

Houck LG, Jenner JF and Mackey BE, 1990, *J. Hort. Sci.*, **65**, 611.

Howard LO, 1896, *U.S. Dept. Agr. Division Entomol. Bull.*, **6**, 13.

ICGFI (International Consultative Group on Food Irradiation), 1988, Provisional guideline for the irradiation of cereal grain for cereal disinfestation. Vienna, Joint FAO/LAEA Division.

Jackson K and Webley D, 1994, in: Highley E, Wright EJ, Banks HJ and Champ BR, (eds), Proc. 6th Int. Wkg Conf. Stored Prod. Prot., April 1994, Canberra, Australia, CAB International, p 635.

Jakobsen I, 1984, *Soil Biology and Biochemistry*, **16**, 281.

Jang EB, 1986, *J. Econ. Entomol.*, **79**, 700.

Jang EB, 1991, *J. Econ. Entomol.*, **84**, 1298.

Jang EB and Moffitt HR, 1994, in: Sharp JL and Hallman GJ, (eds), *Quarantine Treatments for Pests of Food Plants*, (Westview Press) p 225.

Jay EG, 1971, USDA Agriculture Research Service Bulletin No. 51-46, p 6.

Jay EG, 1986, GASGA Seminar on Fumigation Technology in Developing Countries. Tropical Development and Research Institute, London, p 173.

Jona R and Arzone A, 1979, *J. Hort. Sci.*, **54**, 167.

Kalder AA, 1993, *Postharvest Biology and Technology*, **3**, 182.

Kalder AA and Ke D, 1994, in: Paull RE and Armstrong JW, (eds), *Insect Pests and*

Fresh Horticultural Products: Treatments and Responses (Wallingford, UK: CAB International).

Katan J, 1985, in: Parker CA, Rovira AD, Moore KJ, Wong PTW and Kollmorgen JF, (eds), *Ecology and Management of Soilborne Plant Pathogens*. (The American Phytopathological Society, St. Paul, Minnesota) p 274.

Katan J, 1987, in: Chet I (ed), *Innovative Approaches to Plant Disease Control* (New York: John Wiley & Sons) p 77.

Katan J and De Vay JE, 1991, *Soil Solarization* (Boca Raton, Florida: CRC Press) p 267.

Katan J, Greenberger A, Alon H and Grinstein A, 1976, *Phytopath*, **66**, 683.

Katan J, Grinstein A, Greenberger A, Yarden O and De Vay JE, 1987, *Phytoparasitica*, **15**, 229.

Kaufmann HG and Hackbart W, 1967, *Dtsch. Gartenbau*, **14**, 296.

Keever DW, 1989, *J. Agric. Ent.*, **6**, 43.

Kirkpatrick RL and Cagle A, 1978, *J. Kansas Entomol. Soc.*, **51**, 386.

Kodama T and Fukai T, 1979, *Bull. of the Nara Prefecture Agric. Exp. Sta.*, **10**, 71.

Koidsumi K and Shibata K, 1936, *J. Soc. Tropical Agr. (Japan)*, **8**, 82 (in Japanese).

Kraft JM, 1985, in: Parker CA, Rovira AD, Moore KJ, Wong PTW and Kollmorgen KF, (eds), *Ecology and Management of Soilborne Plant Pathogens*. (The American Phytopathological Society, St Paul Minnesota) p 101.

Krishnamurthy TS, Spratt EC and Bell CH, 1986, *J. Stored Prod. Res.*, **22**, 145.

Kuo LS, Su CY, Hseu CY, Chao YF, Chen HY, Liao JY and Huang WC, 1987, Vapor heat treatment for elimination of *Dacus dorsalis* and *Dacus cucurbitae* infesting mango fruits. Taiwan Bureau of Commodity and Quarantine, Ministry of Economic Affairs, Taipei.

Kuo LS, Su CY, Hseu CY, Chao YF, Chen HY, Liao JY and Chu CF, 1989, Vapor heat treatment for elimination of *Dacus dorsalis* infesting carambola fruits. Taiwan Bureau of Commodity Inspection and Quarantine, Ministry of Economic Affairs, Taipei.

Labowsky HJ, 1990, *Landtechnik*, **45**, 270.

Landolt PJ, Chambers DL and Chew V, 1984, *J. Econ. Entomol.*, **77**, 285.

Larson AO and Simmons P, 1924, *J. Agric. Res.*, **27**, 99.

Lasseran JC and Fleurat-Lessard F, 1991, in: Fleurat-Lessard F and Ducom P (eds), Proc. 5th Int. Wkg Conf. Stored Prod. Prot., September 1990, Bordeaux, France, vol II, p 1221.

Latta R, 1932, *J. Econ. Entomol.*, **25**, 1020.

Le Goupil P, 1932, *Revue Path. Veg. Ent. Agric. Fr.*, **19**, 169.

Le Patourel GNJ, 1986, *J. Stored Prod. Res.*, **22**, 63.

Le Torc'h J-M and Fleurat-Lessard F, 1991, in: Fleurat-Lessard F and Ducom P, (eds), Proc. 5th Int. Wkg Conf. Stored Prod. Prot., September 1990, Bordeaux, France, vol II, p 847.

Lee RE, 1991, in: Lee RE and Denlinger DL, (eds), *Insects at Low Temperature* (New York: Chapman and Hall) p 17.

Lemanova NB, Sultonova OD and Zemshman AYa, 1983, *Sadovodstvo, Vinogradarstvo i Vinodelie Moldavii*, **8**, 38.

Lester PJ, Dentener PR, Petry RJ and Alexander SM, 1995, *Postharvest Biology and Technology*, **6**, 349.

Liegel LH, 1986, *Turrialba*, **36**, 11.

Locatelli DP and Daolio E, 1993, *J. Stored Prod. Res.*, **29**, 81.

Locatelli DP and Traversa S, 1989, *Italian J. Food Sci.*, **1**, 53.

Loschiavo S, 1978, *J. Econ. Entomol.*, **71**, 888.

Lounsbury CP, 1907, *Agric. J. Cape Good Hope*, **31**, 186.

Maier DE, 1994, in: Highley E, Wright EJ, Banks HJ and Champ BR, (eds), Proc. 6th Int. Wkg Conf. Stored Prod. Prot., April, Canberra, Australia, p 311.

Mangan RL and Ingle SJ, 1992, *J. Econ. Entomol.*, **85**, 1859.

Mangan RL and Ingle SJ, 1994, *J. Econ. Entomol.*, **87**, 1578.

Matrod L, Faddoul J, Elmeamar A and Al Chaabi S, 1991, *FAO Plant Production and Protection Paper*, **109**, 118.

McFarlane JA, 1989, *Tropical Sci.*, **29**, 75.

McGaughey WH and Akins RG, 1989, *J. Stored Prod. Res.*, **25**, 201.

McGuire RG and Reeder WR, 1992, *J. American Soc. Hortic. Sci.*, **117**, 90.

McSorley R, Dickson DW, de Brito JA, Hewlett TE and Frederick JJ, 1994, *Journal of Nematology* **26**, 175.

Miller WR, McDonald RE, Hallman GJ and Ismail M, 1990, *Pro. Fla State Hortic. Soc.*, **103**, 238.

Moffitt HR, 1990, *Proc. Washington State Hortic. Association (1989)*, **86**, 211.

Moffitt HR and Burditt AK, 1989a, *J. Econ. Entomol.*, **82**, 1379.

Moffitt HR and Burditt AK, 1989b, *J. Econ. Entomol.*, **82**, 1679.

Moore WD, 1949, *Phytopathology*, **39**, 920.

Mordkovich YB, Menshikov NS and Luzan NK, 1985, *EPPO Bulletin*, **15**, 5.

Moreira C, 1923, *Characus e Quint*, **27**, 17.

Morris SC and Jessup AJ, 1994, in: Paull RE and Armstrong JW, (eds), *Insect Pests and Fresh Horticultural Products: Treatments and Responses* (Wallingford, UK: CAB International) p 163.

Mueller DK, 1994, *Fumigants and Pheromones*, **33**, 1.

Muller PJ and van Aartrijk J, 1989, *Acta Horticulturae*, **255**, 261–4.

Muller PJ, van Beers Th and de Rooy M, 1992, Flooding, a non chemical soil treatment to control the root-lesion nematode *Pratylenchus penetrans*. Netherlands. Bulb Research Centre, Lisse.

Munnecke DE and Lindgren DL, 1954, *Phytopathology*, **44**, 605.

Nagji M and Veljovic VM, 1994a, Halozone Technologies Inc. Report, 16 February, 1994.

Nagji M and Veljovic VM, 1994b, Ann. Int. Conf. Methyl Bromide Alternatives and Emission Reductions, November 1994.

Nataredja YC and Hodges RJ, 1990, in: Champ BR, Highley E and Banks HJ, (eds), Fumigation and Controlled Atmosphere Storages of Grain. ACIAR Proceedings No. 25, p 152.

Nataredja YC and Hodges RJ, 1990, in: Champ BR, Highley E and Banks HJ, (eds), Fumigation and Controlled Atmosphere Storages of Grain. ACIAR Proceedings No. 25, p 197.

Nation JL and Burditt AK, 1994, in: Paull RE and Armstrong JW, (eds), *Pests and Fresh Horticultural Products: Treatments and Responses*. (Wallingford, UK: CAB International) p 85.

Navarro S and Calderon M, 1974, *J. Stored Prod. Res.*, **10**, 237.

Nederpel L, 1979, in: Mulder D (ed) *Soil Disinfestation. Developments in Agricultural and Managed-Forest Ecology, 6* (Amsterdam: Elsevier Scientific) p 29.

Nelson SO, 1972, *J. Microwave Power*, **7**, 231.

Nelson SO, 1973, *Bull. Ent. Soc. Am.*, **19**, 157.

Nesheim K, 1984, *Restaurator*, **6**, 147.

Newton J, 1993, Wildey KB and Robinson WH, (eds), Proc. Int. Conf. Invertebrate Pests in the Urban Environment, Cambridge, 1993, BPCC Wheatons Ltd, Exeter, p 329.

Ng PKW, Bushuk W and Borsa J, 1989, *Can. Inst. Food Sci. Technol. J.*, **22**, 173.

Nikitin Yu P and Nikitin P Yu, 1990, *Mekhanizatsiya i Elektrifikatsiya sel'skogo*

Khozyaistva, **12**, 16.

Nothel H, 1968, in: *Isotopes and Radiation in Entomology* (Vienna: IAEA) p 87.

Opperman CH, Taylor CG and Conkling MA, 1994, *Science*, **263**, 22.

Parker WB, 1915, *Bull. U.S. Dep. Agric.*, **235**, 15.

Parker CA, Rovira AD, Moore KJ, Wong PTW and Kollmorgen JF, 1985, *Ecology and Management of Soilborne Plant Pathogens* (The American Phytopathological Society, St. Paul, Minnesota) 358 pp.

Pattee HE and Young CT, 1982, *Peanut science and technology* (Yoakum, Texas: American Peanut Research and Education Society) p 825.

Paull RE and McDonald RE, 1994, in: Paull RE and Armstrong JW, (eds), *Insect Pests and Fresh Horticultural Products: Treatments and Responses.* (Wallingford, UK: CAB International) p 191.

Paull RE and Armstrong JW, 1994, *Insect Pests and Fresh Horticultural Products: Treatments and Responses*, (Wallingford, UK: CAB International) p 1.

Piriyathamrong S, Chouvalitvongporn P and Sudathit B, 1985, in: Moy JH, (ed), Radiation of Food and Agricultural Products, Proceedings of an International Conference, University of Hawaii, p 222.

Plattner AJ, 1989, United States Patent No. 4, 873, 789, p 5.

Poock G, 1910, *Chemikerztg*, **34**, 1127.

Porter IJ and Merriman PR, 1983, *Soil Biology and Biochemistry*, **15**, 39.

Prozell S and Reichmuth C, 1991, in: Fleurat-Lessard F and Ducom P (eds), Proc. 5th Int. Conf. Stored Prod. Prot., September 1990, Bordeaux, France, vol. II, p 911.

Pullman GS and DeVay JE, 1981, *Phytopathology*, **71**, 1285.

Pullman GS, DeVay JE and Garber RH, 1981, *Phytopathology*, **71**, 959.

Pullman GS, DeVay JE, Garber RH and Weinhold AR, 1979, in: Schippers B and Gams S (eds), *Soil-borne Plant Pathogens*, (New York: Academic Press) p 439.

Punja ZK, 1985, *Ann. Rev. Phytopathology*, **23**, 97.

Quarles W and Grossman J, 1995, *The IPM Practitioner*, **XVII (8)**, 1.

Quentin ME, Spencer JL and Miller R, 1991, *Ent. Exp. Appl.*, **60**, 105.

Rattink H, 1982, *Vakblad voor de Bloemisterij*, **37**, 36 and 44.

Rhoades HL, 1983, *Nematropica*, **13**, 9.

Reichmuth C, 1993, *Merkblatt 71 der Biologische Bundesanstalt für Land und Forstwirtschaft edn*, (Saphir Verlag: Ribbesbüttel, Germany) **42**, 38 pp.

Rhodes AA, 1986, Agricultural Research Service and Economic Research Service to United States Department of Energy, Energy Technologies Division, Interagency Agreement Number DE-A104-83AL24327.

Richardson HH and Johnson AC, 1935, *U.S. Dep. Agric. Tech. Bull.*, **1935**, 853.

Riherd C, Nguyen R and Brazzel JR, 1994, in: Sharp, JL and Hallman GJ (eds), *Quarantine Treatments for Pests of Food Plants.* (Westview Press) p 213.

Rodionova OP, Troshina GA, Fedorova IG and Shvartsman MM, 1990, *Tekhnika v Sel'skom Khozyaistve*, **1**, 62.

Rodríguez-Kábana R and Canullo GH, 1992, *Phytoparasitica*, **20**, 211.

Rodríguez-Kábana R and Hollis JP, 1965, *Science*, **148**, 524.

Roorda-van-Eysinga JPNL, 1984, *Acta Horticulturae*, **145**, 262.

Rovira AD and Venn NR, 1985, in: Parker CA, Rovira AD, Moore KJ, Wong PTW and Kollmorgen KF (eds), *Ecology and Management of Soilborne Plant Pathogens*, (The American Phytopathological Society, St Paul Minnesota) p 101.

Rozvaga RI and Bakhishev GN, 1982, in: Mordkovich YaB, (ed), Disinfestation of Plant Products against Quarantine and other Dangerous Pests. Moscow, All-Union Scientific Technical Institute for Quarantine and Plant Protection, 58.

Ruijs MNA, 1992, Proceedings of the International Workshop on Alternatives to

Methyl Bromide for Soil Fumigation. U.N. Environment Programme, 19–21 October 1992, Rotterdam; 22–23 October 1992, Rome/Latina.

Runia WT, 1983, *Acta Horticulturae (Soil Disinfestation)*, **152**, 195.

Ryan L, 1995, *Post-harvest Tobacco Infestation Control* (Chapman & Hall) p 155.

Sabio GC, Alvindia D, Julian DD, Murillo Jr R and Sambrano MS, 1990, in: Champ BR, Highley E and Banks HJ (eds), Fumigation and Controlled Atmosphere Storage of Grain. ACIAR Proceedings No. 25, p 180.

Sadasivan TS, 1965, Effect of mineral nutrients on soil microorganisms and plant disease, in: Baker KF and Snyder WC, (eds), *Ecology of Soil-Borne Plant Pathogens, Prelude to Biological Control*, (Berkeley: University of California Press) p 460.

Schippers B, Geels FP, Hoekstra O, Lamers JG, Maenhout CAAA and Scholte K, 1985, in: Parker CA, Rovira AD, Moore KJ, Wong PTW and Kollmorgen KF (eds), *Ecology and Management of Soilborne Plant Pathogens* (The American Phytopathological Society, St Paul Minnesota) p 127.

Schmidt J, 1983, *Die Zuckerrübe*, **32**, 169.

Scott PR and Hollins TW, 1985, in: Parker CA, Rovira AD, Moore KJ, Wong, PTW and Kollmorgen JF (eds) *Ecology and Management of Soilborne Plant Pathogens*. (The American Phytopathological Society; St Paul Minnesota) p 157.

Sein Jr F, 1935, *J. Agr. University of Puerto Rico*, **19**, 105.

Seo ST, Chambers DL, Komura M and Lee CVL, 1970, *J. Econ. Entomol.*, **63**, 1977.

Sharp JL, 1985, *Proc. Fla State Hortic. Soc.*, **98**, 78.

Sharp JL, 1992, *J. Econ. Entomol.*, **85**, 2302.

Sharp JL, 1993, *J. Econ. Entomol.*, **86**, 462.

Sharp JL, 1994, in: Sharp JL and Hallman GJ, (eds), *Quarantine Treatments for Pests of Food Flants*, p 133–147.

Sharp JL and Spalding DH, 1984, *Pro. Fla State Hortic. Soc.*, **97**, 355.

Sheppard KO, 1984, in: Baur FJ, (ed), *Insect Management for Food Storage and Processing* (American Association of Cereal Chemists, St Paul, Minnesota) p 194.

Shipton PJ, 1977, *Ann. Rcv. Phytopathol.*, **15**, 387.

Skaife SH, 1918, *Bull. Dep. Agric. S. Afr.*, **12**, 32.

Sonneveld C, 1979, in: Mulder D, (ed), *Soil Disinfestation. Developments in Agricultural and Managed-Forest Ecology, 6*. (Amsterdam: Elsevier) p 39.

Soderstrom EL, Brandl DG and Mackey B, 1990, *J. Econ. Entomol.*, **83**, 472.

Soderstrom EL, Brandl DG and Mackey B, 1991, *J. Stored Prod. Res.*, **27**, 95.

Sparling GP and Berrow ML, 1985, *J. Agric. Sci. UK.*, **104**, 223.

Speir TW, Cowling JC, Sparling GP, West AW and Corderoy DM, 1986, *Soil Biol. Biochem.*, **18**, 377.

Spratt E, Dignan G and Banks HJ, 1985, *J. Stored Prod. Res.*, **21**, 41.

Stankiewicz Z and Schreiner H, 1993, *Trans. Inst. Chem. Eng.*, **71**, Part B, 134.

Stover RH, 1979, Flooding of Soil for Disease Control, in: Mulder D (ed), *Soil Disinfestation*. (Amsterdam, Oxford and New York: Elsevier) p 19.

Strang TJK, 1995, Proceedings 3rd International Conference on Biodeterioration of Cultural Property. Bangkok, Thailand, p 199.

Stratil H, Wohlgemuth R, Bolloing H and Zwingelberg H, 1987, *Getreide, Mehl und Brot*, **41**, 294.

Sugimoto T, Furusawa K and Mizobuchi M, 1983, *Res. Bull. Plant Pro. Service (Japan)*, **19**, 81 (in Japanese).

Sunagawa K, Kume K and Iwaizumi R, 1987, *Res. Bull. Plant Pro. Service (Japan)*, **24**, 13 (in Japanese).

Sutherland JW, Thorpe GR and Fricke PW, 1986, in: Proc. Conf. Agric. Eng.,

Adelaide. Canberra, Institution of Engineers, Melbourne, Australia, p 419.

Sutherland JW, Fricke PW and Hill RJ, 1989, *J. Agric. Eng. Res.*, **44**, 113.

Sylvia D and Schenck NC, 1984, *Soil Biol. Biochem.*, **16**, 675.

Szmidt RAK, Hitchon GM and Hall DA, 1989, *Acta Horticulturae*, **255**, 197.

Tarr C, Hilton SJ, Graver J van S and Clingeleffer PR,, 1994, in: Highley E, Wright EJ, Banks HJ and Champ BR, (eds), Proc. 6th Int. Wkg Conf. Stored Prod. Prot., April, Canberra, Australia, p 204.

Thompson RH, 1966, *J. Stored Prod. Res.*, **1**, 353.

Thompson JP, 1990, *Nematologica*, **36**, 123.

Thorpe GR, 1987, *J. Agric. Eng. Res.*, **37**, 27.

Thorpe GR, Evans DE and Sutherland JW, 1984, in: Ripp BE, Banks HJ, Bond EJ, Calverley DJ, Jay G and Navarro S, (eds), *Controlled Atmosphere and Fumigation in Grain Storages*, (Amsterdam: Elsevier) p 617.

Thurston HD, Smith M, Abawi G and Kearl S, 1994, (eds) *Slash/mulch: How farmers use it and what researchers know about it*. (Ithaca, N.Y., Cornell Institute for Food, Agriculture and Development (CIIFAS), Cornell University) 302 pp.

Tilton EW, Burkholder WE and Cogburn RR, 1966a, *J. Econ. Entomol.*, **59**, 944.

Tilton EW, Burkholder WE and Cogburn RR, 1966b, *J. Econ. Entomol.*, **59**, 1363.

Tilton EW and Brower JH, 1987, *Cereal Foods World*, **32**, 330.

Tjamos EC, Vaso Karapapa and Bardas D, 1989, *Acta Horticulturae*, **255**, 139.

Toba HH and Mofitt HR, 1989, in: Reid DS (preparator), International Conference on Technical Innovations in Freezing and Refrigeration of Fruits and Vegetables. Paris: International Institute Refrigeration, p 213.

Toba HH and Mofitt HR, 1991, *J. Econ. Entomol.*, **84**, 1316.

Treherne RC, 1918, *Agric. Gaz. Can.*, **55**, 668.

Trivedi PC and Barker KR, 1986, *Nematropica*, **16**, 213.

Trudgill DL, 1991, *Ann. Rev. of Phytopath*, **29**, 167.

UNEP, 1995, 1994 Report of Methyl Bromide Technical Options Committee for the 1995 Assessment by the UNEP Montreal Protocol on Substances that Deplete the Ozone Layer.

Urbain WM, 1986, *Food Irradiation*. (New York: Academic Press) p 351.

USDA-APHIS (1976) Plant Protection and Quarantine Treatment Manual. Section IV. Treatment facilities. Part 2. Atmosphere fumigation chambers - construction and performance standards—fittings for pressure-leakage test and fumigant concentration sampling. Revised January 1976. 4 pp.

Valentin N, 1993, *International Biodetermination and Biodegradation*, **32**, 263.

Van Assche C and Uyttebroeck P, 1983, *Bulletin-OEPP*, **13**, 491.

Van Berkum JA and Hoestra H, 1979, Practical Aspects of the Chemical Control of Nematodes in Soil, in: Mulder D (ed), *Soil Disinfestation*. (Amsterdam, Oxford and New York: Elsevier Scientific) p 53.

Van de Vrie M, 1986, Food Irradiation Newsletter. Joint FAO/IAEA Division of Isotope and Radiation Applications of Atomic Energy for Food and Agricultural Development.

Van Meekeren A and Van-Meekeren A, 1990, *Groeten-en-Fruit*, **46**, 28.

Van Wambeke E, 1983, *Acta Horticulturae (Soil Disinfestation)*, **152**, 137.

Vela GR, Wu JF and Smith D, 1976, *Soil Science*, **121**, 44.

Wadi JA and Easton GD, 1985, in: Parker CA, Rovira AD, Moore KJ, Wong PTW and Kollmorgen JF, (eds), *Ecology and Management of Soilborne Plant Pathogens* (The American Phytopathological Society, St. Paul, Minnesota) p 134.

Walter DC, Underwood BC and Nelson SR, 1974, *Comp. and Gen. Pharmacol.*, **5**, 165.

Washburn FL, 1905, *Ann. Rep. Minn. Ent.*, No. **9**, 17.

Watters FL, 1984, in: Baur FJ, (ed), *Insect Management for Food Storage and Processing* (American Association of Cereal Chemists, St. Paul, Minnesota) p 267.

Watters FL and MacQueen KF, 1967, *J. Stored Prod. Res.*, **3**, 223.

Weddell JA, 1931, *Queensland Agric. J.*, **36**, 141.

Wilkin DR and Nelson G, 1987, in: Lawson TJ, (ed), *Stored Products Pest Control*, BPCA Monograph No. 37, 247.

Williams-Linera G and Ewel JJ, 1984, *Plant and Soil*, **82**, 263.

Williamson MR and Winkelman PM, 1989, 1989 International Winter Meeting, American Society of Agricultural Engineers at Quebec, Canada. ASAE, St. Joseph, Missouri.

Wohlgemuth R, 1973, *Z. Angew. Ent.*, **74**, 7.

Wontner-Smith T and Chakrabarti B, 1994, *Int. Pest Control*, **36**, 15.

Yang JE, Skogley EO and Schaff BE, 1990, *Soil Science Society of America Journal*, **54**, 1646.

Yassin MY and Ismail AE, 1994, *Anz. Schadlinsk. Plschutz*, **67**, 176.

Yokoyama VY and Miller GT, 1989, *J. Econ. Entomol.*, **82**, 1152.

Yokoyama VY and Miller GT, 1987, *J. Econ. Entomol.*, **80**, 641.

Zachariassen KE, 1985, *Physiol. Rev.*, **65**, 799.

Zakladnoi GA, Men'shenin AI, Pertsovskii ES, Salimov RA, Cherepkov VG and Krssheminskii VS, 1982, *Soviet Atomic Energy*, **52**, 74.

Species Index

Subject Index